INNOVATIVE TECHNOLOGIES FOR THE TREATMENT OF INDUSTRIAL WASTEWATER

A Sustainable Approach

INNOVATIVE TECHNOLOGIES FOR THE TREATMENT OF INDUSTRIAL WASTEWATER

A Sustainable Approach

Edited by

Shirish H. Sonawane, PhD

Y. Pydi Setty, PhD

T. Bala Narsaiah, PhD

S. Srinu Naik, PhD

AAP APPLE ACADEMIC PRESS

Apple Academic Press Inc.
3333 Mistwell Crescent
Oakville, ON L6L 0A2 Canada

Apple Academic Press Inc.
9 Spinnaker Way
Waretown, NJ 08758 USA

Exclusive worldwide distribution by CRC Press, a member of Taylor & Francis Group
No claim to original U.S. Government works
Printed in the United States of America on acid-free paper
International Standard Book Number-13: 978-1-77188-497-6 (Hardcover)
International Standard Book Number-13: 978-1-315-36572-5 (eBook)

Library and Archives Canada Cataloguing in Publication

Innovative technologies for the treatment of industrial wastewater : a sustainable approach / edited by Shirish H. Sonawane, PhD, Y. Pydi Setty, PhD, T. Bala Narsaiah, PhD, and S. Srinu Naik, PhD.
Includes bibliographical references and index.
Issued in print and electronic formats.
ISBN 978-1-77188-497-6 (hardcover).--ISBN 978-1-315-36572-5 (PDF)
1. Sewage--Purification--Technological innovations. 2. Factory and trade waste--Purification--Technological innovations. 3. Sewage--Purification--Environmental aspects. 4. Factory and trade waste--Purification--Environmental aspects. I. Sonawane, Shirish, editor II. Setty, Y. Pydi, editor III. Narsaiah, T. Bala, editor IV. Naik, S. Srinu, editor
TD745.I56 2017 628.3 C2017-905675-1 C2017-905676-X

Library of Congress Cataloging-in-Publication Data

Names: Sonawane, Shirish H., editor.
Title: Innovative technologies for the treatment of industrial wastewater : a sustainable approach / editors: Shirish H. Sonawane, PhD, Y. Pydi Setty, PhD, T. Bala Naesaiah, PhD, S. Srinu Naik, PhD.
Description: Toronto ; [Waretown] New Jersey : Apple Academic Press, 2018. | Includes bibliographical references and index.
Identifiers: LCCN 2017038090 (print) | LCCN 2017043323 (ebook) | ISBN 9781315365725 (ebook) | ISBN 9781771884976 (hardcover : alk. paper)
Subjects: LCSH: Sewage--Purification. | Sustainable engineering.
Classification: LCC TD745 (ebook) | LCC TD745 .I564 2018 (print) | DDC 628.1/683--dc23
LC record available at https://lccn.loc.gov/2017038090

Apple Academic Press also publishes its books in a variety of electronic formats. Some content that appears in print may not be available in electronic format. For information about Apple Academic Press products, visit our website at **www.appleacademicpress.com** and the CRC Press website at **www.crcpress.com**

CONTENTS

LIST OF CONTRIBUTORS

P. Augusta
Department of Chemical Engineering, National Institute of Technology, Tiruchirappalli, Tamil Nadu, India

A. Arunagiri
Department of Chemical Engineering, National Institute of Technology, Tiruchirappalli, Tamil Nadu, India

S. Nawaz Bahamani
Student M.Tech, Chemical Engineering, SV University, Tirupati, A.P., India

Bhaskar Bethi
Department of Chemical Engineering, National Institute of Technology, Warangal – 506004, Telangana State, India

S. L. Bhagat
Department of Chemical Engineering, Pravara Rural Engineering College, Loni – 413736, Ahmednagar, (MS), India

Bhavana
Student M.Tech, Chemical Engineering, SV University, Tirupati, A.P., India

Ch. Shilpa Chakra
Assistant Professor, Centre for Nano Science and Technology, Institute of Science and Technology, JNTUH, Telangana State, India

Dipak K. Chandre
Department of Chemical Engineering, SVIT, Chincholi, Sinnar, Nashik – 422103, India, Tel.: +919423068528, E-mail: dipakchandre@gmail.com

P. B. N. Lakshmi Devi
Department of Chemical Engineering, National Institute of Technology, Warangal – 506004, Tel: +91-870-2462611; Fax: +91-870-2459547

Ana Sofia Figueiredo
CEFEMA, Instituto Superior Técnico, Universidade de Lisboa, Lisbon, Portugal, Departamental Area of Chemical Engineering, Instituto Superior de Engenharia de Lisboa, Instituto Politécnico de Lisboa, Lisbon, Portugal

R. W. Gaikwad
Department of Chemical Engineering, Pravara Rural Engineering College, Loni – 413736, Ahmednagar, (MS), India, E-mail: rwgaikwad@yahoo.com

M. Clara Gonçalves
Department of Chemical Engineering, Instituto Superior Técnico, Universidade de Lisboa, Lisbon, Portugal, E-mail: clara.goncalves@ist.utl.pt

Joseph Govha
Lecturer, Harare Institute of Technology, Chemical and Process Systems Engineering Department, Harare, Zimbabwe

Chandrakant R. Holkar
Department of Chemical Engineering, Institute of Chemical Technology (ICT) Mumbai – 400019, India, Tel.: +919975274304, E-mail: cr.holkar@gmai.com

Ananda J. Jadhav
Department of Chemical Engineering, Institute of Chemical Technology (ICT) Mumbai – 400019, India, Tel.: +918928530865, E-mail: anandjjadhaviitr@gmail.com

M. Jaipal
Department of Chemical Engineering, JNTUA College of Engineering, Anantapuramu – 515002, Andhra Pradesh, India

P. Kalaichelvi
Department of Chemical Engineering, National Institute of Technology, Tiruchirappalli, Tamil Nadu, India, Tel. +91 431 2503110; Fax: +91 431 2500133;
E-mail: kalai@nitt.edu

P. Rajesh Kumar
Department of Chemical Engineering, NIT Warangal, Telangana, India

Milton Manyangadze
Lecturer, Harare Institute of Technology, Chemical and Process Systems Engineering Department, Harare, Zimbabwe, Mobile: +263 775 691 485, E-mail: manyangadzem@gmail.com

S. Srinu Naik
Department of Chemical Engineering, National Institute of Technology, Warangal, 506004, India

T. Bala Narsaiah
Professor, Department of Chemical Engineering, JNTUA College of Engineering, Anantapur – 515002, A.P., India, Tel.: +91-9440245965; E-mail: balutumma@gmail.com

Zar Zar Nwe
Department of Chemical Engineering, National Institute of Technology, Warangal, 506004, India

Aniruddha B. Pandit
Department of Chemical Engineering, Institute of Chemical Technology (ICT) Mumbai – 400019, India, Tel.: +912233612012, E-mail: ab.pandit@ictmumbai.edu.in

Maria Norberta De Pinho
CEFEMA, Instituto Superior Técnico, Universidade de Lisboa, Lisbon, Portugal, Department of Chemical Engineering, Instituto Superior Técnico, Universidade de Lisboa, Lisbon, Portugal, E-mail: marianpinho@ist.utl.pt

Dipak V. Pinjari
Department of Chemical Engineering, Institute of Chemical Technology (ICT) Mumbai – 400019, India, Tel.: +912233612032, E-mail: dv.pinjari@ictmumbai.edu.in

P. Dinesh Sankar Reddy
Department of Chemical Engineering, JNTUA College of Engineering, Anantapuramu – 515002, Andhra Pradesh, India, E-mail: pdsreddy@gmail.com

Abanti Sahoo
Chemical Engineering Department, National Institute of Technology, Rourkela – 769008, Orissa, India

Y. Pydi Setty
Department of Chemical Engineering, National Institute of Technology, Warangal – 506004, Tel: +91-870-2462611; Fax: +91-870-2459547, E-mail: psetty@nitw.ac.in

K. N. Sheeba
Department of Chemical Engineering, National Institute of Technology, Tiruchirappalli, Tamil Nadu, India

Shirish H. Sonawane
Department of Chemical Engineering, National Institute of Technology, Warangal – 546004, India, Tel: +91-870-2462626, E-mail: shirishsonawane09@gmail.com

P. Akhila Swanthanthra
Assistant Professor, Department of Chemical Engineering, SV University, Tirupati, A.P., India, Mobile: +91 9492549980, E-mail: drakhilap@gmail.com

Antonine Terrien
ETU, University of Nantes, France

A. R. Warade
Department of Chemical Engineering, Pravara Rural Engineering College, Loni – 413736, Ahmednagar, (MS), India

LIST OF ABBREVIATIONS

AC	activated carbon
AD	anaerobic digestion
AFR	air flow rate
$Al_2(SO_4)_3$	aluminum sulphate
ANN	artificial neural network
AOP	advanced oxidation process
ARX	auto-regressive with exogenous
AS	activated sludge
ASFF	aerated submerged fixed-film
ASS	activated sludge system
BAS	biofilm airlift suspension
BG	brilliant green
BI	biodegradation index
BOD	biochemical oxygen demand
CA	cellulose acetate
$CaCl_2$	calcium chloride
CaO	calcium oxide
CCD	central composite design
CMC	critical miscelle concentration
COD	chemical oxygen demand
COD/N	chemical oxygen demand to nitrogen ratio
CPCB	Central Pollution Control Board
CSTR	continuous stirred-tank reactor
CV	crystal violet
DAF	dissolved air flotation
DO	dissolved oxygen
DTA	differential thermal analysis
EF	electro-flotation
EGSB	expanded granular sludge blanket
ELM	emulsion liquid membrane

FA	free ammonia
FBR	fluidized bed reactors
$Fe_2(SO_4)_3$	ferric sulphate
$FeCl_3$	ferric chloride
FIR	finite impulse response
FNA	free nitrous acid
HAOP	hybrid advanced oxidation process
HC	hydrodynamic cavitations
HF	hydrogen fluoride
HRT	hydraulic retention time
HTCO	high temperature catalytic oxidation
IAF	air flotation
IC	inorganic carbon
IC	internal circulation
ICT	Institute of Chemical Technology
IEPA	Iran Environmental Protection Agency
IFBBR	inverse fluidized bed biofilm reactor
IFBR	inverse fluidized bed reactor
LEM	liquid emulsion membrane
MABR	membrane-aerated biofilm reactor
MAT	microwave assisted technique
MB	methylene blue
MBBR	moving bed bioreactor
MBR	membrane bioreactor
MF	microfiltration
MW	microwave
NDIR	non-dispersive infrared
NF	nanofiltration
NH_3	ammonia
NNPLS	neural network partial least squares
NSHS	nano silica hollow spheres
OAX	absorbable organic halides
OLR	organic loading rate
PAC	powdered activated carbon
PAH	polycyclic aromatic hydrocarbons
PES	polyethersulfone

PFS	polyferrichydroxysulphate
PH	peanut husk
PPy	polypyrrole
RAS	return activated sludge
RBC	rotating biological contactor
RO	reverse osmosis
RSM	response surface methodology
SBR	sequencing batch reactor
SEM	scanning electron microscopy
SHARON	single high ammonia removal over nitrite
SRT	sludge residence times
SS	suspended solids
TDS	total dissolved solids
TAN	total ammonia nitrogen
TEM	transmission electron microscope
TGA-DTA	thermal gravimetric differential thermal analyzer
TOC	total organic carbon
TS	total solid
TSS	total suspended solids
TTIP	titanium tetra isopropoxide
UASB	anaerobic up flow sludge blanket
UAT	ultrasound assisted technique
UF	ultrafiltration
US	ultrasound
USEPA	United States Environmental Protection Agency
VOC	volatile organic compounds
VSS	volatile suspended solids
WAS	waste activated sludge
WHO	World Health Organization
XRD	x-ray diffraction

ACKNOWLEDGMENTS

This book, *Innovative Technologies for the Treatment of Industrial Wastewater,* has come from the original work carried out by the various authors and the review of work done in the past. We have received around 20 chapters, and out of those we have selected and collected 12 chapters to publish in book chapter form. We got our inspiration from Professor T. Srinivasa Rao, Director, National Institute of Technology, Warangal, and the Eminent Scientist Dr. B. D. Kulkarni, National Chemical Laboratory, Pune, who have inspired us to take up the task of writing a book.

This book covers the innovative technologies in the area of wastewater treatment. Some of the important and recent areas mainly include nanotechnology, cavitation-based wastewater treatment, liquid emulsion membrane, inverse fluidization technology, biological nitrification in a batch gas–liquid–solid bioreactor, and adsorption. We feel very happy for bringing out such a special type of book.

All the chapters went through a crucial process of peer review and were cleared through the addition of the experts' comments and suggestions. Furthermore, we would also like to acknowledge Mr. Ashish Kumar and the Apple Academic Press team for their prompt and supportive attention to all our queries related to editorial assistance. Special acknowledgment to my research scholar Mr. Bhaskar Bethi, who was looking after uploading the documents and sending reminders to the authors to complete the book in time.

With all humility, we acknowledge the initial strength derived for this book from Professor A. B. Pandit, ICT Mumbai; Professor S. Mishra; and Professor R. D. Kulkarni from UICT Jalgaon along with the unwavering encouragement.

Dr. Shirish Sonawane acknowledges his wife Kiran and son Satwik and Vedika who suffered seclusion and neglect due to involvement with this book for last six months.

—***Shirish Hari Sonawane, PhD***
Y. Pydi Setty, PhD
T. Bala Narsaiah, PhD
S. Srinu Naik, PhD

ABOUT THE EDITORS

Shirish Hari Sonawane, PhD
Associate Professor, Department of Chemical Engineering, National Institute of Technology, Warangal, Telangana State, India

Shirish Sonawane, PhD, is currently an Associate Professor in the Chemical Engineering Department at the National Institute of Technology Warangal, India. His research interests are focused on the synthesis of hybrid nanomaterials, cavitation-based inorganic particle synthesis, sonochemical synthesis of nanolatex, process intensification, and microreactors for nanoparticles production. Dr. Sonawane is the recipient of a fast-track young scientist project in 2007 from the Department of Science and Technology, India. He has industrial experience from reputed chemical industries such as Bayer Polymers (R&D). He also worked in the Chemical Engineering Department and Process Control Laboratory at the University of Dortmund, Germany, in 2002, on the emulsion polymerization process control modeling. He has published more than 50 research papers in reputed journals and seven book chapters and holds six Indian patents applications. He was a recipient of prestigious BOYSCAST Fellowship from the Department of Science and Technology from the government of India in 2009. He is a visiting academic and worked in the Particle Fluid Processing Center, University of Melbourne, Australia, and the Chemical Engineering Department at the Instito Superio Technico in Lisbon, Portugal. He is a member of several professional associations and is a reviewer for several international journals.

Y. Pydi Setty, PhD
Professor, Department of Chemical Engineering, National Institute of Technology, Warangal, Telangana, India

Y. Pydi Setty, PhD, is currently a Professor in the Chemical Engineering Department at the National Institute of Technology Warangal, India. His research interests are focused on hydrodynamics, residence time distribution and drying studies in a bubbling fluidized bed dryer and circulating fluidized bed dryer, biological wastewater treatment using fluidized bed bioreactor and packed-bed bioreactor, waste to energy using microbial fuel cell, bioethanol production, modeling, and simulation and optimization of chemical engineering processes. He has completed an ISRO project for preparation of nano Al particles. He has also completed a consultancy project on pilot scale production of nano additives arganate phase-I. He is also Co-developer for the MHRD-sponsored NMEICT project on "Novel Separation Techniques." He has published several research articles in professional journals, has two patents filed, and has presented many articles at national and international conferences. He has received several awards for his papers at various professional conferences as well as an Outstanding Faculty Award in 2015 by Venus International Foundation, Chennai, India. He has organized two short-term training programs, sponsored by AICTE-ISTE, and he also was the convenor for the International Conference on Chemical and Bioprocess Engineering, India.

T. Bala Narsaiah, PhD
Professor, Department of Chemical Engineering at Jawaharlal Nehru Technological University, College of Engineering, Anantapur, Andhra Pradesh, India

T. Bala Narsaiah, PhD, is an Professor in the Department of Chemical Engineering at Jawaharlal Nehru Technological University, College of Engineering, Anantapur, Andhra Pradesh, India. He has 16 years of teaching along with seven years of research experience. He has published and presented

30 papers in international and national conferences and journals. He also worked as Chairman, Board of Studies, Chemical Engineering at JNTUH. He has received several distinctions, including being nominated by the government of Andhra Pradesh to be a member of the executive council at Jawaharlal Nehru Technological University Hyderabad; a Best Teacher Award from the Swami Ramananda Tirtha Institute of Science and Technology, Nalgonda, in the year 2006, and nominated as an Associate Fellow of AP Akademi of Sciences (APAS). His research interest areas are gas-solid and liquid-solid circulating fluidized bed technology; wastewater treatments; rheological studies of complex fluids; process control, synthesis and application of nanomaterials; and process modeling and simulation and heat transfer.

S. Srinu Naik, PhD
Assistant Professor, Department of Chemical Engineering, University College of Technology, Osmania University, Hyderabad, India

S. Srinu Naik, PhD, is an Assistant Professor in Chemical Engineering Department at the National Institute of Technology in Warangal, Andhra Pradesh, India. He is presently on academic lien and working at the University College of Technology at Osmania University, Hyderabad. He has visited the University of Massachusetts at Amherst, USA, and trained in electrospun cellulose nanofiber fabrication. He has published several articles in international journals and conference proceedings and has conducted an International Conference on Chemical and Bioprocess Engineering, India. He has guided several postgraduate students.

PREFACE

This book, *Innovative Technologies for the Treatment of Industrial Wastewater,* is written based on the potential use of the advanced sustainable methodologies in wastewater treatment and the use of intensified chemical processes in order to make a cleaner environment. There are number of books that give the introductory background of the unit operations used in wastewater treatment processes and fundamental aspects of the water treatment processes. However, to use the new advanced technologies to minimize the waste and recovery of the valuable products from the waste is a challenge in this century. Conventional wastewater treatment processes require large amounts of energy, involve a number of equipment, and entail significant capital and maintenance costs. The disposal of the sludge and decomposition of the sludge is problem in the conventional methods. While sustainable wastewater treatment technologies involve the generation of the energy, economically feasible treatment methods require less equipment cost, require less energy, and are sludge-free processes or generate no solid waste.

Industrial wastewater consists of a variety of discharges based on the type of industry, such as dairy/food industries that consist of more fats and high BOD value with variation in the pH value, while the electroplating industries may consist of more inorganic matter and dissolved solids. The oil extraction industries will have more solvents contained in the effluent. Dyes and textile and speciality chemicals consist of higher organic load with high TDS. Hence, every type of manufacturing industry will have a different method for treatment of the effluent. Hence, the major changes will be brought out by the advanced treatment techniques in the effluent treatment processes. Hence, this book will cover all possible types of the advanced treatment techniques to be used in the effluent treatment processes.

Complete mineralization is one of the important issues in the chemical treatment processes of wastewater. Out of all the treatment processes, one of the important treatment processes is biological wastewater treatment. Hence, in Chapter 1, we have introduced biological nitrification in a batch bioreactor. Ammonia is the most commonly occurring nitrogenous pollutant in municipal sewage and in agricultural and industrial wastewater. Biological nitrification in a batch reactor is discussed in detail.

There are number of reports obtained to formulate a smaller droplet size liquid emulsion membrane and use it for the removal of the heavy metals. However, very few attempts are reported for the removal of dyes using the liquid emulsion membrane. Hence, in Chapter 2, a new method has been reported for formation of stable emulsion using "ultrasound" and stable emulsion, used for removal of the dye molecule.

Removal of the fluoride and arsenic from surface water and to use the water as potable water is a challenging task for providing drinking water in rural places. There are very few viable technologies available to treat fluoride-containing water. Hence, Chapter 3 deals with the use of the adsorption method for the removal of fluoride using aluminum iron mixed oxide.

Similarly Chapter 4 also deals with the removal of the fluoride using low-cost adsorbent. Chapter 4 deals with adsorption kinetics of *Azadirachta indica* (Neem) and *Ficus religiosa* (Pipal) for removal of fluoride from drinking water.

Chapter 5 deals with the investigation of removal of cadmium using low-cost adsorbents like maize corn leaves powder and *Syzygiumcumini* leaves powder (popularly known as jamun). In this chapter the authors report the effect of different parameters such as contact time, pH, initial concentration of cadmium, temperature and adsorbent dosage on the adsorption of cadmium.

Chapter 6 deals with external mass transfer studies on adsorption of methylene blue on *Psidium guava* leaves powder. In this chapter, the author has carried out a detailed survey of the effect of different parameters such as contact time, pH, concentration of the dye onto the mass transfer coefficient and finally adsorption process. Chapter 7 details the components of wastewater and the major industries that are contributing organic wastewater. This chapter also reports on the inverse fluidization process and its significant importance to environmental, biochemical engineering, and oil-water separation. This chapter gives a detailed overview of the inverse fluidization technology along theory and mechanism, hydrodynamics of inverse fluidization and detailed design, and some case studies of steel plant wastewater treatment has also been reported.

Pollution caused by distillery effluent stream is one of the most critical environmental issues worldwide. Distillery industries are one of the most polluting industries worldwide. Chapter 8 deals with use of hydrodynamic cavitations for distillery wastewater treatment. In this chapter, the authors have reported the detailed study of the effect of inlet pressure for mineralization, effect of dilution for mineralization, effect of cavitations for color reduction, effect of cavitations for biodegradability for distillery waste, etc.

The treatment of emulsified oil in wastewater is the need in terms of an economic perspective but also in terms of reducing the pollution of water resources. Hence, in Chapter 9, the author reports on microwave- and ultrasound-assisted surfactant treated adsorbent for the efficient removal of emulsified oil from wastewater. Both (ultrasound and microwave) energy-based intensification techniques are used to prepare the corn husk as adsorbent and used for the removal of the oil.

There are many studies that were reported for the degradation of mixtures of dyes using photocatalysis and other advanced oxidation processes. However, very little literature was found on degradation of the mixture of dyes and degradation using hydrodynamic cavitation. Hence, Chapter 10 deals with a performance evaluation of tertiary dye mixture degradation by hybrid advanced oxidation process, which consists of hydrodynamic cavitation and other AOPs for mixed dye degradation.

Chapter 11 deals with preparation of the mixed matrix membranes and the effect of their chemical properties, and water permeation properties are studied as function of the TiO_2 nanoparticles. The influence of TiO_2 nanoparticles size and shape on the CA asymmetric membrane composites permeability has been reported.

Finally, Chapter 12 deals with application of the nano adsorbents for the removal of the heavy metals from wastewater. This chapter also intends to establish the kinetics of the adsorption of one of the toxic heavy metals, Pb^{2+}, by employing nanoparticles of silica (SiO_2) as an adsorbent in wastewater treatment.

Finally, the authors want to highlight the use of the advanced separation techniques and wastewater treatment methodologies that is the need of this time. The authors feel that this book will definitely the stimulate interest and use of novel technologies in wastewater treatment processes.

—Shirish Hari Sonawane, PhD
Y. Pydi Setty, PhD
T. Bala Narsaiah, PhD
S. Srinu Naik, PhD

CHAPTER 1

REMOVAL OF FLUORIDE IN WATER USING AMORPHOUS NANO METAL OXIDES

JOSEPH GOVHA,[1] T. BALA NARSAIAH,[2] and CH. SHILPA CHAKRA[2]

[1]*Department of Chemical and Process Systems Engineering, School of Engineering and Technology, Harare Institute of Technology, Ganges Road, Belvedere, Harare, Zimbabwe*

[2]*Department of Chemical Engineering, JNTUA College of Engineering, Anantapur – 515002, A.P., India, E-mail: balutumma@gmail.com; Tel.: +91-9440245965*

[3]*Centre for Nano Science and Technology, Institute of Science and Technology, Jawaharlal Nehru Technological University, Hyderabad – 500085, India*

CONTENTS

ABSTRACT

The co-precipitation method was used to synthesis Fe-Al mixed oxide. The morphology of the nanoparticles was characterized by XRD, TGA-DTA, and SEM Analysis. The nanoparticles synthesized where spherical in nature. Fe-Al mixed oxide is amorphous. Fe-Al mixed oxide fluoride removal efficiency and its capacity for fluoride removal was investigated through batch experiments. The effect of different parameters like, contact time, pH of solution, adsorbent dosage, initial fluoride concentration and temperature were investigated to determine the adsorption capacity of Fe-Al mixed oxide. It was observed that 81.4% fluoride removal efficiency was achieved in 3 hours at a pH of 7 and adsorbent dose of 2 g/L. The maximum adsorption capacity of the adsorbent was found to be 8.196 mg/g. The data obtained was found to best fit the Langmuir isotherm and pseudo-second-order kinetic model.

1.1 INTRODUCTION

The effect of fluoride on health is concentration dependent, at low concentration it has been found to reduce dental carries while at high concentration above 2 mg/L of F, it has been found to cause dental and skeletal fluorosis [1, 2]. As per World Health Organization (WHO) recommendations a maximum of 1.5 mg/L is the maximum allowable concentration of safe drinking water [3]. The fluoride is present in water due to both natural and anthropogenic activities. The high fluoride level is attributed mainly to geological formation of the area. The fluoride that contaminates water is released due to geochemical reaction [5]. High fluoride concentrations occur in calcium-poor aquifers and in areas where fluoride-bearing minerals are common [4, 5].

There are over 200 million people in the world who are affected by drinking water with high fluoride, hence there is need to find a solution that will reduce the impacts of excess fluoride concentration in water as in most cases it's the only available source of water.

The endemic nature of fluoride has led to the development of various defluoridation methods which are ion exchange, precipitation-coagulation, membrane techniques (reverse osmosis, nanofiltration, dialysis and electro-dialysis) and adsorption [1, 8]. Adsorption is widely used technique for fluoride removal in water because it is simple and there is a wide range of adsorbents [6, 9, 10]. Bulk adsorbents have low adsorption capacity as compared to Nanoadsorbents, hence the need to prepare and test them for efficiency. The high surface to volume ratio and presence of large number of active sites on surface, results in high adsorption capacity and fast kinetics.

Nano-composites are composed of two or more different materials in the Nano scale range. Composites are produced in an effort to harness best properties of the individual materials and come up with an adsorbent with superior and effective adsorbent properties. In an effort to take advantage of synergistic behavior of materials a number of composite materials have been produced. A number of synthesis routes are available for producing iron aluminum oxide with specific properties. The methods are wet-chemical methods like coprecipitation, hydrothermal, sonochemical reactions, sol-gel method, combustion and, micro emulsion [7, 11, 13, 14]. The co-precipitation method is simple and cost effective, but it is pH sensitive. Sol–gel technique is sophisticated and requires stringent drying conditions and expensive alkoxide precursors.

In the present work co-precipitation at room temperature was employed to produce Iron aluminum oxide, since it is simple and cost effective. The resulting properties, size, structure and morphology, of the synthesized Iron aluminum oxide were characterized using X-ray diffraction (XRD), scanning electron microscopy (SEM) and thermal gravimetric-differential thermal Analyzer (TGA-DTA).

1.2 EXPERIMENTAL SECTION

1.2.1 MATERIALS

$Al_2(SO_4)_3.16H_2O$, $FeSO_4.7H_2O$, deionized water, sodium hydroxide (NaOH), Whatman filter paper, conical flasks, beakers, magnetic stirrer, burette, burette stand, pistol, mortar, drying oven, and crucibles.

1.2.2 SYNTHESIS OF FE-AL OXIDE

Prepare a mixture of 0.1 M $FeSO_4.7H_2O$ and 0.1 M $Al_2(SO_4)_3.16H_2O$, using distilled water. Stir the mixture on a magnetic stirrer at 450 rpm for 30 min. Use a burette to add NaOH drop wise to induce precipitation while stirring. Continue adding NaOH until pH gets to 9.5. Continue stirring for 1 h. Allow the precipitate to age overnight. Filter and wash three times with distilled water. Dry at 80°C for 2 hours and Calcine at 300°C for 3 h.

1.2.3 CHARACTERIZATION

The XRD diffraction (Shimadzu, 700 N), were recorded using Cu- Kά (1.5406Å) monochromatic radiation source, operating voltage and current maintained at 40 kV and 30 mA, respectively in the 2θ range 10–80°. Thermo-gravimetric analysis and differential thermal analysis (TGA-DTA SII EXSTAR 6300R, Japan) was performed using nitrogen at a heating rate of 10°C min^{-1}. Scanning electron microscopy (SEM) Analysis (HITACHI S-3700N) was used to study morphology of the particles. The fluoride concentration in this work was analyzed by UV-Vis spectrophotometer at wavelength of 570 nm.

1.2.4 BATCH ADSORPTION

Sodium fluoride was used for preparation of standard fluoride (1000 mg/L) in deionized water. The required concentration of fluoride solution was obtained by diluting the standard solution to required concentration. The time dependent adsorption reactions were conducted by mixing 300 mL of fluoride solution with 0.6 g of adsorbent at pH 7, 28°C and 500 (±10) rpm. At intervals of 30 min. 10 mL of reaction mixture was withdrawn by a pipette and filtered immediately after withdrawal. The residual fluoride concentration was then Analyzed using spectrophotometer

1.3 RESULTS AND DISCUSSIONS

1.3.1 CHARACTERIZATION OF THE NANOADSORBENTS

1.3.1.1 X-Ray Analysis

The XRD pattern obtained for iron aluminum oxide is shown in Figure 1.1. It has four broad peaks which indicate amorphous nature of the synthesized mixed oxide.

1.3.1.2 SEM Analysis

The morphology of the mixed oxides was investigated by using a SEM and Figure 1.2 shows the images of the iron aluminum oxide oxides. The nanoparticles of the mixed oxide are spherical and they show tendencies to agglomerate. The size particles range obtained for Fe-Al mixed oxide was 73–114 nm.

1.3.1.3 TGA-DT Analysis

Thermo-gravimetric analysis is a vital tool for evaluating the relative stability of the material with reference to temperature change. The TGA-DTA

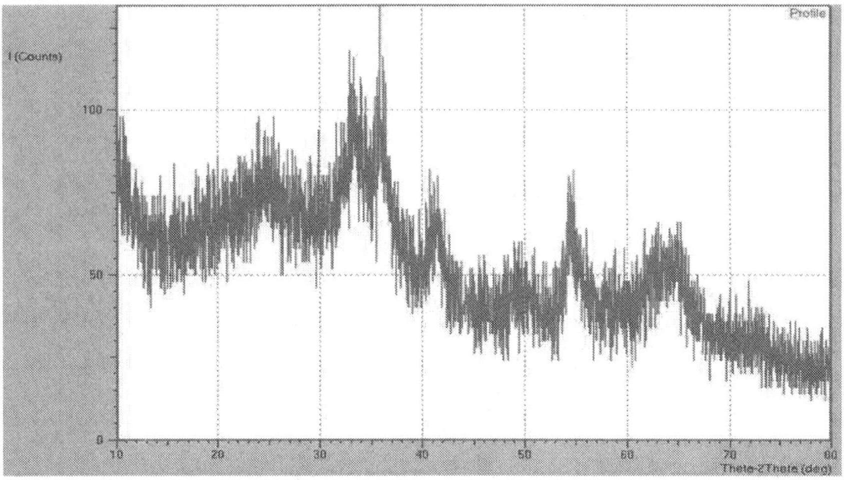

FIGURE 1.1 XRD for synthesized nano-composites.

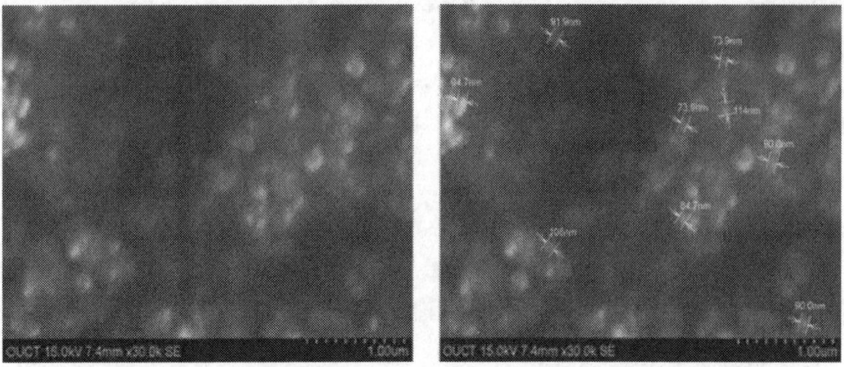

FIGURE 1.2 Scanning electron photomicrographs of Fe-Al mixed oxide.

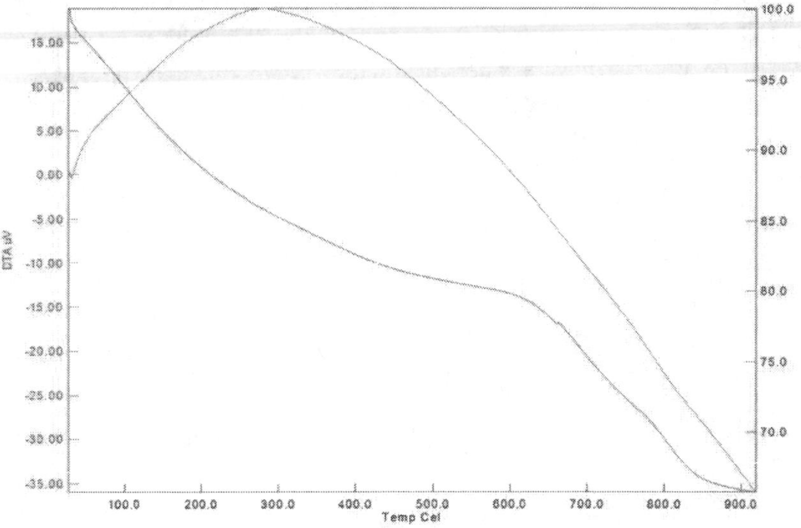

FIGURE 1.3 TGA-DTA Analysis for Fe-Al mixed oxide.

curves in the range 15 to 950°C are shown in Figure 1.3. The weight loss is from Fe-Al mixed oxide with a loss of 34.242%. The loss in weight is due to loss of physically adsorbed water and dehydroxylation. Fe-Al mixed oxide has a high number of hydroxyl groups and hence the high weight loss. Differential thermal analysis (DTA) for Fe-Al, shows an exothermic peak around 290°C. These results indicate the hydrous nature of the mixed oxides and the peaks could be due to hydroxylation [14].

1.3.1.4 Adsorption Kinetic Studies

The efficiency of an adsorbent is measured by the rate of adsorption as one of the most important factors. The small size of nano-composites results in high surface to volume ratio and this results in the adsorbate easily accessing the numerous active adsorption sites as pore diffusion is reduced. Rate of adsorption varies with surface area and pore properties of the sorbent. The results for batch studies conducted for Fe-Al mixed oxide are shown in Figure 1.4. Fe-Al mixed oxide successfully removed 81.4% of fluoride within 3 h. The change in the rate of adsorption is due to the depletion of vacant adsorption sites as the active sites become occupied by the adsorbent (fluoride). The decrease in concentration gradient with time also contributes to the lowering of adsorption rate.

The kinetics of adsorption is better explained by Analysing kinetic data using reaction kinetic models. The kinetics of adsorption was computed using different mathematical models which are pseudo-first-order, pseudo-second-order and intraparticle diffusion. The estimation of kinetic parameters was carried out by linearizing the kinetic models.

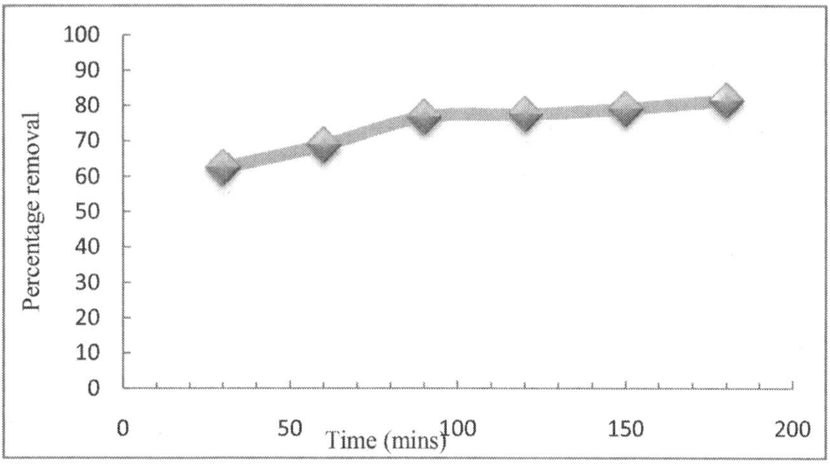

FIGURE 1.4 Percentage removal of fluoride vs Time (initial concentration = 10 mg/L; adsorbent dose = 2 g/L; temperature = 28 ± 2(°C).

The plot of $\log(q_e - q_t)$ *vs.* t should give a straight line from which k_1 and q_e can be calculated from the slope and intercept, respectively. The Lagergren pseudo-first order plot of $\log(q_e - q_t)$ *vs.* t is shown in Figure 1.5.

The plot for pseudo-second order is shown in Figure 1.6 it is a linear plot of t/q_t versus t, and provides the value of k_2 and q_e from the intercept and straight line.

The correlation coefficients and kinetic parameters for the two models (pseudo-first-order, pseudo-second-order) were calculated from plots of $\log(q_e - q_t)$ *vs.* t and t/q_t *vs.* t, respectively. The values of R^2 and the kinetic parameters are tabulated in Table 1.1. The values for R^2 are both above 0.85 for the two models which indicate that the adsorption kinetics can be explained by both models but pseudo-second-order provides the best fit.

1.3.1.5 Adsorption Isotherm Analysis

Adsorption capacity is one of basic parameters that is used for design of adsorption systems. The equilibrium data was analyzed by linear regression of isotherm models of Langmuir and Freundlich. A linear plot of

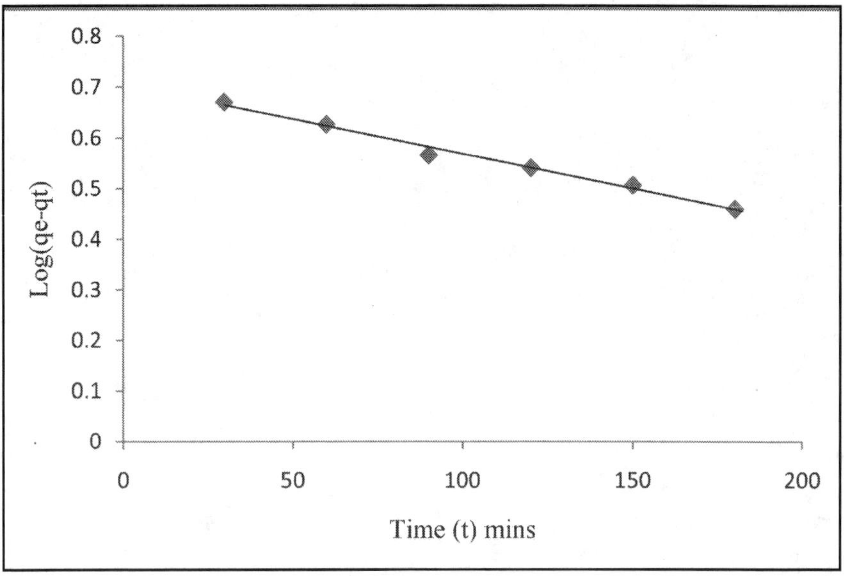

FIGURE 1.5 Adsorption kinetics, $\log(q_e - q_t)$ vs t (initial concentration = 10 mg/L; adsorbent dose = 2 g/L; temperature = 28 ± 2°C.

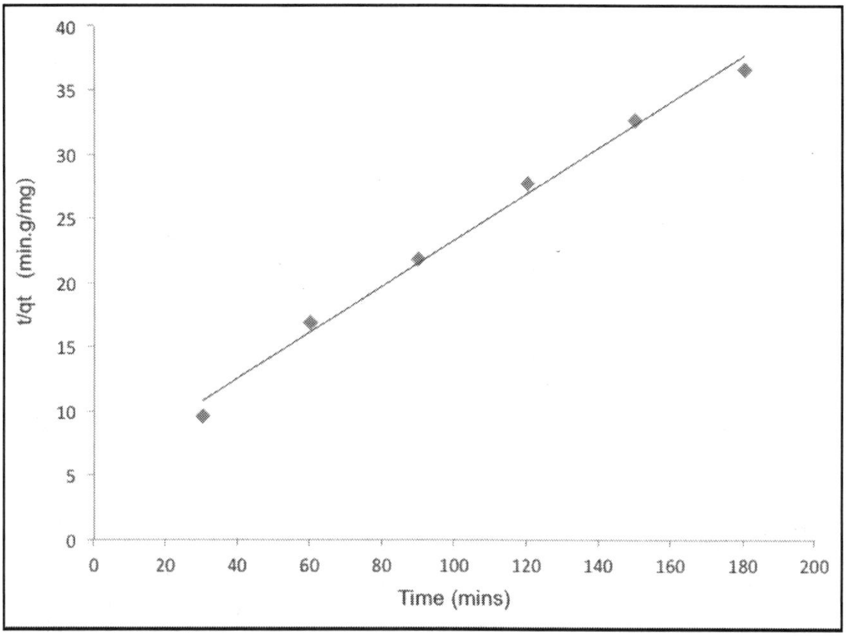

FIGURE 1.6 Adsorption kinetics, t/q_t vs t (initial concentration = 10 mg/L; adsorbent dose = 2 g/L; temperature = 28 ± 2(°C).

TABLE 1.1 Kinetic Parameters for Fluoride Adsorption onto Fe-Al Mixed Oxide

Initial Conc. (mg/L)	q_{exp}	Pseudo-first-order k$_1$ (L/min)	q$_e$ (mg/g)	R^2	Pseudo-second-order k$_2$ (mg/g min$^{(1/2)}$)	q$_e$ (mg/g)	R^2
10.00	4.92	0.0032	5.08	0.988	0.01	5.57	0.992

C_e/q_e versus C_e for Langmuir isotherm is shown in Figure 1.7 and that of log q_e versus log C_e for Freundlich isotherm is shown in Figure 1.8.

The results of fitting Langmuir and Freundlich isotherms are summarized in Table 1.2. The obtained data fit the Langmuir isotherm well as compared to Freundlich. The fitting of data to Langmuir isotherm means that adsorption is homogeneous and it is monolayer. The monolayer adsorption capacity (q_m) obtained for the Fe-Al binary mixed oxide

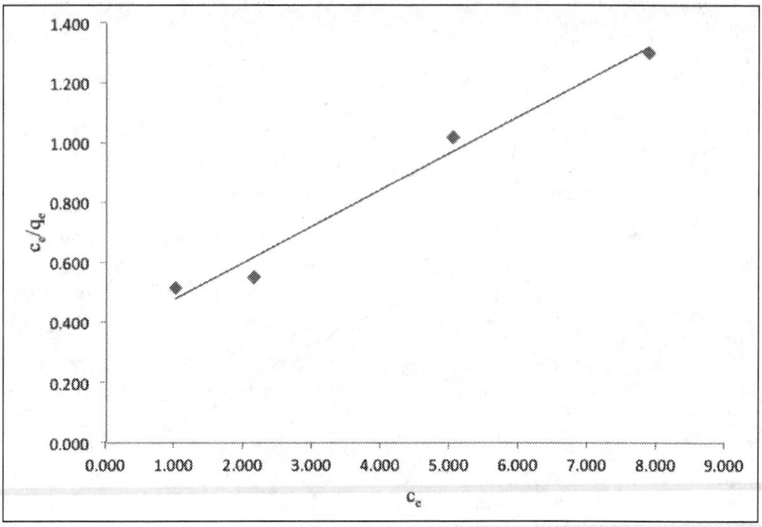

FIGURE 1.7 Linearized Langmuir isotherm, C_e/q_e versus C_e.

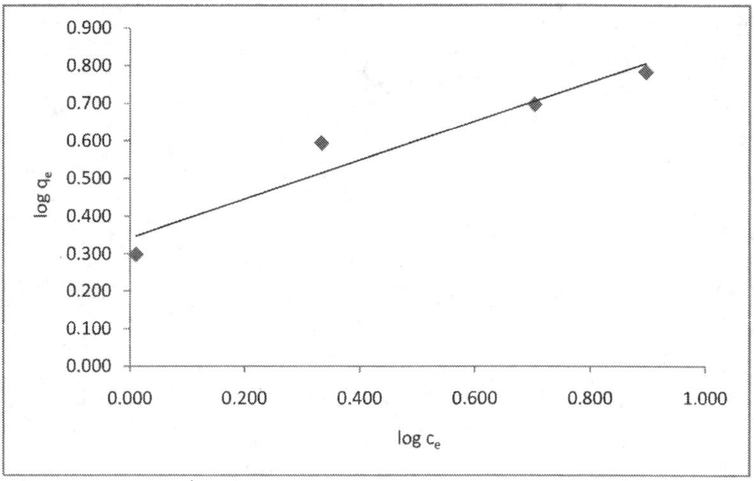

FIGURE 1.8 Linearized Freundlich isotherm, log q_e against log C_e.

TABLE 1.2 Langumuir and Freundlich for Fe-Al Binary Oxide

Isotherm models					
Langmuir Model			***Freundlich Model***		
q_m *(mg/g)*	*b (L/mg)*	r^2	K *(mg$^{1-(1/n)}$L$^{1/n}$g^{-1})*	n	r^2
8.196	2.89	0.98	2.20	1.94	0.93

is 8.196 which is between the adsorption of Iron (III) $q_m = 7.50$ mg/g and aluminum (III) $q_m = 11.42$ mg/g.

The value of R_L is used to indicate the shape of the isotherm: where unfavorable ($R_L > 1$), linear ($R_L = 1$), favorable ($0 < R_L < 1$) or irreversible ($R_L = 0$). The value obtained for this work is 0.03 for initial concentration of 10 mg/L suggesting favorable adsorption of fluoride onto the studied binary mixed oxide for the conditions used in this work.

1.3.1.6 Effect of pH

The adsorption of ions on the oxide surface is affected by initial pH of solution. The effect of pH on the adsorption of fluoride is shown in Figure 1.9. The results indicate that as the solution pH increases the amount of fluoride adsorbed on the mixed oxide also increases. The increase continues until it reaches a maximum at pH 8 where 80.9% of fluoride is removed and then starts to decrease at pH 9. The decrease in fluoride adsorption at low pH is due to formation of hydrogen fluoride (HF) which reduces the available amount of F-ions to be adsorbed. At high pH it is believed that the reduction in pH is due to competing OH- which will result in the surface of adsorbent having a negative charge

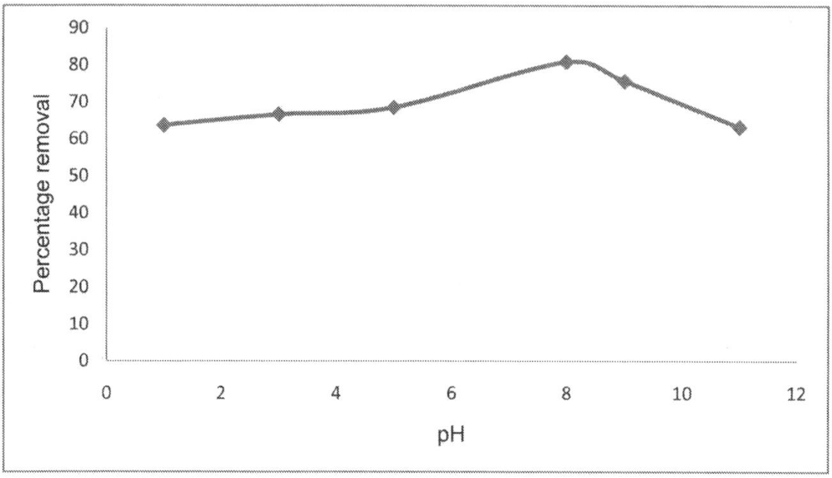

FIGURE 1.9 Effect of pH on fluoride removal; (initial concentration = 10 mg/L; adsorbent dose = 2 g/L; temperature = 28 ± 1°C.

which will result in the F-ions being repelled there by reducing the fluoride removal efficiency. This adsorbent is effective in the normal pH range of drinking water hence it can be applied for defluoridation of drinking water.

1.3.1.7 Effect of Adsorbent

The effect of adsorbent dose on the removal of fluoride was studied under the following conditions: initial concentration = 10 mg/L; pH = 7 ± 0.2; temperature = 28 ± 1°C and contact time = 90 mins. Figure 1.10 shows the findings and it is evident that removal of fluoride increased from 69.27% to 81.18% for 0.05–1 g/50 mL of Fe-Al mixed oxide. The percentage of fluoride removal increases with increase in the amount of adsorbent this is due to the availability of more active sites as amount of adsorbent increases. The rate at which fluoride is removed decrease as the adsorbent increases due to decrease in the available fluoride that can be adsorbed on the active sites.

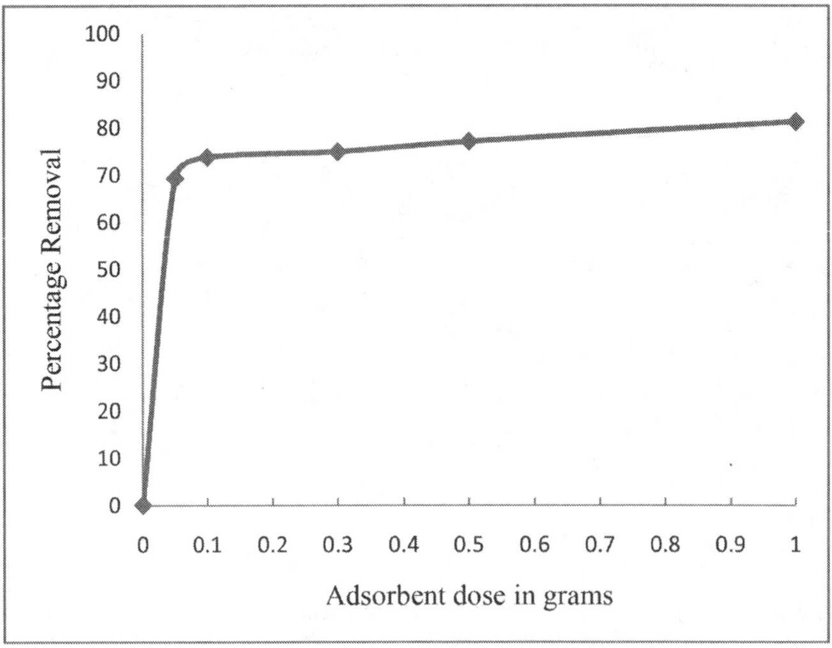

FIGURE 1.10 Plot of percentage removal against adsorbent dose.

1.3.1.8 Effect of Temperature on Adsorption

The influence of temperature on fluoride removal when using Fe-Al mixed oxide adsorbent was studied under the following conditions: initial concentration = 10 mg/L; adsorbent dose = 2 g/L; pH = 7 ± 0.2 and contact time = 90 mins. As the temperature is increased the percentage of fluoride removed also increases as shown in Figure 1.11. This means that fluoride adsorption on Fe-Al mixed oxide is endothermic.

Thermodynamic parameters were evaluated by assuming that ΔH° and to be constant within temperature range of study. The values were calculated from the slope and intercept of a plot of In $(q_{e/}c_e)$ vs 1/T which is shown in Figure 1.12. This relationship is obtained from equation:

$$\ln K = (\Delta S^o/R) - (H^0/RT)$$

where $K = q_e/C_e$.

The thermodynamic parameters are presented in Table 1.3. The values indicate that the adsorption process is endothermic as the value of enthalpy obtained is positive. The positive value of entropy indicates that there is increasing randomness during the sorption process. The negative value of ΔG° at all the temperature shows adsorption process is spontaneous.

1.3.1.9 Fluoride Removal Mechanism

The possible mechanism of fluoride adsorption by Fe-Al mixed oxide can be attributed to attraction of the fluoride ion onto the positively charged adsorbent

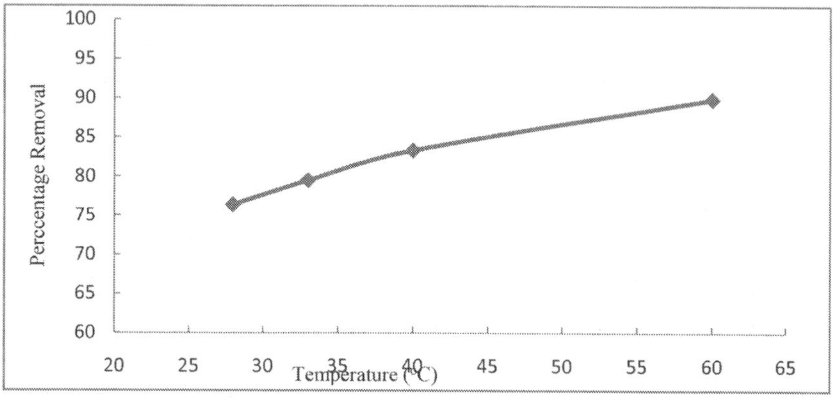

FIGURE 1.11 Plot of percentage removal against temperature.

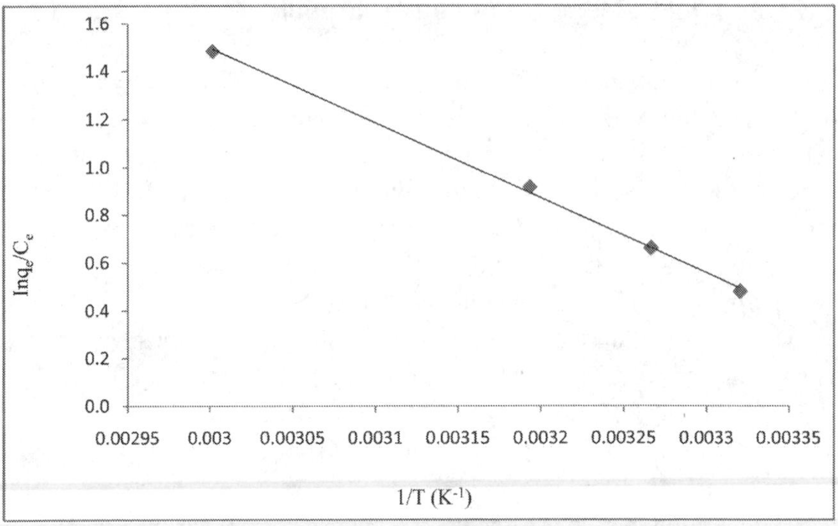

FIGURE 1.12 Plot of In (q_e/C_e) vs $1/T$.

Table 1.3 Summary Thermodynamic Parameters

Temp (°C)	ΔH° (kJ mol^{-1})	ΔΣ° (J mol^{-1} K^{-1})	ΔΓ° (kJ mol^{-1})
28	26.031	90.539	–1.203
33			–1.685
40			–2.386
60			–4.107

surface this mainly occurs in pH that is less than 8. In alkaline pH the adsorbent surface becomes negatively charged and it repels the negatively charged fluoride ions. Thus, fluoride is adsorbed through electrostatic attraction.

1.4 CONCLUSION

The amorphous Fe-Al mixed oxide has a high adsorption capacity. The maximum adsorption capacity of Fe-Al mixed oxide was found to be 8.196 mg/g. The adsorption of fluoride on to the mixed oxide is pH dependent with optimum fluoride removal being achieved at pH 7 for solute concentration of 10 mg/g. The mixed oxide has fast kinetics with an equilibrium time of 90 mins. Isotherm data fit well to the Langmuir isotherm hence adsorption process is monolayer and the adsorbent surface contains homogeneous adsorption sites. Thermodynamic parameters adsorption process is endothermic ($\Delta H^0 = 26.031$ kJ mol^{-1}) and spontaneous ($\Delta G^\circ = -1.203$ kJ mol^{-1} at 28°C). The kinetic data is best described by pseudo-second order model. Fe-Al mixed oxide can be used for fluoride removal in drinking water since its adsorption capacity is comparable to other adsorbents and it operates in the pH range of drinking water. The challenge in using this adsorbent result from it not being able to reduce the fluoride concentration to the WHO standard of 1.5 mg/L in one stage thus two or more adsorption stages are required to achieve a residual concentration of less than 1.5 mg/L. The nanoparticles are too small and they have poor hydraulic properties hence they cannot be used in packed columns or continuous flow processes hence there is need to look for a carrier that can support this adsorbent if it's to be commercially applied.

ACKNOWLEDGMENTS

We would like to thank Head, Centre for Chemical Sciences and Technology and Centre for Nano Science and Technology, IST, JNTUH, Hyderabad for the invaluable support in conducting our experiments. We also thank Osmania University College of Technology, Hyderabad for providing SEM and XRD analysis.

KEYWORDS

- **adsorption**
- **amorphous**
- **co-precipitation**
- **fluoride**
- **isotherms**
- **nanometal oxides**

REFERENCES

1. Amit Bhatnagar, E. K. (2011). Fluoride removal from water by adsorption: A review. *Chemical Engineering Journal*, 811–840.
2. Fakhri, A. (2013). Application of response surface methodology to optimize the process variables for fluoride ion removal using maghemite nanoparticles. *Journal of Saudi Chemical Society,* 18, 340–347
3. Gerrard Eddy Jai Poinern, M. K. J. (2011). Defluoridation behavior of nanostructured hydroxyapatite synthesized through an ultrasonic and microwave combined technique. *Journal of Hazardous Materials*, 185(1), 29–37.
4. Fawell, J., & Bailey K. (2006). Fluoride in Drinking water. IWA Publishing.
5. Krishna Biswas, S. K. (2007). Adsorption of Fluoride from Aqueous solution by synthetic Iron (III)-Aluminium(III) mixed oxide. *Ind. Eng. Chem. Res.*, 5346–5356.
6. Kumar, V. T. (2013). A critical study on efficiency of different materials for fluoride removal from aqueous media. *Chemistry Central Journal*, *7*, 51.
7. Lin Chen, He, B. Y., He, S., Wang, T. J., Su, C. L., Jin, Y. (2012). Fe-Ti oxide nanoadsorbent synthesized by co-precipitation for fluoride removal from drinking water and its adsorption mechanism. *Powder Technology,227*, 3–8.
8. Mohapatra, M., Anand, S., Mishra, B.K., Dion, E. G., Singh, P. (2009). Review of fluoride removal from drinking water. *Journal of Environmental Management,* 91, 67–77.
9. Meenkashi, Maheshwari, R. C. (2006). Fluoride in drinking water and its removal. *Journal of Hazardous Materials B,* 137, 456–463.
10. Patricia Miretzky, A. F. (2011). Fluoride removal from water by chitosan derivatives and composites: A review. *Journal of Fluorine Chemistry,* 231–240.
11. Piao Xu, G. M. (2012). Use of iron oxide nanomaterials in wastewater treatment: A review. *Science of the Total Environment,* 1–10.
12. Swain, S.K., Tanushree, P., Patnaik, P.C., Usha, J., Dey, R.K. (2013). Development of new alginate entrapped Fe(III)–Zr(IV) binary mixed oxide. *Chemical Engineering Journal,215–216*, 763–771.
13. Wang, S.G., Yue, M., Shi, Y. J., Gong, W. X. (2009). Defluoridation performance and mechanism of nano-scale aluminum oxide hydroxide in aqueous solution. *J. Chem. Technol. Biotechnol.,* 1043–1050.

CHAPTER 2

BIOLOGICAL NITRIFICATION IN A BATCH GAS–LIQUID–SOLID BIOREACTOR

P. B. N. LAKSHMI DEVI and Y. PYDI SETTY

Department of Chemical Engineering, National Institute of Technology, Warangal, 506004, India, Tel.: +91-870-2462611, Fax: +91-870-2459547, E-mail: psetty@nitw.ac.in

CONTENTS

2.1 INTRODUCTION

Ammonia is a colorless gas that dissolves readily in water. It is commonly present in most water supplies in trace quantities as ammonium ion due to degradation of nitrogenous organic matter. Biological nitrification process can be used to treat nitrogenous wastewater even for high ammonium concentrations. Biological treatment processes are cheaper compared to chemical, physical or combined chemical-physical treatment methods [7]. Ammonia can also be removed by inorganic zeolites such as chabazite, mordenite and clinoptilolite. Some zeolites are more highly selective for ammonia and are able to remove it throughout the acceptable pH range for potable water (6 to 9) without the danger of dumping.

Ammonium hydroxide is formed when ammonia contacts with moisture present in skin, eyes, oral cavity, respiratory tract, etc. Presence of ammonium hydroxide leads to cellular destruction as it causes disruption of cell membrane lipids. Water released during cell protein breakage further causes damage. Ammonia present in air gives unpleasant odor even at low concentrations. It harms vegetation at high concentrations. Though low concentration of ammonia is necessary for plants, over-fertilization can lead to excessive concentration resulting leaching to water bodies [26]. Ammonia plays a role in transportation and enhanced deposition of acidic pollutants in soil. This leads to acidification of ground water bodies harming plant and animal life.

Ammonia is the most commonly occurring nitrogenous pollutant in wastewater. Municipal sewage, agricultural and industrial wastewater are the main sources of ammoniacal nitrogen. The main polluting industrial sources of ammoniacal nitrogen include oil refineries, coal gasification units, dairy industry, distilleries, fertilizer and pharmaceutical industries, glass production units, cellulose, pulp and paper industries. European Association has given approximately 0.5 mg/L as maximum limit of ammonia for drinking water and guide level is given as 2.5 and 1.0 mg/L for discharge to surface waters and wells, respectively according to Iran Environmental Protection Agency (IEPA) [16, 36]. It causes irritation to eyes, skin and respiratory problems to human beings as well as animals.

Ammonia exists as NH_3 and NH_4^+ in aqueous solution. Unionized ammonia (NH_3) is more toxic than NH_4^+ at low concentrations [29]. The proportion of NH_3 to NH_4^+, depends on temperature, pH value and salinity [44]. Hence, the key limiting water quality parameter in aquaculture systems design and operation is total ammonia nitrogen (TAN) concentration instead of ammonia nitrogen [25]. The classical solution to the problem of ammonia removal is biological nitrification.

Biological nitrification can be accomplished by suspended and attached growth of microorganisms. In a suspended system, microorganisms move freely in the liquid. This provides direct contact between cells and water, whereas a biofilm is formed that is attached to the solid support in an attached system. The latter is called biofilm process. In this case the micro-organisms are immobilized.

Reactions (1) and (2) show the basic reactions that occur in a nitrification process [45, 48]. Here the first reaction shows that ammonia is oxidized to nitrite (NO_{2-}) by autotrophic bacteria like *Nitrosomonas*. Nitrite is unstable and gets oxidized to nitrate (NO_{3-}) by other bacteria like *Nitrobacter*. Nitrate is less toxic than nitrite.

$$NH_4^+ + 1.5\ O_2 \rightarrow 2H^+ + H_2O + NO_2^- \quad (1)$$

$$NO_2^- + 0.5\ O_2 \rightarrow NO_3^- \quad (2)$$

These bacteria drive energy from the above reactions for their survival. They consume oxygen, produce hydrogen ions and also produce nitrite as an intermediate product [8].

2.1.2 *TREATMENT METHODS*

Some of the common treatment methods used for biological nitrification of wastewater are given below:

2.1.2.1 Convectional Activated Sludge

These processes can be classified as complete-mix, plug flow and step feed treatment methods. They have lay out of an aeration basin and secondary clarifier with return and waste sludge pumps. Time required for these processes are longer than BOD removal.

2.1.2.1.1 Advantages

These processes are capable of treating different types of wastewater. They are less complex and easy to operate when compared to other treatment processes. Processes designed for BOD removal can be modified for biological nitrification.

2.1.2.1.2 Disadvantages

These processes require high capital cost. Aeration basin and clarifiers are constructed with concrete. They use equipments like blowers, pumps, etc. These processes are susceptible to bulking sludge from filamentous organisms. This may require an addition of anoxic step.

2.1.2.2 Extended Aeration

These processes are similar to convectional activated sludge treatment processes. They include aeration basins, clarifiers, return activated sludge waste activated sludge. They have higher hydraulic and solids residence times in the process. Twenty-four hours is the hydraulic residence time and the sludge residence time is more than 20 days at design flow rates and organic loadings. These processes include Convectional extended aeration and oxidation ditches. With enough air, nitrification occurs easily in extended aeration processes.

2.1.2.2.1 Advantages

These processes provide high quality effluent for different types of wastewater as the solids residence time and hydraulic residence time are higher. Extended aeration processes are less compel for operation compared to convectional activated sludge treatment processes. They provide high level of biological nitrification with enough oxygen.

2.1.2.2.2 Disadvantages

The main disadvantage is to choose the proper size of the equipment that gives higher solids resistance time and hydraulic residence time. They are costlier because the aeration basins and clarifiers are constructed with concrete and require mechanical equipment.

2.1.2.3 Sequencing Batch Reactor

The difference between earlier two processes and sequencing batch reactors (SBR) is that in sequencing batch reactor both aeration and clarification processes take place in the same reactor basin. The Sequencing batch reactor has four steps in the reactor basin: fill, react/aeration, settle and decant. Wastage usually occurs during the react/aeration step. Hydraulic and solids residence times can be manipulated suitably to enhance nitrification with enough air.

2.1.2.3.1 Advantages

Size of the system is compact as both aeration and clarification occur in a single unit unlike convectional activated sludge process or extended aeration systems. No return activated sludge (RAS) pumps or separate clarification equipments are necessary. The processes can be manipulated based on the time allocated for each step so that desired quality of effluent can be obtained.

2.1.2.3.2 Disadvantages

Ability to manipulate by changing he time allocated for each step may lead to difficulties in operation of the plant. So the operating personnel should be well trained. Otherwise, the effluent may not have desired quality.

2.1.2.4 Fixed Film

There are several types of fixed film treatment processes. They are filter/activated sludge treatment process, rotating biological contactor (RBC) or moving bed bioreactor (MBBR). They can be used for BOD removal as well as for biological nitrification. In the equipments used in these processes, the microorganisms are attached to the solid material instead of being suspended as seen in the equipments used in the earlier cited treatment processes. The solid carrier materials used are usually plastic media. The microorganisms grow on these plastic media in a tower where wastewater is present for treatment. In continuous towers used for treatment the wastewater is sent continuously. The trickling filter is usually followed by activated sludge process. The required equipments are fans, blowers, clarifiers, RAS and waste activated sludge (WAS) pumps. RBCs consist of circular discs, which come into contact with air supplied during rotation. They are followed by clarifiers, RAS and WAS equipment. MBBR technology is the integrated fixed film activated sludge treatment process, where the medium is suspended in the aeration basin and BOD removal and biological nitrification occur in the same basin.

2.1.2.4.1 Advantages

The combined advantages of both trickling filters as well as activated sludge treatment processes can be obtained in these processes. Trickling filters are energy efficient. The activated sludge process prevents sloughing material creating poor effluent quality. These three types of systems reduce the footprint required as compared to convectional activated sludge treatment processes.

2.1.2.4.2 Disadvantages

The disadvantages of fixed film treatment technologies are high solids retention time, high pumping energy and sloughing from RBCs. The MBBR treatment process requires higher levels of dissolved oxygen. MBBR basins require dissolved oxygen levels up to 7 mg/L.

2.1.2.5 Membrane Bioreactor

These processes consist of: (i) anoxic basins, (ii) pre-aeration basins, and (iii) the MBR basins. Prior to the anoxic basin raw wastewater is to be passed through a fine screen. Mixed liquor then flows into the pre-aeration basins and then into MBR basins. The membranes are present in the MBR basins where wastewater is passed through the membranes. Effluent is delivered by permeate pumps into the disinfection process and then to discharge. In this process the secondary clarification equipment is not necessary while it is necessary in other treatment processes. RAS and WAS pumping are required.

2.1.2.5.1 Advantages

This process produces effluent water of high quality. Hence, additional clarification and filtration is not needed, thus reducing the cost of these equipments. This process provides biological nitrification and also total nitrogen removal can be achieved. It can fit into a smaller area.

2.1.2.5.2 Disadvantages

Cost of construction of the equipment is high. Maintenance cost also is high as frequent replacement of membranes is necessary. Hence, the operation and maintenance cost high. It requires more power and also more operator attention.

A detailed review on biological nitrification was given by Lakshmi [18] which covered review on effect of different parameters like substrate concentration, dissolved oxygen, temperature, pH, alkalinity, salinity, turbulence, SND process, heterotrophic nitrification. The authors also reviewed on Immobilization and nitrification kinetics. Apart from this, additional review of earlier work is given in the following section.

2.1.3 REVIEW OF EARLIER WORK

2.1.3.1 Review of Work Done in Various Reactors

Kugaprasatham et al. [17] studied the effect of hydraulic conditions on nitrifying biofilm grown under a low ammonia nitrogen concentration (about 1 g/m^3) in a cylindrical reactor. Terada et al. [43] evaluated the oxygen supply rate by calculating the oxygen transfer rate from the membrane to the bulk for a membrane-aerated biofilm reactor (MABR). Mosquera-Corral et al. [30] studied on ammonia removal from wastewater by varying initial ammonium ion and inorganic carbon concentrations during the operational time in a Single High Ammonia Removal Over Nitrite (SHARON) reactor. Different concentrations of sodium acetate ($CH_3COONa.3H_2O$, 0–1.7 g/L) and sodium chloride (NaCl, 0–30 g/L) were added to study the effects of the presence of organic matter and salts over the process in different operational periods. An equi-molar composition (NH_4^+-N/NO_2-N: 1/1) was obtained at pH 7.16 and HRT of 1.1 days when the ammonia influent concentration was 1000 mg NH_4^+-N/L. Ciudad et al. [10] performed experimental work in a Sequencing Batch Reactor (SBR) for batch mode of operation and also in a continuous operation using a biofilm rotating disk reactor at different pH values ranging from 7.5 to 8.6. Ruiz et al. [37] have performed experimental work on nitrification to study the effect of pH over the range 6.45 and 8.95. Ruiz et al. [37] reported that nitrification suddenly fell for pH values below 6.45 and above 8.95. Ruiz et al. [37] further reported that the fall in nitrification is due to inhibition of both bacteria. Young Zhen et al. [49] studied the feasibility of partial nitrification of wastewater at ambient temperature by aeration control in a sequencing bench bioreactor (SBR) operated for 8 months, with an input of domestic wastewater. Bougard et al. [6] observed the effect of temperature in an inverse turbulent bed reactor for nitrification of high synthetic wastewater (250–2000 mg N-NH_4^+/L) with a granular floating solid at temperatures of 30 and 35°C for 120 days. Zhu and Chen [51] reported that the impact of temperature on a fixed film nitrification rate was less significant than that predicted by the van't Hoff-Arrhenius equation. Sandu et al. [40] in his experimental work used three ratios of bed-height to diameter for fluidized bed nitrification filters. They used 2 x 4 ABS plastic beads with specific gravity of 1.06. The columns of diameters, 12.7, 15.2 and 17.8 cm have been used for the experimental study.

In general, the reactor types applicable for the flow of particulate biofilms in wastewater treatment processes are categorized into anaerobic upflow sludge blanket (UASB), fluidized bed reactors (FBR), expanded granular sludge blanket (EGSB), biofilm airlift suspension (BAS), and internal circulation (IC) reactors [11]. Fluidized bed bioreactors have been carried out for all of the basic secondary and tertiary processes and have shown many advantages over other technologies such as conventional suspended growth including: a large specific surface for attached bio-logical growth of 800–1200 m^2/m^3, high biomass concentrations of 8000–12,000 mg/L for nitrification and 30,000–40,000 mg/L for denitrification [11, 42], long sludge residence times (SRT) and low observed yields which reduce sludge management costs and may result in elimination of secondary clarification requirements.

TABLE 2.1 Summary of Earlier Work on Kinetics and Modeling

Details	Results	Reference
Inlet COD concentration was in the range of 0.057–1.8 g COD.l^{-1} and NH_{3}_N concentration was in the range of 0.0152–0.48 g NH_{3}_N.l^{-1} with hydraulic loading ranging from 0.028 to 0.112 $m^3.m^{-2}.d^{-1}$.	Kinetic parameters were evaluated for combined carbon-nitrogen removal in a four-stage aerated submerged fixed-film (ASFF) bioreactor of 50 l volume using synthetic wastewater at 20°C.	[15]
Kinetic model was derived to describe the kinetics of the first stage of biological nitrification.	Mechanism including the reaction step of formation and decomposition of an intermediate complex between ammonia, bacteria and oxygen was reported.	[33]
Mass transfer and kinetics of nitrification was investigated using free and alginate gel immobilized *Nitrosomonas* cells.	Effective diffusivity of oxygen in alginate gel was estimated to be around 8.7×10^{-9} $m^2.s^{-1}$ for assumed first order kinetics at 25°C and the activation energy for nitrification with free *Nitrosomonas* cells was around 73 kJ.mol^{-1}.	[5]
The effectiveness of the nitrification process was evaluated by nitrification kinetics in a recirculating aquaculture system under different water quality and operating conditions using various types of fixed film biofilters.	It was reported that the impact of parameters such as substrate and dissolved oxygen concentrations, organic matter, temperature, pH, alkalinity, salinity and turbulence level depend upon nitrification kinetics for predicting the performance of a biofilter.	[8]
A simplified mathematical model was developed to describe the biological reactions taking place in the process, in order to determine the optimum conditions for stable nitrite accumulation without nitrate production.	This model represents several features, such as the inhibition of ammonia-oxidizing bacteria by its substrate, NH3 and product HNO_2, and the inhibition of nitriteoxidizing bacteria by free ammonia (NH3). Authors stated that the model correctly describes this nitritation process in sequencing batch reactor.	[34]
A kinetic model was derived for simultaneous inhibition by free ammonia (FA) and free nitrous acid (FNA).	The model predicted that nitrite oxidation should be affected more seriously than ammonium oxidation by the simultaneous inhibition, which would accelerate the accumulation of nitrite in a strong nitrogenous wastewater treatment.	[35]

TABLE 2.1 (Continued)

Details	Results	Reference
A mathematical model was developed for simulation of ammoniacal nitrogen from basic mass balances.	The model was validated well with the experimental results which was supported by the R^2value of 0.79.	[3]
Biological aerated filtration model was evaluated using GPS-X.	The model was validated with the experimental data obtained in a semi-industrial pilot scale submerged biofilter operated at tertiary nitrification stage, which receives the effluent from a medium loaded activated sludge process developed for municipal wastewater.	[46]
Stoichiometric and kinetic parameters were determined for biological processes operating with fixed-bed reactors through mass balances.	The authors reported that in-situ pulse respirometry was a useful tool to characterize fixed-bed reactors as an alternative to other methods, such as mass balance.	[2]
An enhanced Activated Sludge Model No. 1 (ASM1) with two step nitrification/denitrification. Process was developed and implemented in the Benchmark Simulation Model No.1 wastewater treatment plant (WWTP).	The two control models mentioned were implemented for WWTP and concluded that the operational costs were reduced. The authors further reported that effluent quality index can be improved by MPC model.	[14]
Three kinetic models were developed and evaluated for predicting the removal of nitrogen and organics in vertical flow wetlands. These models were developed by combining first-order kinetics, Monod kinetics and multiple Monod kinetics with continuous stirred-tank reactor (CSTR) flow pattern.	The validity of combining Monod and multiple Monod kinetics with CSTR flow pattern has been reported for modeling and design of vertical flow wetland systems	[38]
Five models have been developed for biodegradation of nitrogen and organics removal in vertical and horizontal flow wetland systems.	First order kinetics, Monod and multiple Monod kinetics have been considered in the model development. These kinetics considered was combined with continuous-stirred tank reactor (CSTR) or plug flow pattern for inlet and outlet concentrations of each key pollutants across a single wetland.	[39]

TABLE 2.1 (Continued)

Details	Results	Reference
Using a single reactor high activity ammonia removal over nitrite (SHARON) process, the static behavior was studied, experimental work was performed and validated with a model for nitrification process.	The model includes mass balances of total ammonium, total nitrite, ammonium oxidizers, and nitrite oxidizers. The effect of operating conditions (dilution rate and ammonium feed concentration) and kinetic parameters on the performance of the bioreactor was reported.	[28]
A three phase fluidized bed biofilm reactor has been used to make a study on the removal of $NH4^+$, COD (Chemical Oxygen Demand) and TN (Total Nitrogen) and developed a model for simultaneous nitrification and denitrification (SND).	$NH4^+$, COD, NO_3 and TN profiles along the height of the bioreactor can be predicted using the model.	[41]
A three-dimensional three-phase fluid model has been developed to simulate the hydrodynamics, oxygen mass transfer, carbon oxidation, nitrification and denitrification in an oxidation ditch.	The authors reported that the test results were in good agreement with laboratory experimental data. They also reported the transport and biological process models developed, making reasonable predictions for: (1) the biochemical kinetics of dissolved oxygen, chemical oxygen demand (COD) and nitrogen variation, and (2) the physical kinematics of sludge sedimentation.	[24]

Table 2.1 gives a summary of earlier work on nitrification kinetics and modeling. The review on response surface methodology and artificial neural networks is in the following subsections.

2.1.3.2 Review on RSM and ANN

Choi and Park [9] reported the hybrid ANN technique. The technique showed enhancement of prediction capability and reduction of over fitting problem of neural networks. The results indicate that information from noisy data can be extracted using the hybrid ANN technique. Also nonlinearity of complex wastewater treatment processes can be described using this technique.

Lee et al. [22] developed a neural network partial least squares (NNPLS) combined with finite impulse response (FIR) and auto-regressive with exogenous (ARX) inputs to model a full-scale biological wastewater treatment plant and reported that NNPLS with ARX inputs is capable of modeling the dynamics of the nonlinear wastewater treatment plant. The model also gave improved prediction performance compared to the conventional linear PLS model.

Machon et al. [27] worked on coke wastewater biodegradation containing high concentrations of ammonium, thiocyanate, phenols and other organic compounds in a laboratory-scale activated sludge system (ASS). They developed a feed-forward neural network for estimating the effluent ammonium concentration of the treatment plant. The approach was found to be satisfactory. The methodology involves several tests varying the size of the hidden layer and subset of input variables. The model developed, can be used for simulations under varying conditions of the influent stream and to estimate the effluent ammonium concentration. It is observed from their study that the neural network gives better results than classical mathematical models for biological wastewater treatment.

Nasrin et al. [32] performed experimental work on semi-aerobic leachate biodegradation aerobically using powdered activated carbon (PAC) and also without using PAC. Two 16 L laboratory-scale activated sludge reactors have been used in parallel for the experimental work. The work was done at room temperature and pH value was adjusted to 6.5 ± 0.5. PAC of 75–150 μm size has been used in one of the reactors to observe its effect on semi-aerobic leachate biodegradation. They studied on factorial design considering the parameters like hydraulic retention time (0.92, 1.57 and 2.22 d) and influent COD concentration (750, 1800 and 2850 mg/L). Higher COD and ammoniacal nitrogen removal has been achieved due to PAC addition.

Lee et al. [23] developed a real-time remote monitoring system for wastewater treatment plants (WWTPs) at the optimum operational conditions. Both operation data and measurement values were taken from a novel mobile multi-sensor system, which were transmitted on-line by a telecommunication system. The developed multivariate statistical process controls. Neural network software sensor techniques were applied to supervise local WWTPs. They stated that remote monitoring system makes it possible to monitor the WWTPs and to support the operation of local wastewater systems.

Dogan et al. [12] investigated on BOD estimation and improving its accuracy using artificial neural network (ANN). The available data set was partitioned into a training set and a test set according to network. Nodes of 2, 3, 5, and 10 were tested to obtain optimum number of nodes in the hidden layer. It is reported that the ANN architecture having 8 inputs and 1 hidden layer with 3 nodes gave the best choice. They also compared the data with the results of ANN model and stated that the ANN model gave satisfactory estimates for the BOD prediction.

Bashir et al. [4] investigated on optimization of ammoniacal nitrogen ($NH_{3-}N$) removal from Malaysian semi-aerobic landfill stabilized leachate using synthetic cation exchange resin. The experimental design was carried out based on central composite design (CCD) with response surface methodology (RSM). They reported that the optimum conditions obtained were 24.6 cm^3 resin dosage, 6.00 min contact time, and 147.0 rpm shaking speed, which results in 94.2% removal of $NH_{3-}N$ as obtained from the predicted model, which fitted well with the laboratory results (i.e., 92%).

Akhbari et al. [1] performed experimental work on the removal of carbon, nitrogen and phosphorus from a synthetic wastewater in an integrated rotating biological contactor (RBC) and activated sludge (AS) system. The experiments were conducted based on a central composite design (CCD) and Analyzed using response surface methodology (RSM). To analyze the process, four significant independent variables, hydraulic retention time (HRT), chemical oxygen demand to nitrogen ratio (COD/N), internal recirculation from aerobic to anoxic zone (R) and discs rotating speed (r) and six dependent process and quality parameters as the process responses were studied. The ranges of the variables are taken as HRT (9.7, 18.7 h), COD/N (4, 20), R (1, 4) and r (5, 15 rpm). They reported that the most effective factors on the process performance are HRT and R in terms of COD and TN removal. They also reported that the optimal region was at the range of HRT (18, 18.7 h) and r (5, 15 rpm), at the medians of the COD:N:P (1000:83:30) and R (2.5).

Mousavi et al. [31] performed experimental work on sequential nitrification and denitrification of nitrogen-rich wastewater under different operating conditions in a twin-chamber upflow bio-electrochemical reactor with palm shell granular activated carbon as biofilm carrier and third electrode. The dependent variables such as electric current and hydraulic retention time are optimized with independent variables on NH_4^+-N removal, using response surface methodology. The operating range of electric intensity is 20–50 mA and hydraulic retention time is 6–24 h. They reported that the optimum condition for ammonium oxidation (90%) was obtained with an electric intensity of 32 mA and hydraulic retention time of 19.2 h.

2.1.4 PRESENT WORK

The present work has been carried out to study the effect of airflow rate, pH, temperature, amount of solids (polypropylene beads) and different substrates in a gas–liquid–solid bioreactor. For comparison experiments were also performed in a gas-liquid bioreactor. Also optimization of the parameters has been made using RSM to develop a design equation for the ammonium ion concentration. It is seen from the literature that the reported work on application of RSM to biological nitrification of wastewater is little. The available work in the literature shows that the biological ammonia removal from wastewater using gas–liquid–solid bioreactor is also very little. Hence, the present work using gas–liquid–solid and gas-liquid bioreactors on microbial ammonia removal is significant.

2.2 MATERIALS AND METHODS

2.2.1 CULTURE PREPARATION

Nitrosomonas (NCIM No-5076) and *Nitrobacter* (NCIM No-5062) were obtained from National Chemical Laboratory, Pune, India. The subcultures were prepared according to the procedure given below:

(i) *Nitrosomonas* medium: Solution-I, with the composition of K_2HPO_4- 1.0 g, $MgSO_4.7H_2O$-0.2 g, $CaCl_2.2H_2O$-20.0 mg, $FeSO_4.7H_2O$-50.0 mg, $MnCl_2.4H_2O$ 2.0 mg, $Na_2MoO_4.2H_2O$-1.0 mg per one liter of distilled water was prepared. The pH of the medium was adjusted to 8.5 with 5N NaOH, then 0.5% $CaCO_3$ was added to the medium. Solution-II, with the composition of NH_4Cl-3.0 g and distilled water-100.0 mL was prepared. Both were sterilized at 121°C for 15 min and then to 100.0 mL of solution-I, 5.0 mL solution-II was added aseptically.

(ii) *Nitrobacter* medium: Solution-I, with the composition of $MgSO_4.7H_2O$-0.2 g, K_2HPO_4–1.0 g, $FeSO_4.7H_2O$-50.0 mg, $CaCl_2.2H_2O$-20.0 mg, $MnCl_2.4H_2O$-2.0 mg, $Na_2MoO_4.2H_2O$-1.0 mg per one liter of distilled water was prepared. The medium was adjusted to pH 8.5 with NaOH and sterilized 121°C for 15 min. Solution-II, with the composition of $NaNO_2$–6.0 g and distilled water-100.0 mL was prepared. Both were sterilized at 121°C for 15 min and then to 100.0 mL of solution-I, 5.0 mL of solution-II was added aseptically.

The subcultures were preserved in a refrigerator at a temperature of 4°C.

2.2.2 PREPARATION OF INOCULUM [19, 20]

The liquid broth having the composition $(NH_4)_2CO_3$ 0.303 g, $NaHCO_3$ 25 g, $MgSO_4.7H_2O$ 0.257 g, Na_2HPO_4–1.135 g, KH_2PO_4 1.092 g, $FeCl_3.6H_2O$ 0.035 g, Sucrose 6.174 g per liter [50] was prepared. By varying the amount of ammonium carbonate proportionately, the composition was adjusted to 100 ppm of initial ammonium concentration. HCl and NaOH were used for adjusting the pH value of the solution to 7.5. The broth was sterilized in an autoclave for 15 min at 15 psi pressure and 120°C to kill the undesirable microorganisms. After cooling the broth to room temperature (28°C) the medium of *Nitrosomonas* and *Nitrobacter* were introduced into the broth. Further, the medium was kept for the growth in an incubator for 24 h for the formation of biofilm over polypropylene beads (with 2 mm diameter and density of 880 kg/m^3) for attached growth of microorganisms where the temperature was maintained at 30°C. This is used as a medium in the bioreactor. 50 mL of the product was collected at an interval of every 2 h and was used for the analysis of final ammonium ion and nitrate ion concentration. The analysis for ammonium ion and nitrate ion were carried out using Orion Ion Selective Electrode.

The ammonium and nitrate electrodes were calibrated with the solvents provided my Thermo Scientific Instruments Limited, USA and then analysis of samples was carried out using electrodes. The Dissolved Oxygen (DO) was also measured with DO electrode provided by Thermo Scientific Instruments Limited, USA.

2.2.3 EXPERIMENTAL SETUP AND PROCEDURE [21]

The gas–liquid–solid bioreactor consists of a glass column of 0.5 m height and 100 mm outer diameter with a capacity of 3.4 liters as shown in Figure 2.1. The setup is provided with a glass jacket of 122 mm outer diameter, for temperature control in the reactor at the set point. Provision was made for the supply of air/N_2/O_2 as per requirement. The liquid broth was prepared as mentioned in Section 2.2 and transferred into the reactor. The air was circulated continuously throughout the process into the reactor using an air pump. For uniform distribution of air, a gas sparger was provided at the base of the column. Due to air circulation in the medium

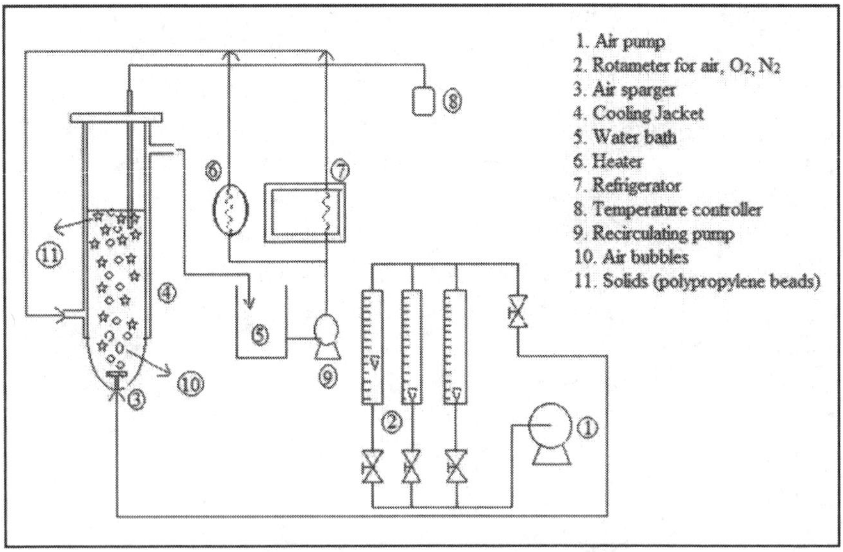

FIGURE 2.1 Schematic diagram of a gas–liquid–solid bioreactor.

the polypropylene beads get suspended throughout the medium in the experiments of gas–liquid–solid bioreactor. In case of experimental work using gas-liquid bioreactor, liquid broth used does not contain any carrier material. Samples were collected at different time intervals. They were analyzed for ammonium ion and nitrate ion concentrations.

2.3 RESULTS AND DISCUSSION

The effect of major parameters such as airflow rate, pH, temperature, amount of solids and different substrates were studied and observed the optimal values for each parameter. Also the effect of various nutrients like concentration of magnesium sulphate, concentration of di-sodium hydrogen phosphate, concentration of potassium di-hydrogen phosphate, concentration of sodium hydrogen carbonate, concentration of sucrose and concentration of ferric chloride on ammonia removal were studied and the optimal values have been observed. The results of the experimental study are given in detail below.

2.3.1 EFFECT OF AIR FLOW RATE

2.3.1.1 Effect of Air Flow Rate on Ammonium Ion Concentration Using Gas-Liquid Bioreactor

To assess the effect of airflow rate on the initial ammonium concentration different airflow rates ranging from 1 to 3 lpm were selected. Samples of fermentation media were prepared as per the composition given in Section 2.2. The conditions for all the samples were maintained at a temperature of 30°C, pH of 7.5 and initial concentration of ammonium ion of 100 ppm in a gas-liquid bioreactor. The samples were filtered with 0.22 micron filter paper in order to measure biomass concentration and then sample was analyzed for ammonium and nitrate ions for all parameters studied in the present work.

The concentration of ammonium and nitrate ions were plotted against fermentation time and the optimal value for maximum removal of initial ammonium ion was observed. Figures 2.2 and 2.3 represent

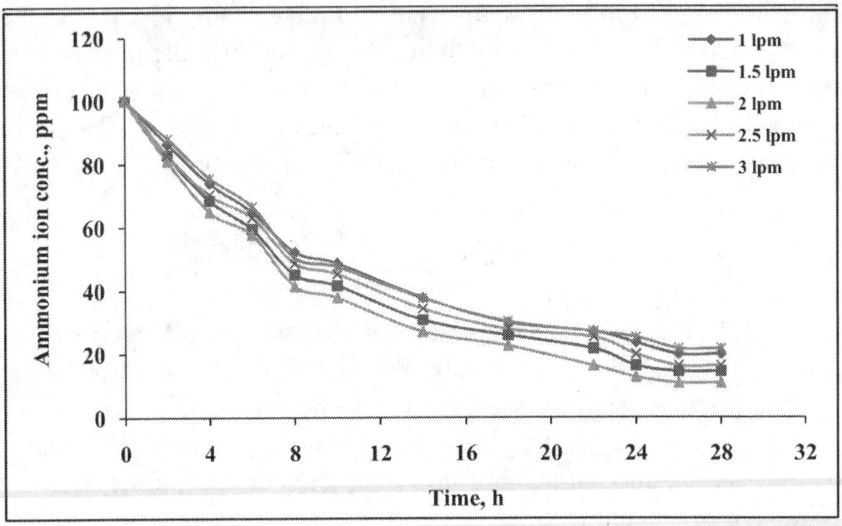

FIGURE 2.2 Effect of airflow rate on ammonium ion concentration using gas-liquid bioreactor.

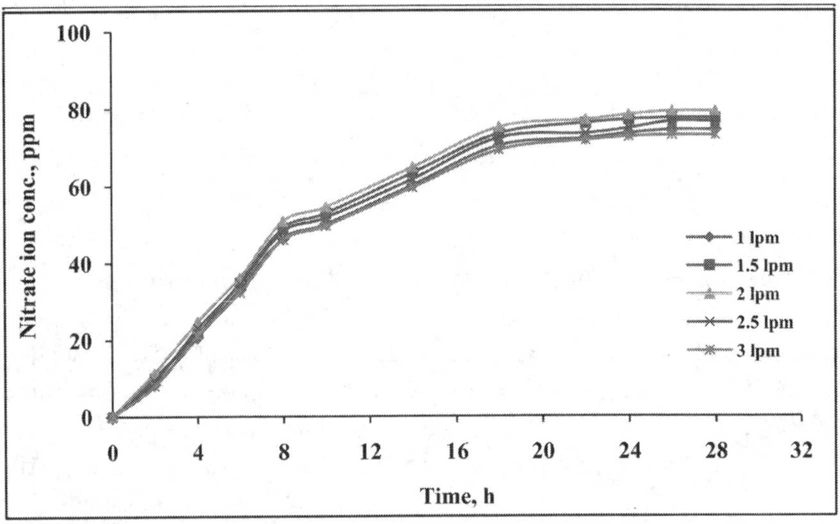

FIGURE 2.3 Effect of airflow rate on nitrate ion concentration using gas

the concentration of ammonium ion and nitrate ion in the experiments using gas-liquid bioreactor. From Figures 2.2 and 2.3, it is observed that the optimal airflow rate among above considered flow rates occurs at 2 lpm in the gas-liquid bioreactor. The ammonium ion concentration has dropped to 10.8 ppm and the nitrate ion concentration was found to be 79.27 ppm within 28 h.

2.3.1.2 Effect of Air Flow Rate on Ammonium Ion Concentration Using Gas–Liquid–Solid Bioreactor

Samples of fermentation media were prepared as per the composition reported in Section 2.2. The conditions for all the samples were maintained at a temperature of 30°C, pH of 7.5 and initial concentration of ammonium ion of 100 ppm in the experimentation using gas–liquid–solid bioreactor. The concentration of ammonium and nitrate ions were plotted against fermentation time and the optimal value for maximum removal of ammonium ion was observed. Figures 2.4 and 2.5 represent the ammonium ion and nitrate ion concentrations in the gas–liquid–solid bioreactor. It is observed from Figures 2.4 and 2.5 that the optimal airflow rate among above considered flow rates occurs at 2 lpm in the gas–liquid–solid bioreactor, with maximum removal of ammonium ion concentration being 89.2% within 22 h. Maximum nitrate ion concentration is found to be 81.04 ppm which is higher than that found in gas-liquid bioreactor indicating less accumulation of nitrite ion in gas–liquid–solid bioreactor. Thus, gas–liquid–solid bioreactor took less time to reach the same removal level

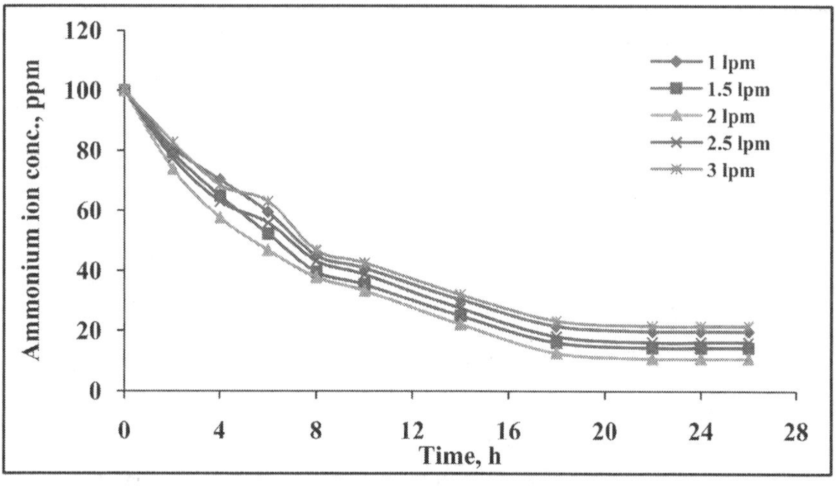

FIGURE 2.4 Effect of airflow rate on ammonium ion concentration using

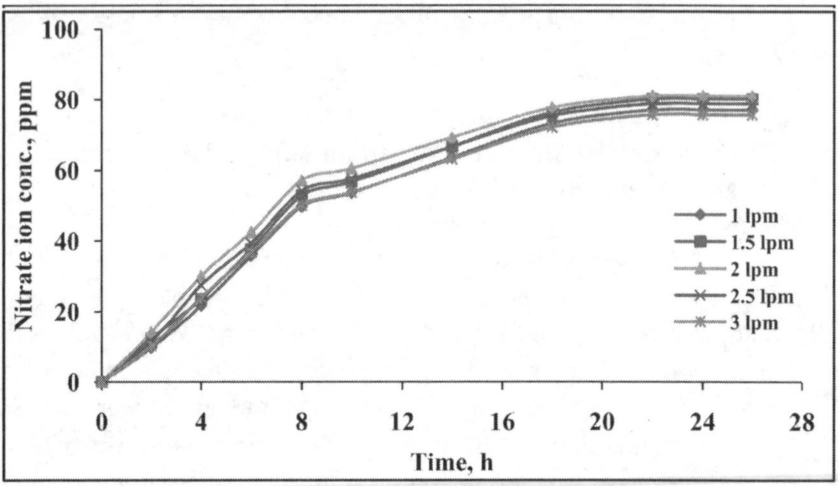

FIGURE 2.5 Effect of airflow rate on nitrate ion concentration using gas–liquid–solid bioreactor.

when compared to gas-liquid bioreactor. Hence, further experiments were carried out in the gas–liquid–solid bioreactor and the optimum airflow rate used was 2 lpm. It is also observed that with increase in airflow rate the removal ammonium ion increased upto 2 lpm and then decreased which may be due to less contact time between the biofilm and air bubbles at higher airflow rates.

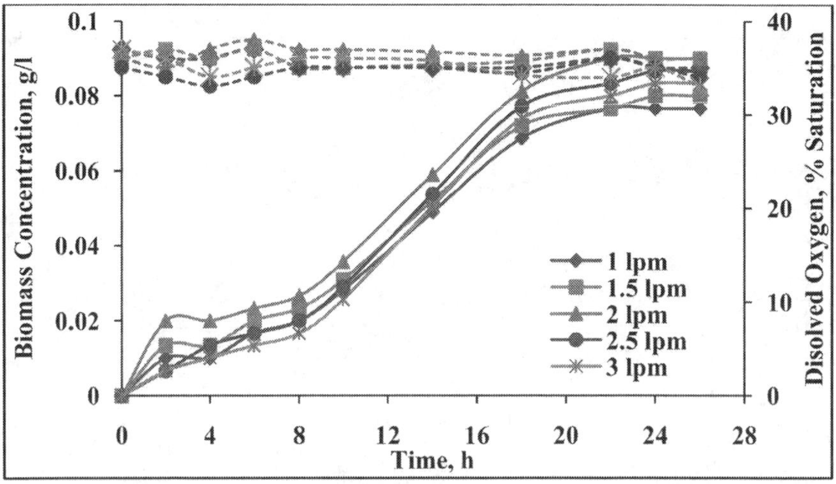

FIGURE 2.6 Effect of airflow rate on biomass and DO concentration using gas–liquid–solid bioreactor.

Figure 2.6 represents the biomass concentration and DO against fermentation time. It is observed that biomass concentration increases with time where as DO remains constant throughout the experiment due to continuous flow of air into the reactor. The straight line represents the concentration of Biomass where as dotted line represents DO concentration.

2.3.2 EFFECT OF PH

To assess the effect of pH on the ammonium ion concentration, experiments with different pH values ranging from 6 to 8.5 were performed. Samples of fermentation media were prepared as per the composition given in Section 2.2. The conditions for all the samples were maintained at a temperature of 30°C, airflow rate 2 lpm and initial concentration of ammonium ion of 100 ppm. The concentration of ammonium ion was plotted against fermentation time and the optimal value for maximum removal of ammonium ion was observed. Figure 2.7 represents the effect of pH on removal of ammonium ion in a gas–liquid–solid bioreactor. It is observed from Figure 2.7 that the optimal pH among above considered values occurs at 7 using gas–liquid–solid bioreactor. Maximum removal of ammonium ion is 92.8%. Hence, pH of 7 was continued for further

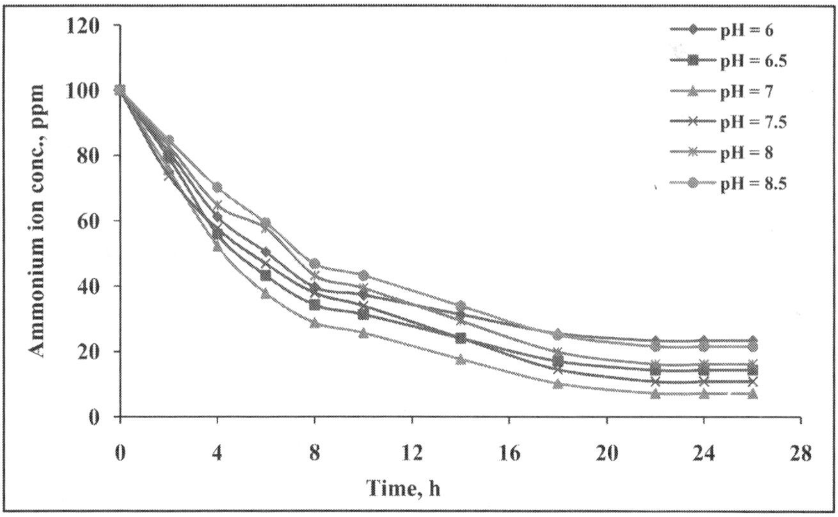

FIGURE 2.7 Effect of pH on ammonium ion concentration using gas–liquid–solid bioreactor

experimental work. In the present work maximum nitrification occurs at pH of 7, with less accumulation of nitrite for the given concentration of ammonium ion of 100 ppm. At other pH values in the experimental study nitrification is found to be less, which may be due to inhibition of bacteria as reported by Ruiz et al. [37].

2.3.3 EFFECT OF TEMPERATURE

To assess the effect of temperature on the ammonium concentration, experiments with temperatures ranging from 24°C to 39°C were performed. Samples of fermentation media were prepared as per the composition given in Section 2.2. The conditions for all the samples were maintained at an airflow rate 2 lpm, pH of 7 and initial concentration of ammonium ion of 100 ppm. The concentration of ammonium ion was plotted against fermentation time and the optimal value for maximum removal of ammonium ion was observed. Figure 2.8 represents the effect of temperature on removal of ammonium ion for gas–liquid–solid bioreactor. It is observed from Figure 2.8 that the optimum temperature among above considered values occurs at 30°C using gas–liquid–solid bioreactor, with maximum removal of ammonium ion being 92.8%. Hence, further experiments were

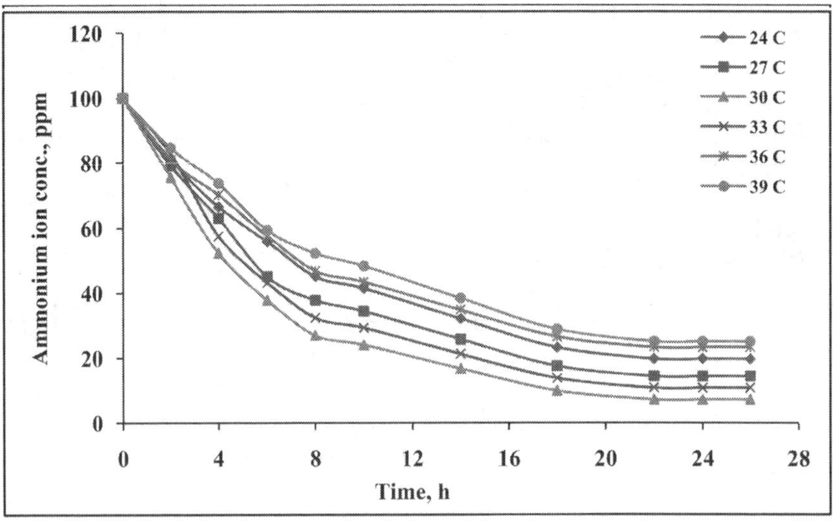

FIGURE 2.8 Effect of temperature on ammonium ion concentration using gas–liquid–solid bioreactor.

carried at optimum temperature of 30°C. It is seen that nitrification is less at temperatures below 30°C and above 30°C which may be due to varying saturation level of DO concentration as temperature varies as reported by Zhu and Chen [51].

2.3.4 EFFECT OF AMOUNT OF POLYPROPYLENE BEADS

To assess the effect of amount of polypropylene beads on the ammonium concentration, experiments with different weights of solids with concentrations of 5 g/L, 10 g/L, 15 g/L, 20 g/L and 25 g/L were performed. Samples of fermentation media were prepared as per the composition given in Section 2.2. The conditions for all the samples were maintained at an airflow rate of 2 lpm, pH of 7, temperature of 30°C and initial concentration of ammonium ion at 100 ppm. The concentration of ammonium ion was plotted against fermentation time and the optimal value for maximum removal of ammonium ion was observed. Figure 2.9 represents the removal of ammonium ion for gas–liquid–solid bioreactor for different amounts of solids. It is observed from Figure 2.9 that the optimal amount of solids among above considered values occurs at 10 g/L using gas–liquid–solid bioreactor, with maximum removal of ammonium ion

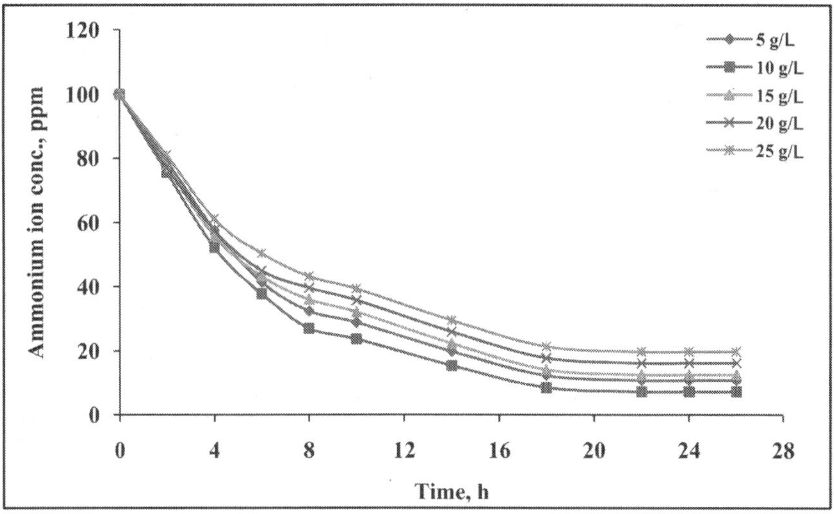

FIGURE 2.9 Effect of amount of solids on ammonium ion concentration using gas–liquid–solid bioreactor.

concentration with composition reduced to 7.2 ppm. Hence, the amount of solids for gas–liquid–solid bioreactor of 10 g/L was used for further experimental work.

2.3.5 EFFECT OF DIFFERENT SUBSTRATES

To assess the effect of different substrates (ammonia sources) on the ammonium concentration, three substrates namely ammonium carbonate, ammonium sulphate and ammonium chloride were studied. Samples of fermentation media were prepared as per the composition mentioned in Section 2.2. The conditions for all the samples were maintained at an airflow rate of 2 lpm, pH of 7, temperature of 30°C and initial concentration of ammonium ion at 100 ppm. The concentration of ammonium ions was plotted against fermentation time and the optimal value for maximum removal of ammonium ion was observed. Figure 2.10 represents the removal of ammonium ion for gas–liquid–solid bioreactor for different substrates. It is observed from Figure 2.10 that the suitable substrate among above considered substrates was ammonium carbonate using gas–

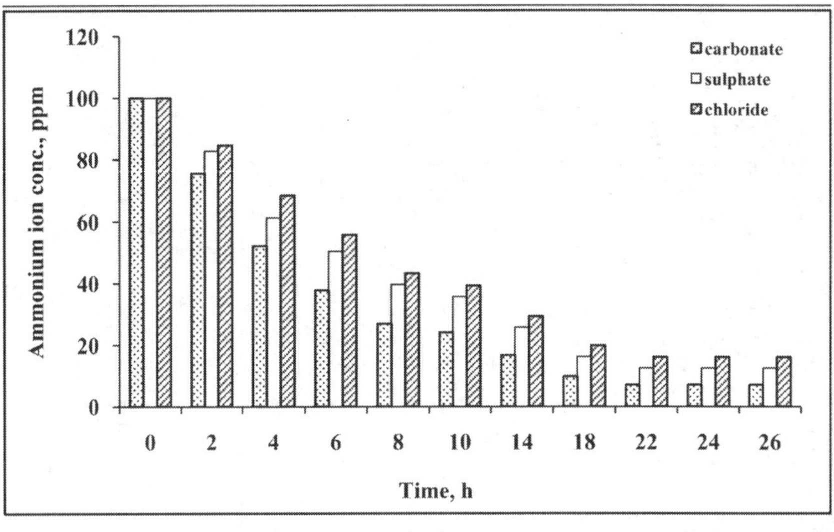

FIGURE 2.10 Effect of substrates on ammonium ion concentration using gas–liquid–solid bioreactor.

liquid–solid bioreactor, with maximum removal of ammonium ion concentration and has dropped to 7.2 ppm. The study shows that ammonium carbonate suits best for the two microorganisms considered in the study.

2.3.6 EFFECT OF CONCENTRATION OF MAGNESIUM SULPHATE ON AMMONIUM ION CONCENTRATION IN GAS–LIQUID–SOLID BIOREACTOR

The effect of magnesium sulphate concentration on ammonium ion concentration experiments at different concentrations varying from 0.1 g/L to 0.5 g/L was studied. Samples of media were prepared as per the composition given in Section 2.2. The conditions for all the samples were maintained at an air flow rate of 2 lpm, pH of 7, temperature of 30°C, amount of solids of 10 g/L and initial concentration of ammonium ion of 100 ppm. The concentration of ammonium ion was plotted against time and the optimal value for maximum removal of initial ammonium ion for different values of concentrations has been observed. Figure

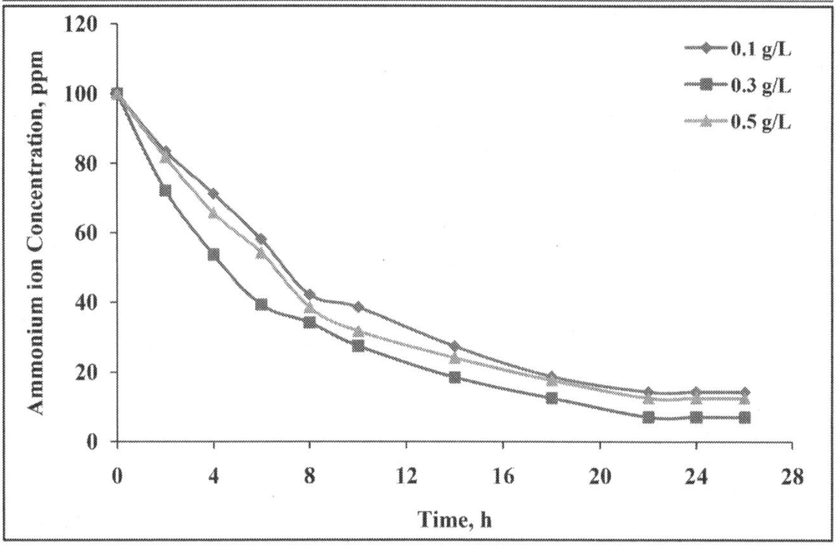

FIGURE 2.11 Effect of concentration of magnesium sulphate on ammonium ion concentration in gas–liquid–solid bioreactor.

2.11 represents the concentration of ammonium ion with time in gas–liquid–solid bioreactor. It is observed from Figure 2.11 that the optimal concentration of magnesium sulphate among above considered values occurs at 0.3 g/L in gas–liquid–solid bioreactor with maximum removal of ammonium ion that is present initially.

2.3.7 EFFECT OF CONCENTRATION OF DI-SODIUM HYDROGEN PHOSPHATE ON AMMONIUM ION CONCENTRATION IN GAS–LIQUID–SOLID BIOREACTOR

To study the effect of di-sodium hydrogen phosphate, different concentrations ranging from 1 g/L to 3 g/L were used. Samples were prepared as per the composition given in Section 2.2. The conditions for all the samples were maintained at an air flow rate of 2 lpm, pH of 7, temperature of 30°C, amount of solids of 10 g/L and initial concentration of ammonium ion of 100 ppm. The concentration of ammonium ion was plotted with time. Figure 2.12 represents the variation

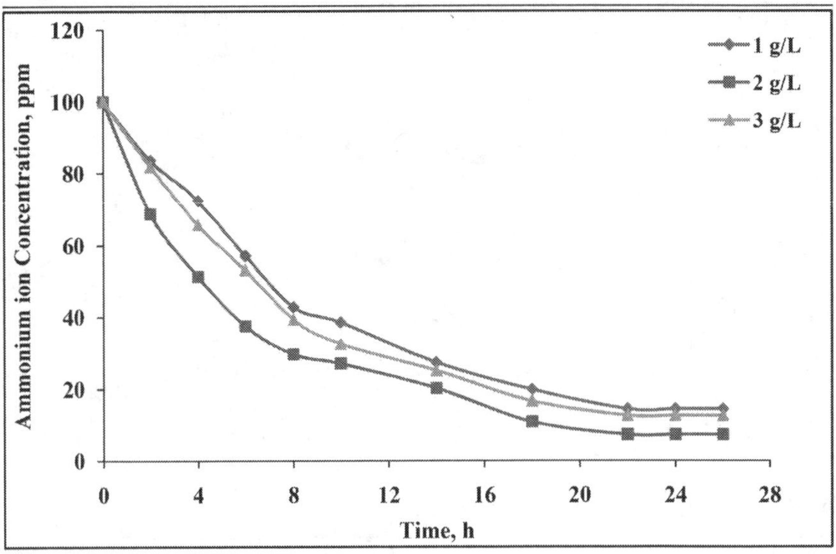

FIGURE 2.12 Effect of concentration of di-sodium hydrogen phosphate on ammonium ion concentration in gas–liquid–solid bioreactor.

of ammonium ion concentration with time in gas–liquid–solid bioreactor. It is observed from Figure 2.12 that the optimal concentration of di-sodium hydrogen phosphate is 2 g/L with maximum removal of initial ammonium ion present.

2.3.8 EFFECT OF CONCENTRATION OF POTASSIUM DI-HYDROGEN PHOSPHATE ON AMMONIUM ION CONCENTRATION IN GAS–LIQUID–SOLID BIOREACTOR

To study the effect of potassium di-hydrogen phosphate, different concentrations ranging from 1 g/L to 5 g/L were used. Samples were prepared as per the composition given in Section 2.2. The conditions for all the samples were maintained at an air flow rate of 2 lpm, pH of 7, temperature of 30°C, amount of solids of 10 g/L and initial concentration of ammonium ion of 100 ppm. The concentration of ammonium ion was plotted with time. Figure 2.13 represents the variation of ammonium ion concentration with time in gas–liquid–solid bioreactor. It is observed from Figure 2.13 that the optimal concentration of

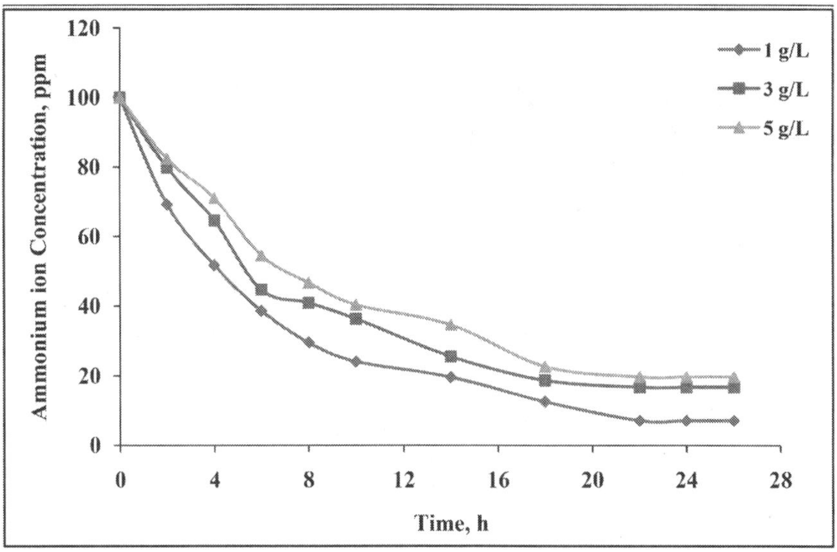

FIGURE 2.13 Effect of concentration of potassium di-hydrogen phosphate on ammonium ion concentration in gas–liquid–solid bioreactor.

potassium di-hydrogen phosphate is 1 g/L with maximum ammonium ion removed.

2.3.9 EFFECT OF CONCENTRATION OF SODIUM HYDROGEN CARBONATE ON AMMONIUM ION CONCENTRATION IN GAS–LIQUID–SOLID BIOREACTOR

To observe the effect of sodium hydrogen carbonate, different concentrations ranging from 20 g/L to 40 g/L were used. Samples were prepared as per the composition given in Section 2.2. The conditions for all the samples were maintained at an air flow rate of 2 lpm, pH of 7, temperature of 30°C, amount of solids of 10 g/L and initial concentration of ammonium ion of 100 ppm. The concentration of ammonium ion was plotted against time. Figure 2.14 represents the variation of concentration of ammonium ion with time in gas–liquid–solid bioreactor. It is observed from Figure 2.14 that the optimal concentration of

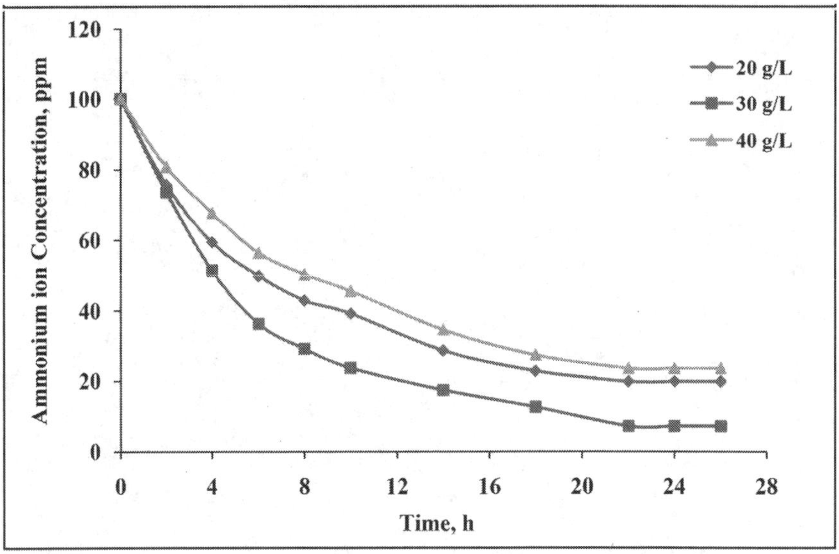

FIGURE 2.14 Effect of concentration of sodium hydrogen carbonate on ammonium ion concentration in gas–liquid–solid bioreactor.

sodium hydrogen carbonate is 30 g/L with maximum initial ammonium ion removed.

2.3.10 EFFECT OF CONCENTRATION OF SUCROSE ON AMMONIUM ION CONCENTRATION IN GAS–LIQUID–SOLID BIOREACTOR

To observe the effect of sucrose, different concentrations ranging from 5 g/L to 9 g/L were used. Samples were prepared as per the composition given in Section 2.2. The conditions for all the samples were maintained at an air flow rate of 2 lpm, pH of 7, temperature of 30°C, amount of solids of 10 g/L and initial concentration of ammonium ion of 100 ppm. The concentration of ammonium ion was plotted against time. Figure 2.15 represents the variation of concentration of ammonium ion with time in

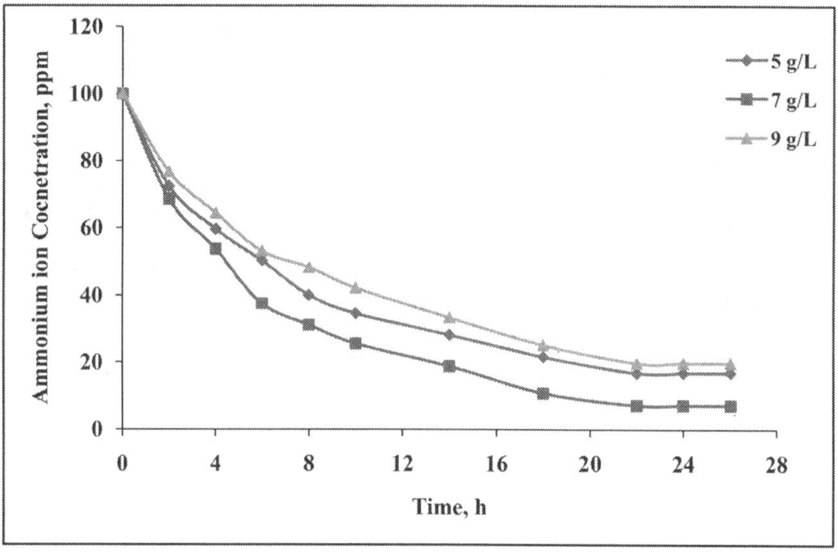

FIGURE 2.15 Effect of concentration of sucrose on ammonium ion concentration in gas–liquid–solid bioreactor.

gas–liquid–solid bioreactor. It is observed from Figure 2.15 that the optimal concentration of sucrose is 7 g/L with maximum initial ammonium ion removed.

2.3.11 EFFECT OF CONCENTRATION OF FERRIC CHLORIDE ON AMMONIUM ION CONCENTRATION IN GAS–LIQUID–SOLID BIOREACTOR

To observe the effect of ferric chloride, different concentrations ranging from 0.03 g/L to 0.05 g/L were used. Samples were prepared as per the composition given in Section 2.2 The conditions for all the samples were maintained at an air flow rate of 2 lpm, pH of 7, temperature of 30°C, amount of solids of 10 g/L and initial concentration of ammonium ion of 100 ppm. The concentration of ammonium ion was plotted with time. Figure 2.16 represents the concentration of ammonium ion with time in gas–liquid–solid bioreactor. It is observed from Figure 2.16 that the optimal concentration of ferric chloride is 0.04 g/L with maximum initial ammonium ion removed.

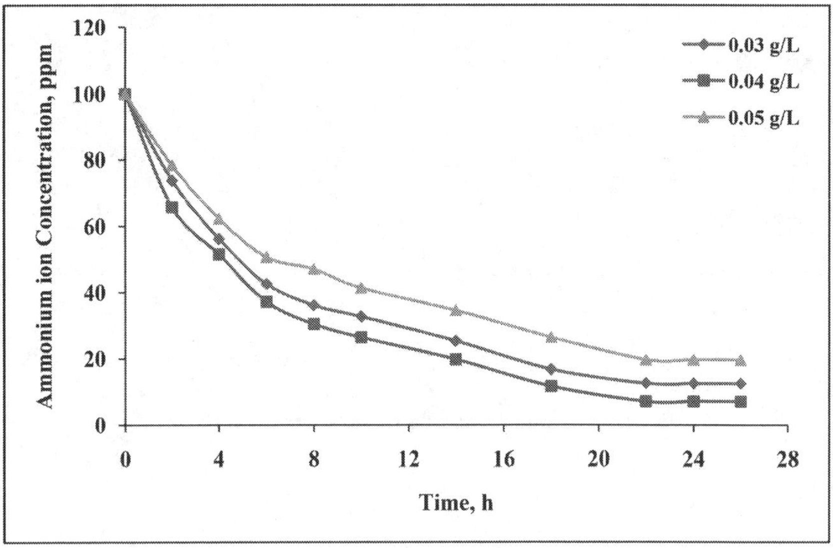

FIGURE 2.16 Effect of concentration of ferric chloride on ammonium ion concentration in gas–liquid–solid bioreactor.

2.3.12 ESTIMATION OF KINETIC PARAMETERS USING BRIGGS–HALDANE APPROACH

From the experimental data of all optimal parameters, a plot was drawn for (1/μ) vs. (1/C_s). The kinetic parameters were calculated using equation

$$1/\mu = K_M/(\mu_{\max} C_S) + 1/\mu_{\max}$$

From Figure 2.17, it is observed that μ_{max} = 7.812 h^{-1} and K_M = 11.062 mg/L.

2.3.13 RESPONSE SURFACE METHODOLOGY

The Response Surface Methodology is an economical applied mathematical technique for optimization of multiple parameters with minimum range of experiments and it uses a collection of designed experiments to get best response [13, 47]. The response surface methodology was developed using design expert software and the parameters were optimized using central composite method. The parameters like air flow rate, pH, temperature and amount of solids were optimized using response surface methodology. The ranges of parameters were given in Table 2.2.

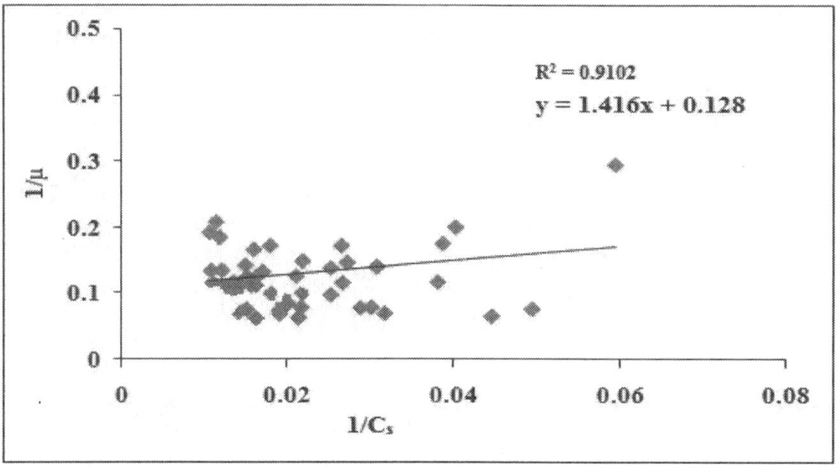

FIGURE 2.17 Estimation of kinetic parameters.

For the four parameters the central composite method designed 30 number of runs for best combinations. Experiments were performed according to the design and final ammonium ion concentration was analyzed. Table 2.3 gives design of experiments for the four parameters. The quadratic model was derived from the ANOVA results (Table 2.4). From Table 2.4, it was observed that the model is significant as seen from the value of R^2 which is 0.9328. Values of "p-value" less than 0.0500 indicate that model terms are significant. Values greater than 0.1000 indicate that the model terms are not significant.

The response surface plots for the effect of different parameters on ammonium ion concentration are represented in Figure 2.18:

(a) airflow rate and pH with actual factors of temperature, 30°C and amount of solids, 10 g/L;
(b) airflow rate and temperature with actual factors of pH 7 and amount of solids, 10 g/L;
(c) airflow rate and amount of solids with actual factors of temperature, 30°C and pH;
(d) pH and temperature with actual factors of airflow rate, 2 lpm and amount of solids, 10 g/L;
(e) pH and amount of solids with actual factors of airflow rate, 2 lpm and temperature, 30°C;
(f) temperature and amount of solids with actual factors of airflow rate, 2 lpm and pH 7.

The final concentration of ammonium ion (ppm) can be calculated with the design equation of central composite design as:

$$\text{Ammonium concentration} = 8.319 + 1.24{*}A + 0.497{*}B + 0.063{*}C + 0.594{*}D - 1.395{*}A{*}B + 0.655{*}A{*}C - 0.181{*}A{*}D - 0.47{*}B{*}C - 0.706{*}B{*}D - 0.956{*}C{*}D + 5.5{*}A^2 + 1.184{*}B^2 + 3.434{*}C^2 + 0.914{*}D^2$$

where A = airflow rate, B = pH, C = temperature, and D = amount of solids.

The optimal values for the parameters observed from the response surface method are air flow rate of 1.89 lpm, pH of 6.76, temperature of 31.84°C and amount of solids (beads), 10.3 g/L and the ammonium ion concentration is seen reduced to 7.98 ppm. Whereas from experimental

TABLE 2.2 Ranges of the Parameters for Central Composite Design

S. No	Parameter	Low value	High value
1	Air flow rate	1 lpm	3 lpm
2	pH	6	8
3	Temperature	27°C	33°C
4	Amount of solids	5 g/L	15 g/L

TABLE 2.3 Design of Experiments for Four Parameters

S. No	Air flow rate (L/min)	pH	Temperature (°C)	Amount of solids (g/L)	Final ammonium ion concentration (ppm)
1	1	6	27	15	19.8
2	2	7	30	10	7.2
3	2	7	30	10	7.2
4	3	6	33	5	23.6
5	1	6	33	15	14.4
6	3	6	27	5	16.8
7	3	6	27	15	21.2
8	2	7	30	10	7.2
9	3	8	33	5	19.8
10	1	7	30	10	14.4
11	1	8	27	5	19.8
12	3	8	27	15	21.6
13	1	6	33	5	14.4
14	2	7	30	10	7.2
15	2	5	30	10	12.6
16	3	8	27	5	19.8
17	2	7	30	5	10.8
18	2	7	24	10	21.6
19	1	6	27	10	16.8
20	2	7	36	10	23.4
21	1	8	33	15	19.8
22	3	7	30	10	16.8
23	1	8	27	15	19.8
24	1	8	33	5	19.8
25	2	7	30	15	12.6
26	2	7	30	10	7.2

TABLE 2.3 (Continued)

S. No	Air flow rate (L/min)	pH	Temperature (°C)	Amount of solids (g/L)	Final ammonium ion concentration (ppm)
27	2	9	30	10	14.4
28	3	8	33	15	16.8
29	2	7	30	10	7.2
30	3	6	33	15	23.4

TABLE 2.4 ANOVA for Response Surface Quadratic Model

Source	Sum of Squares	Df	Mean Square	F Value	p-value	Prob > F
Model	825.9922	14	58.99944	14.87318	< 0.0001	Significant
A-Air flow rate	26.59352	1	26.59352	6.703968	0.0205	
B-pH	5.746297	1	5.746297	1.448585	0.2474	
C-temperature	0.093725	1	0.093725	0.023627	0.8799	
D-amount of solids	5.655188	1	5.655188	1.425618	0.2510	
AB	29.7712	1	29.7712	7.505029	0.0152	
AC	6.558496	1	6.558496	1.653333	0.2180	
AD	0.459981	1	0.459981	0.115957	0.7382	
BC	3.381028	1	3.381028	0.852324	0.3705	
BD	6.999505	1	6.999505	1.764507	0.2039	
CD	12.83454	1	12.83454	3.235463	0.0922	
A^2	117.6738	1	117.6738	29.66443	< 0.0001	
B^2	39.1985	1	39.1985	9.881559	0.0067	
C^2	329.7305	1	329.7305	83.12185	< 0.0001	
D^2	3.130397	1	3.130397	0.789143	0.3884	
Residual	59.5025	15	3.966834			
Lack of Fit	59.5025	10	5.95025			
Pure Error	0	5	0			
Cor Total	885.4947	29				

R-Squared = 0.932803;

Adj R-Squared = 0.870086;

Pred R-Squared = 0.624243.

work performed using the experimental optimal values the concentration is found reduced to 7.2 ppm. When the optimal values like airflow rate of 2 lpm, pH of 7, temperature of 30°C and amount of solids of 10 g/L are substituted in the above design equation the concentration is seen reduced

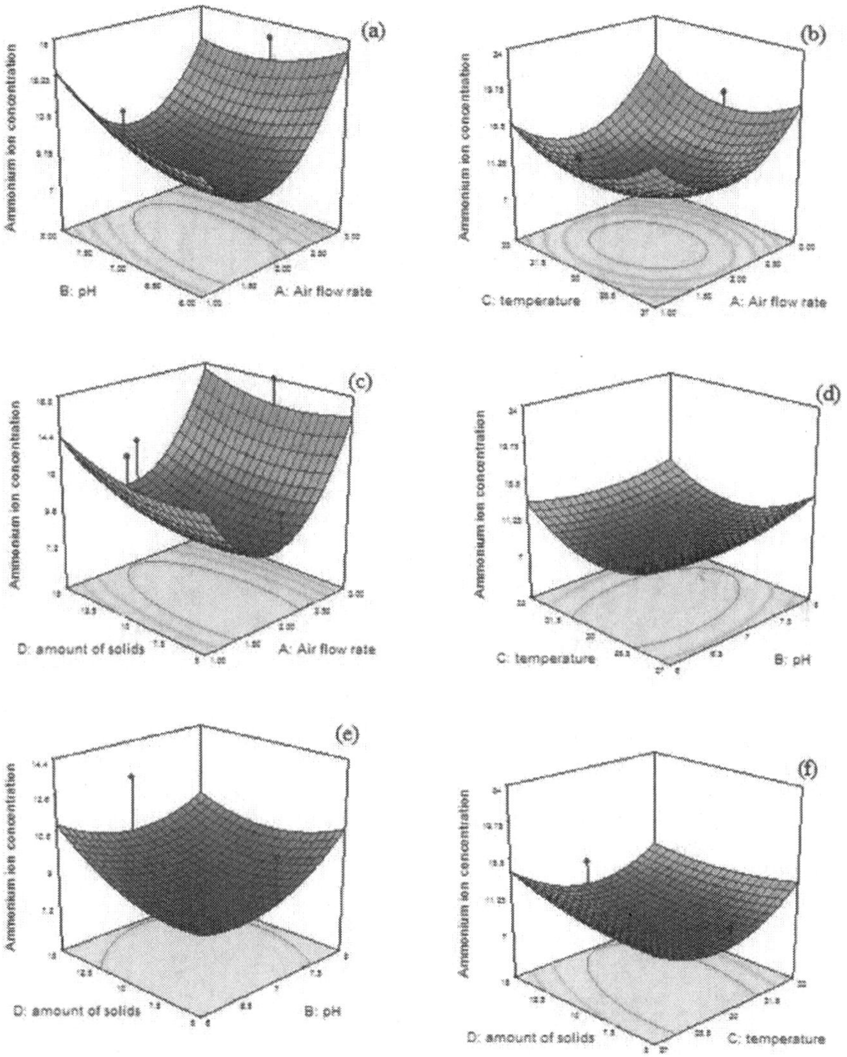

FIGURE 2.18 Response surface plots of ammonium ion concentration with (a) airflow rate and pH; (b) airflow rate and temperature; (c) airflow rate and amount of solids; (d) pH and temperature; (e) pH and amount of solids; and (f) temperature and amount of solids.

to 7.36 ppm. Figure 2.19 represents the predicted and actual Ammonium ion concentration.

2.3.15 Artificial Neural Network

Artificial neural network (ANN) is a computational model that is inspired by the structure and/or functional aspects of biological neural networks. The purpose of a neural network is to learn to recognize patterns in the data.

A typical neural network model consists of three independent layers: input, hidden, and output layers. Each layer is comprised of several processing neurons. While input and output layers perform as boundary between the neural network and the environment, the hidden layer

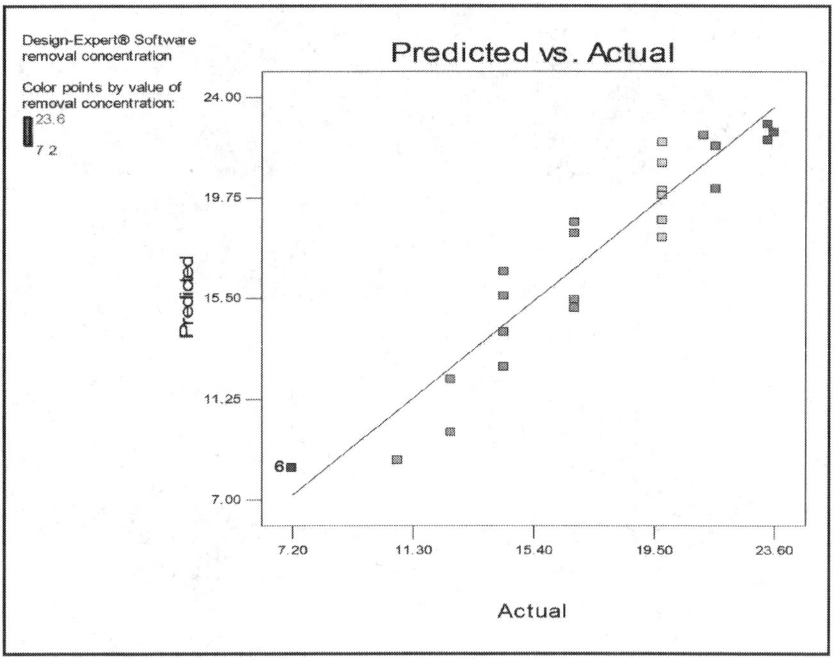

FIGURE 2.19 Predicted vs. actual ammonium ion concentration.

and input/output layers may fully interconnect with each other through the information flow channels between the neurons. The neurons in the input layer store the scaled input values. Each processing neuron in the hidden layer is assumed to have a small amount of local memory and may convert inputs to the neuron outputs. The pattern of hidden layer to be applied in the modeling analysis can be either multiple layers or a single layer.

The experimental data is modeled using artificial neural networks to have more accurate prediction on ammonium ion removal and 50 combinations of experimental data were taken, in which 30 were used for training and remaining 20 were used for testing. In the present study four neurons were considered in input layer (air flow rate (AFR), pH, temperature and amount of solids). Neurons in hidden layer were varied from 1–5, 1–10, 1–15 and one neuron in output layer as response for ammonium ion removal was considered. Sigmoid function is used for the present work and back-propagation algorithm was applied for training the network and for updating the desired weights.

Correlation coefficient (R) can be used to determine how well the network output fits the desired output. Higher value of R is desirable for an effective ANN model. Correlation coefficient values measure the correlation between outputs and targets. An 'R' value of 1 means a close relationship, 0 a random relationship. So a higher value of R is desirable to obtain an effective ANN model. From Figure 2.20, it is observed that R is close to 1 and therefore it can be inferred that the ANN model 4–10–1 is effective for the experimental values. Figure 2.21 shows the optimal neural network topology.

The error is calculated for both the models with experimental data and an error plot was drawn for each run as shown in Figure 2.22. It is observed from the Figure 2.22 that the error has been reduced in ANN when compared to RSM. Table 2.5 gives the experimental and predicted values of Ammonium ion concentration using RSM and ANN.

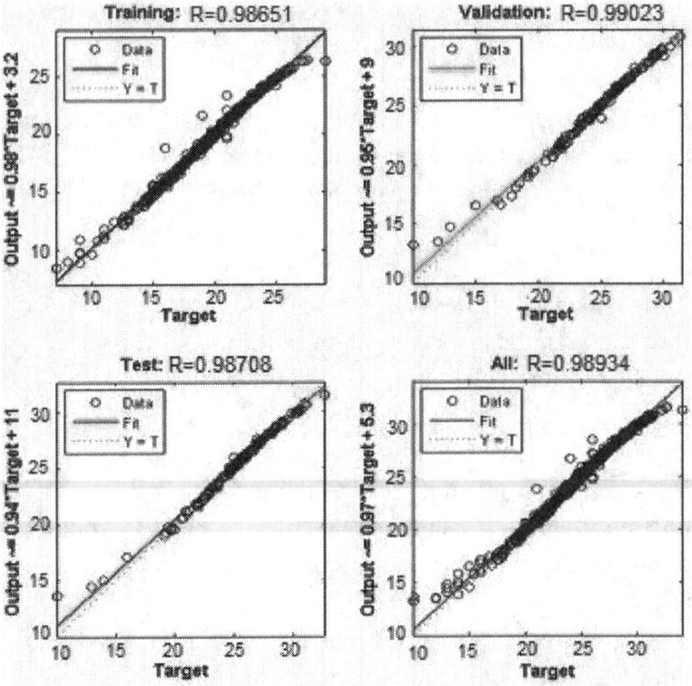

FIGURE 2.20 Regression plots of Output vs. Target.

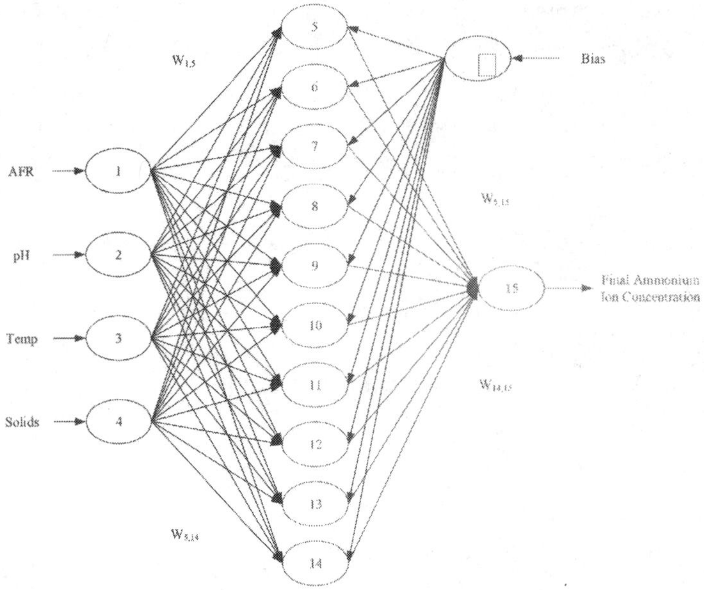

FIGURE 2.21 Optimum Neural Network topology.

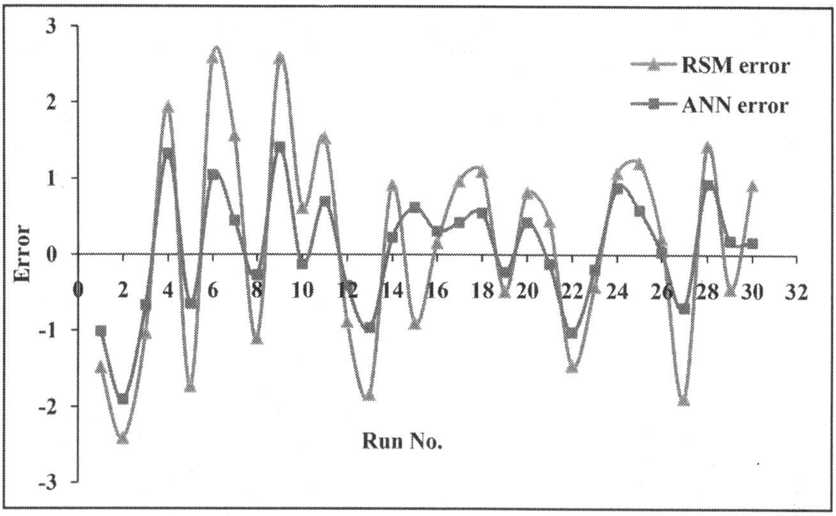

FIGURE 2.22 Error of RSM and ANN.

TABLE 2.5 Experimental and Predicted Values of Ammonium Ion Concentration Using RSM and ANN

S. No.	Experimental value	RSM predicted value	RSM error	ANN predicted value	ANN error
1	19.8	21.281	–1.481	20.812	–1.012
2	7.2	9.612	–2.412	9.106	–1.906
3	7.2	8.235	–1.035	7.864	–0.664
4	23.6	21.659	1.941	22.275	1.325
5	14.4	16.128	–1.728	15.043	–0.643
6	16.8	14.208	2.592	15.756	1.044
7	21.2	19.634	1.566	20.756	0.444
8	7.2	8.301	–1.101	7.469	–0.269
9	19.8	17.216	2.584	18.392	1.408
10	14.4	13.785	0.615	14.526	–0.126
11	19.8	18.267	1.533	19.108	0.692
12	21.6	22.482	–0.882	22.007	–0.407
13	14.4	16.237	–1.837	15.363	–0.963
14	7.2	6.285	0.915	6.976	0.224
15	12.6	13.502	–0.902	11.983	0.617
16	19.8	19.645	0.155	19.492	0.308

TABLE 2.5 (Continued)

S. No.	Experimental value	RSM predicted value	RSM error	ANN predicted value	ANN error
17	10.8	9.835	0.965	10.381	0.419
18	21.6	20.507	1.093	21.053	0.547
19	16.8	17.281	–0.481	17.0286	–0.2286
20	23.4	22.582	0.818	22.975	0.425
21	19.8	19.356	0.444	19.923	–0.123
22	16.8	18.257	–1.457	17.821	–1.021
23	19.8	20.215	–0.415	19.992	–0.192
24	19.8	18.726	1.074	18.925	0.875
25	12.6	11.392	1.208	12.016	0.584
26	7.2	6.984	0.216	7.165	0.035
27	14.4	16.285	–1.885	15.092	–0.692
28	16.8	15.368	1.432	15.872	0.928
29	7.2	7.647	–0.447	7.018	0.182
30	23.4	22.475	0.925	23.235	0.165

2.4 CONCLUSION

The engineering information is required for establishing reliable nitrification performance of biological wastewater treatment systems and an understanding of nitrogen transformation in receiving waters. The goal is to make full-scale use of the nitrification process in order to achieve improved water pollution control. From the present study, it is observed that the removal of ammonium ion in biological nitrification process using gas–liquid–solid bioreactor is more effective than using gas-liquid bioreactor. Also the effect of airflow rate, pH, temperature, amount of solids and different substrates were studied in a gas–liquid–solid bioreactor and the optimum values have been determined. The parameters were also optimized using response surface method and the design equation was developed for the final concentration of ammonium ion and compared with the experimental values. The process parameters were modeled with Artificial Neural Network for the prediction of ammonium ion concentration. The error in ANN is small when compared to RSM with experimental data was observed. Ammoniacal wastewater is the major pollutant in many process industries and hence biological wastewater treatment as seen in the present work can play an important role for treating ammoniacal wastewater.

KEYWORDS

- **artificial neural network**
- **gas-liquid bioreactor**
- **gas–liquid–solid bioreactor**
- **nitrification**
- **response surface methodology**

REFERENCES

1. Akhbari, A., Zinatizadeh, A. A. L., Mohammadi, P., Irandoust, M., & Mansouri, Y., (2011). Process modeling and analysis of biological nutrients removal in an integrated RBC-AS system using response surface methodology. *Chemical Engineering Journal, 168*, 269–279.
2. Alberto, O., Oliveira, C. S., Alba, J., Carrion, M., & Thalasso, F., (2011). Determination of apparent kinetic and stoichiometric parameters in a nitrifying fixed bed reactor by in situ pulse respirometry. *Biochemical Engineering Journal, 55*, 123–130.
3. Arun, K. T., Bhargava, R., & Pramod, K., (2010). Nitrification kinetics of activated sludge biofilm system: A mathematical model. *Bioresource Technology,101*, 5827–5835.
4. Bashir, M. J. K., Aziz, H. A., Yusoff, M. S., & Adlan, M. N., (2010). Application of response surface methodology (RSM) for optimization of ammoniacal nitrogen removal from semi-aerobic landfill leachate using ion exchange resin. *Desalination, 254*, 154–161.
5. Benyahia, F., & Polomarkaki, R., (2005). Mass transfer and kinetic studies under no cell growth conditions in nitrification using alginate gel immobilized *Nitrosomonas*. *Process Biochemistry. 40*, 1251–1262.
6. Bougard, D., Bernet, N., Cheneby, D., & Delgenes, J. P., (2006). Nitrification of a high-strength wastewater in an inverse turbulent bed reactor Effect of temperature on nitrite accumulation. *Process Biochemistry, 41*, 106–113.
7. Canziani, R., Emondi, V., Garavaglia, M., Malpei, F., Pasinetti, E., & Buttiglieri, G., (2006). Effect of oxygen concentration on biological nitrification and microbial ki-

netics in a cross-flow membrane bioreactor (MBR) and moving-bed biofilm reactor (MBBR) treating old landfill leachate. *Journal of Membrane Science*, *286*, 202–212.
8. Chen, S., Ling, J., & Blancheton, J., (2006). Nitrification kinetics of biofilm as affected by water quality factors. *Aquaculture Engineering*, *34*, 179–197.
9. Choi, D. & Park, H., (2001). A hybrid artificial neural network as a software sensor for optimal control of a wastewater treatment process. *Water Research*, *35*(16), 3959–3967.
10. Ciudad, G., Gonzalez, R., Bornhardt, C., & Antileo, C., (2007). Modes of operation and pH control as enhancement factors for partial nitrification with oxygen transport limitation. *Water Research*. *41*, 4621–4629.
11. Cooper, P.F., & Sutton, P.M., (1983). Treatment of wastewaters using biological fluidized beds. *Chemical Engineering*, 392–395.
12. Dogan, E., Sengorur, B., & Koklu, R., (2009). Modeling biological oxygen demand of the melen river in Turkey using an artificial neural network technique. *Journal of Environmental Management*, *90*, 1229–1235.
13. Francis, F., Sabu, A., Nampoothiri, K. M., Ramachandran, S., Ghosh, S., Szakasc, A., et al., (2003). Use of response surface methodology for optimizing process parameters for the production of an amylase by *aspergillus orzae*, *Biochemical Engineering Journal*, *15*, 107–115.
14. George, S. O., Cristea, V. M., & Agachi, P. S., (2011). Cost reduction of the wastewater treatment plant operation by MPC based on modified ASM1 with two-step nitrification/denitrification model. *Computers and Chemical Engineering*, *35*, 2469–2479.
15. Hamoda, M. F. Zeidan, M. O., & Al-Haddad, A. A., (1996). Biological nitrification kinetics in a fixed-film reactor. *Bioresource Technology*, *58*, 41–48.
16. Keyvani, N., (2008). Regulations and Environmental Standards in Human Environment Field. *Iran Enviromental Protection Agancy*, *45*–54.
17. Kugaprasatham, S., Nagaoka, H. & Ohgaki, S., (1991). Effect of short term and long-term changes in hydraulic conditions on nitrifying biofilm. *Water Science and Technology*, *23*, 1487–1494.
18. Lakshmi Devi, P. B. N., & Pydi Setty, Y., (2014a). Biological nitrification of wastewater. *International Journal of Applied Biology and Pharmaceutical Technology*, *5*(4), 120–129.
19. Lakshmi Devi, P. B. N., & Pydi Setty, Y., (2014b). Effect of nutrients on biological nitrification. *British Biotechnology Journal*, *4*(12), 1283–1290.
20. Lakshmi Devi, P. B. N., & Pydi Setty, Y., (2015). Removal of ammonia from wastewater using biological nitrification. Shirish H., Sonawane, Pydi Setty, Y., & Srinu Naik, S., (Eds.), Chemical and Bioprocess Engineering: Trends and Developments, Chapter 29, 309–320.
21. Lakshmi Devi, P. B. N., (2015). Studies on biological nitrification of ammoniacal wastewater. Ph. D. Thesis, National Institute of Technology: Warangal, India.
22. Lee, D. S., Lee, M. W, Woo, S. H., Kim, Y., & Park, J. M., (2006). Nonlinear dynamic partial least squares modeling of a full-scale biological wastewater treatment plant. *Process Biochemistry*, *41*, 2050–2057.
23. Lee, M. W., Hong, S. H., Choi, H., Kim, J., Lee, D. S., & Park, J. M., (2008). Real time remote monitoring of small-scaled biological wastewater treatment plants by a

multivariate statistical process control and neural network-based software sensors. *Process Biochemistry, 43*, 1107–1113.
24. Lei, L. & Ni, J., (2014).Three dimensional three-phase model for simulation of hydrodynamics, oxygen mass transfer, carbon oxidation, nitrification and denitrification in an oxidation ditch. *Water Research, 53*, 200–214.
25. Losordo, T. M. & Westers, H., (1994). System carrying capacity and flow estimation. In: Timmons, M. B., Losordo, T. M., (Eds.), Aquaculture Water Reuse Systems: Engineering Design and Management, 9–60.
26. Mehran, A., George, N., & Zhu, J., (2010). Dynamic testing of the twin circulating fluidized bed bioreactor (TCFBBR) for nutrient removal from municipal wastewater. *Chemical Engineering Journal, 162*, 616–625.
27. Machon, I., Lopez, H., Rodriguez-Iglesias, J., Maranon, E., & Vazquez, I., (2007). Simulation of a coke wastewater nitrification process using a feed-forward neural net. *Environmental Modeling and Software, 22*, 1382–1387.
28. Malik, A., (2012). Study of performance of nitrification process. *Journal of King Saud University-Engineering Sciences, 24*, 53–59.
29. Meade, J. W., (1985). Allowable ammonia for fish culture. *Prog. Fish-Culturist, 47*, 135–145.
30. Mosquera-Corral, A., Gonzalez, F., Campos, J. L., & Mendez, R., (2005). Partial nitrification in a sharon reactor in the presence of salts and organic carbon compounds. *Process Biochemistry, 40*, 3109–3118.
31. Mousavi, S., Ibrahim, S., & Aroua M. K., (2012). Sequential nitrification and denitrification in a novel palm shell granular activated carbon twin chamber upflow bio-electrochemical reactor for treating ammonia-rich wastewater. *Bioresource Technology, 125*, 256–266.
32. Nasrin, A., Hamidi-Bin, A. A., Mohamed, H. I., & Ali, A. Z., (2007). Powdered activated carbon augmented activated sludge process for treatment of semi-aerobic landfill leachate using response surface methodology. *Bioresource Technology, 98*, 3570–3578.
33. Nevov, V., & Kamenski, D., (2004). Model based study of the first stage of biological nitrification (NH_4^+ oxidation to NO_2). *Environmental Modeling and Software, 19*, 517–524.
34. Pambrun, V., Paul, E., & Sperandio, M., (2008). Control and modeling of partial nitrification of effluents with high ammonia concentrations in sequencing batch reactor. *Chemical Engineering and Processing, 47*, 323–329.
35. Park, S., & Bae, W., (2009). Modeling kinetics of ammonium oxidation and nitrite oxidation under simultaneous inhibition by free ammonia and free nitrous acid. *Process Biochemistry, 44*, 631–640.
36. Rahmani, A. R., Nasseri, S., Mesdaghinia, A. R., & Mahvi, A. H., (2004). Investigation of ammonia removal from polluted waters by Clinoptilolite zeolite Inter. *Journal of Environmental Science and Technology, 1*, 125–133.
37. Ruiz, G., Jeison, D., & Chamy, R., (2003). Nitrification with high nitrite accumulation for the treatment of wastewater with high ammonia concentration. *Water Research, 37*, 1371–1377.
38. Saeed, T., & Sun, G., (2011a). The removal of nitrogen and organics on vertical wetland reactors: predictive models. *Bioresource Technology, 102*, 1205–1213.

39. Saeed, T., & Sun, G., (2011b). Kinetic modeling of nitrogen and organics removal in vertical and horizontal flow wetlands. *Water Research, 45*, 3137–3152.
40. Sandu, S. I., Boardman, G. B., Watten, B. J., & Brazil, B. L., (2002). Factors influencing the nitrification efficiency of fluidized bed filter with a plastic bead medium. *Aquaculture Engineering, 26*, 41–59.
41. Seifi, M., & Fazaelipoor, M. H., (2012). Modeling simultaneous nitrification and denitrification (SND) in a fluidized bed biofilm reactor. *Applied Mathematical Modeling, 36*, 5603–5613.
42. Sutton, P. M., & Mishra, P. N., (1990). Fluidized bed biological wastewater treatment: effects of scale-up on system performance. *Water Science and Technology, 22*, 419–430.
43. Terada, A., Yamamoto, T., Igarashi, R., Tsuneda, S. & Hirata, A., (2006). Feasibility of a membrane-aerated biofilm reactor to achieve controllable nitrification. *Biochemical Engineering Journal, 28*, 123–130.
44. Trussell, R. P., (1972). The percent un-ionized ammonia in aqueous ammonia solutions at different pH levels and temperatures. *J. Fish. Res. Board of Canada,29*, 1505–1507.
45. United States Environmental Protection Agency (USEPA), (1984). Methods for the Chemical analysis of Water and Wastewater, EPA-600/4-79-020. U.S. Environmental Protection Agency, Office of Research and Development, Environmental Monitoring and Support Laboratory: Cincinnati, OH, USA.
46. Vigne, E., Choubert, J., Canler, J., Heduit, A., Sorensen, K., & Lessard, P. A (2010). biofiltration model for tertiary nitrification of municipal wastewater. *Water Research, 44*, 4399–4410.
47. Vohra, A., & Satyanarayana, T. (2002). Statistical optimization of medium components by response surface methodology to enhance phytase production by *pichia anomla. Process Biochemistry, 7*, 999–1004.
48. WPCF, (1983). Nutrient Control. Manual of Practice. Publication number FD-7, Water Pollution Control Federation: Washington, DC, USA.
49. Young Zhen, P., Shouyou, G., Shuying, W., & Lu, B. B., (2007). Partial nitrification from domestic wastewater by aeration control at ambient temperature. *Chinese Journal of Chemical Engineering, 15*(1), 115–121.
50. Zhu, S., & Chen, S., (2001). Effects of organic carbon on nitrification rate in fixed film biofilters. *Aquaculture Engineering, 25*, 1–11.
51. Zhu, S., & Chen, S., (2002). The impact of temperature on nitrification rate in fixed film biofilters. *Aquaculture Engineering, 26*, 221–237.

CHAPTER 3

ULTRASOUND ASSISTED FORMATION OF STABLE EMULSION AND ITS APPLICATIONS AS LIQUID EMULSION MEMBRANE IN WASTEWATER TREATMENT

ZAR ZAR NWE, S. SRINU NAIK, SHIRISH H. SONAWANE, and Y. PYDI SETTY

Department of Chemical Engineering, National Institute of Technology, Warangal, 506004, India, Tel: +91-870-2462611; Fax: +91-870-2459547, E-mail: y.pydisetty@gmail.com

CONTENTS

ABSTRACT

In this work stable emulsion droplets have been prepared using ultrasound assisted technique, in which continuous phase was water and droplet phase was hexane. Tween 80 surfactant was used as surface active agent with quantity above Critical Miscelle Concentration (CMC) value. The stock solution of desired concentration was obtained by diluting with distilled water. The blue colored complex was formed very quickly and the solution is stirred for one minute to ensure complete development of the complex prior to measuring the absorbance. Effect of variables such as cavitation time, Energy dissipated concentration of hexane and surfactants and their effect on emulsion stability is reported. The most suitable parameter conditions are at 50 mL distilled water, 4 mL hexane, 0.1 g Tween 80, 1 mL emulsion solution, 50 mL dye solution of 20 ppm concentration and 5 min sonication time. It is observed that maximum dye removal efficiency was obtained at the above said parameter values.

3.1 INTRODUCTION

Textile dye effluents are difficult to treat by conventional and biological processes, as most of the dyes are toxic and non-biodegradable. The process for degradation takes longer time. Major problem with these dye molecules is that they are complex structure, and are difficult to break-down biologically and can't be treated efficiently by only conventional methods. Metals represent one of the problematic groups of pollutants that can be found in industrial wastewater. Dye is one of these elements which can penetrate into ground water and cause serious poisoning. In addition, the epidemiologic studies have proved its carcinogenic effects. Thus, it is important to develop a reasonable method in order to remove and/or recover this element from the aquatic environment [1–4].

Emulsion liquid membranes are of considerable potential as effective tool for wide variety of separations. It is also called surfactant liquid membrane or liquid surfactant membrane, are essentially double emulsion, water/oil/water (W/O/W) systems or oil/water/oil (O/W/O) systems for the W/O/W systems, the liquid membrane is the oil phase that is between two water phases, and the configuration is stabilized by oil soluble surfactant. Emulsion Liquid membranes are usually prepared by first forming an emulsion between two immiscible phases, and then dispersing the emulsion into a third (continuous) phase by agitation for extraction of dyes [5–8]. The membrane phase is the liquid phase that separates the encapsulated, internal droplets in the emulsion from the external, continuous phase. In general, the internal, encapsulated phase and the external, continuous phase are miscible. To maintain the integrity of the emulsion during the extraction process, the membrane phase generally contains some surfactant(s) and additive(s) as stabilizing agents, and it also contains a base material that is a solvent for all the other ingredients. Based on the solvent and surfactant and discrete and continuous phase the liquid emulsion system could be oil in water or water in oil. The droplet of emulsion could be in micrometer to sub micrometer in size. The one of the major advantage of this liquid emulsion membrane is that it has large interfacial area which makes them ideal candidate for extraction and treatment of industrial effluent specifically for dye removal. Making uniform small size droplets is one of the challenging task. The droplet size of discrete phase of the Emulsion Liquid Membrane (ELM) can be controlled based on the surfactant concentration [9, 10].

When exposing liquids to ultrasound the sound waves that propagate into the liquid result in alternating high-pressure and low-pressure cycles. This applies mechanical stress on the attracting forces between the individual particles. Ultrasonic cavitation in liquids causes high speed liquid jets of up to 1000 km/h [11, 12]. When more powerful ultrasound at a lower frequency is applied to a two phase system, it is possible to produce the chemical changes as a result of acoustically generated cavitation. The generation of radicals takes place due to the extreme environment created by acoustic cavitation [11, 12]. In addition, the physical effect of the medium on the wave is referred to low power or high frequency ultrasound. The physical effect in the fields of nanomaterials like thermal heating, mass transfer, emulsification, and surface cleaning are induced by cavitation Ultrasound is found to be one of the most efficient methods for the preparation of stable monomer droplets without any aggregation during the emulsion process [10–12]. The most obvious role is in the dispersion of materials in liquids in order to break particle agglomerates. Another process is the application of ultrasound during emulsion preparation is formation of stable smaller droplet size. Generally, due to micro mixing and micro jets shearing smaller droplet size emulsion is formed with increased size uniformity. Such jets press liquid at high pressure between the particles and separate them from each other. Smaller particles are accelerated with the liquid jets and collide at high speeds. This makes ultrasound an effective means for dispersion [10–12].

The number of attempts has been made for preparation of LEM and use for removal of heavy metals and ions [13–16]. However, there are very few reports found to control the size of emulsion droplet using cavitation technique and use for the water treatment. Hence, through this manuscript, an attempt is made for synthesis of small droplet size LEM and removal dye from the wastewater.

3.2 MATERIALS AND METHODS

3.2.1 MATERIALS

Tween 80 is extra pure reagent used as surfactant. Methylene blue (MB) is the alkaline pigment solution. Hexane and Heptane are used as solvents. All the chemicals used are of AR grade.

3.2. METHODS

3.2.2.1 Preparation of Standard Solution of Dye

The stock solution of MB was prepared (1,000 mg/L), and then solution of desired concentration was obtained by diluting stock solution with distilled water. The blue colored complex is formed very quickly and the solution is stirred for 1 min to ensure complete development of the complex prior to measuring the absorbance. A calibration plot was made in the concentration range of 10 to 500 ppm for determination of the dye concentration. This spectrum was taken with UV-visible spectro-photometer using a 1 cm quartz cuvette filled with 4 mL solution of Methylene Blue in Water. UV-Visible absorption spectrum of Methylene Blue solution shows the peak at 650 nm (Figure 3.1). Therefore, the peak at the wavelength of 650 nm is selected to plot a calibration curve for the Methylene Blue solution.

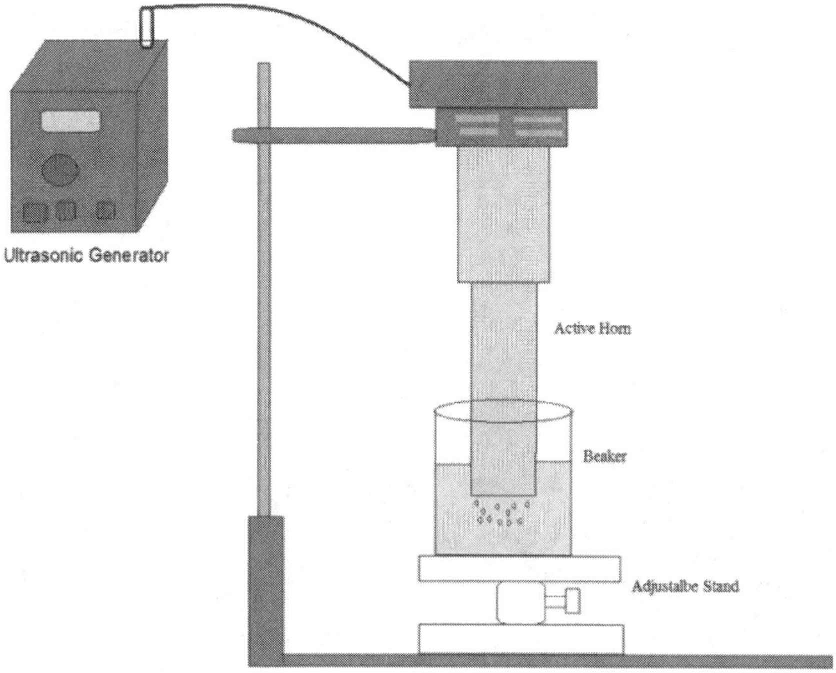

FIGURE 3.1 Schematic diagram of sonication device.

Figure 3.1 shows absorbent intensities of the Methylene Blue solution at the wavelength of 650 nm in response to the given concentration.

3.2.2.2 Liquid Emulsion Membrane Preparation

About 0.1 g of Tween 80 was added to accurately measured 10 mL of distilled water in a beaker. This volume of distilled water is kept constant for all experiments. The mixture was stirred by using ultrasound sonication device shown in Figure 3.2. The system was kept for 5 min in the sonication device. Then 4 mL of hexane was mixed with solution mixture for 5 min sonication time by ultrasound assisted sonication device (Prompt: 1.27 cm dia.). As an operating condition during stirring the solution, 2 sec working sonication time and 1 second off time alternatively (total 5 min sonication time) has been followed. Experimental procedure is shown in Figure 3.3 and Table 3.1 shows the composition and parameters used

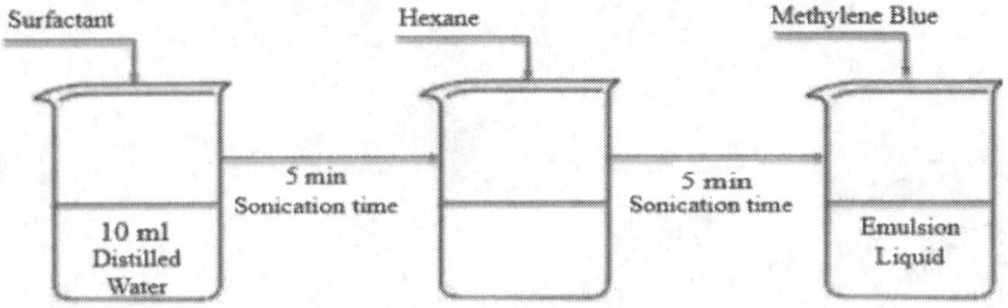

FIGURE 3.2 Schematic diagram of methylene blue removal by liquid emulsion membrane.

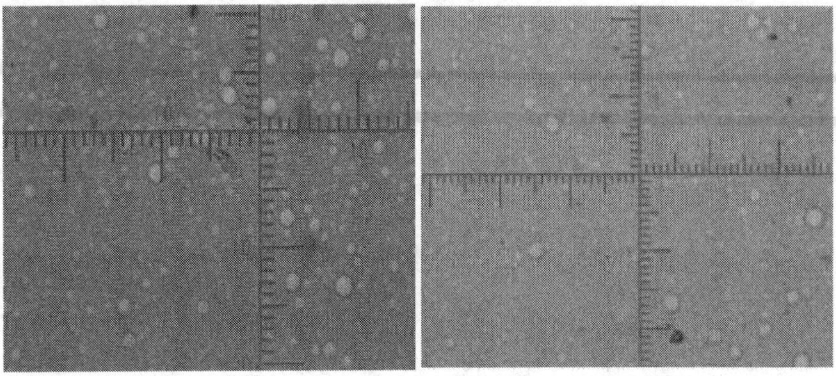

FIGURE 3.3 Determination of emulsion droplet diameter using a microscope.

TABLE 3.1 Composition and Parameters Used for Emulsion Liquid Membrane Preparation

	Operating parameter
Distilled water	10 mL
Tween 80 (surfactant)	0.1 g
Hexane	4 mL
1st sonication time	5 min
2nd sonication time	5 min
Emulsion Solution	5 mL
Dye solution	250 mL
Extraction time interval	10 min
Stirrer (speed and time)	1000 rpm, 10 min

in emulsion liquid membrane. This whole process is energy efficient and takes place at room temperature.

3.3.2.3 Removal of Methylene Blue Dye

Around 250 mL of dye-containing feed was taken in a beaker. Around 5 mL of stable liquid emulsion was added to the feed. Initially dye solution was kept in the separating funnel. Then stable emulsion solution was added to the funnel to react with dye solution. The emulsion slowly began to get dispersed. Samples were drawn at regular intervals up to 10 min and were collected in clean and dry test tubes. Each sample was subjected to gravity settling for about 30 min. The organic phase emulsion solution being lighter got collected at the top and clear solution was present at the bottom in the funnel. The bottom layer which is aqueous phase clear solution was taken for analysis. The removal of dye is evaluated as:

$$\text{Removal efficiency (\%)} = 100 \times (C_o - C_t)/C_o$$

where, C_o is the initial concentration and C_t is the concentration at time t [4].

3.2.2.4 UV-Visible Spectrophotometer Analysis for Measurement of Concentration

The concentration of dye at the clear solution extracted from the feed was determined by UV-Visible spectrophotometer (Thermo Scientific, USA; model: GENESYS 2 and Systronics, India, Model: 2201). The standard method was used to calculate the concentration of Methylene Blue in water.

3.3 RESULTS AND DISCUSSION

When the emulsion is dispersed by agitation in the external, continuous phase during the extraction process, many small globules of the emulsion were formed. These globules were dispersed within the sample solution so that the inner phase never comes into direct contact with the sample. The target compound diffuses from the sample solution through the membrane into the inner phase, sometimes with the aid of an added carrier. Agitation of the emulsion produces many small droplets within the sample so that the contact surface area to analyze is very large, providing efficient transfer to the inner phase. The size of the globules depends on the characteristics and concentration of the surfactant in the emulsion, the viscosity of the emulsion, the intensity and the mode of mixing. The diameter of emulsion droplet as shown in Figure 3.4 was determined using a microscope.

The diluent forms the major proportion of the emulsion. So the performance of several solvents was compared. The main properties to be

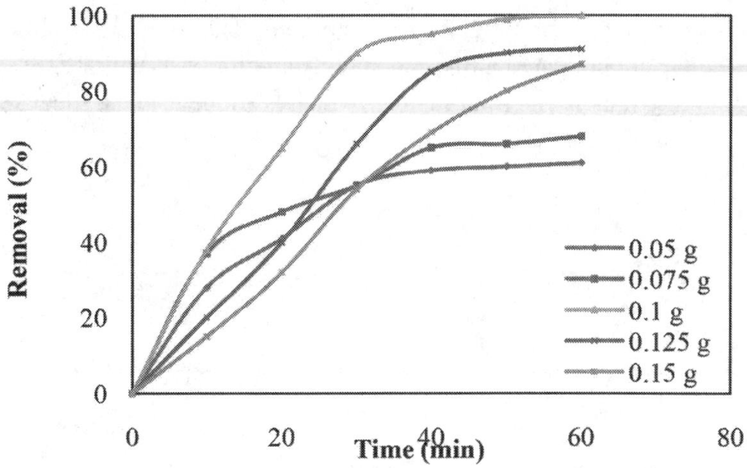

FIGURE 3.4(a) Variation of quantity of Tween 80 with time at 20 ppm MB.

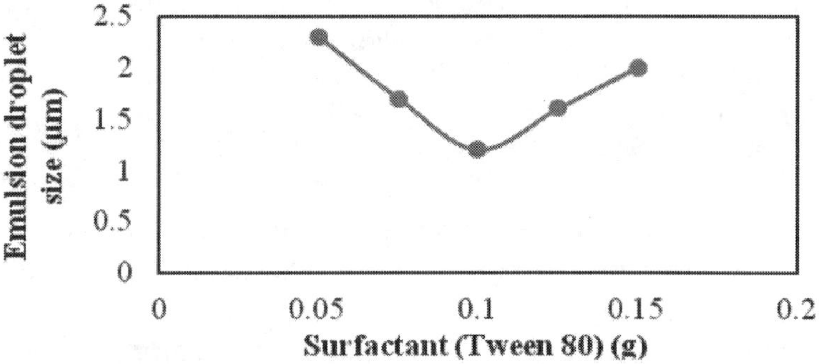

FIGURE 3.4(b) Effect of quantity of Tween 80 on emulsion droplet size.

considered are the density and viscosity, which determine the permeability and thickness of the membrane walls which affect the stability of the emulsion and the extraction time and yield. Hexane performed well, giving recoveries of 99%, after 60 min of contact time at 20 ppm of Methylene Blue. The removal performance was influenced by many factors, such as sonication time, energy dissipated, concentration of diluents and surfactants. In real world, it would be desirable to have an idea of initial concentration of dye and diluting the feed solution and the sonication time. The initial concentration of Methylene Blue in the aqueous solution also had an effect. According to UV-visible spectrophotometer for levels up to 500 ppm, maximum recovery was maintained at low concentrations like 50 ppm but it began to fall away for more concentrated solutions, ranging from about 65–95% for 100–200 ppm methylene blue after 80 min extraction time. According to the result increase in the dye concentration decreases the percentage of removal efficiency.

3.3.1 EFFECT OF QUANTITY OF SURFACTANT

Surfactant concentration is an important factor as it directly affects the stability, swelling and break up of Emulsion Liquid Membrane. Figure 3.5 represents the variation of percentage removal of MB for various Tween 80 quantities. It is observed from Figure 3.5 that the percentage removal of MB increases up to 0.1 g of the amount of Tween 80 and decreases thereafter. At lower surfactant quantities (less than 0.1 g), emulsions break easily leading to poor extraction. At higher surfactant quantities (beyond 0.1 g), although the membrane stability increases, mass transfer resistance also increases due to presence of more surfactant at aqueous–organic phase interface, resulting in less transfer of dye molecules to internal phase. Thus, the dye extraction is reduced. It is observed that dye removal is maximum (~99%) at Tween 80 quantities of 0.1 g. It may be observed from Figure 3.5 that removal of Methylene Blue reaches a constant value after about 45 min, due to reduction in driving force (i.e., the concentration gradient of Methylene Blue between external and internal phases) and then Methylene Blue removal does not fall below the maximum level at any of the surfactant quantity. The best value of surfactant quantity was

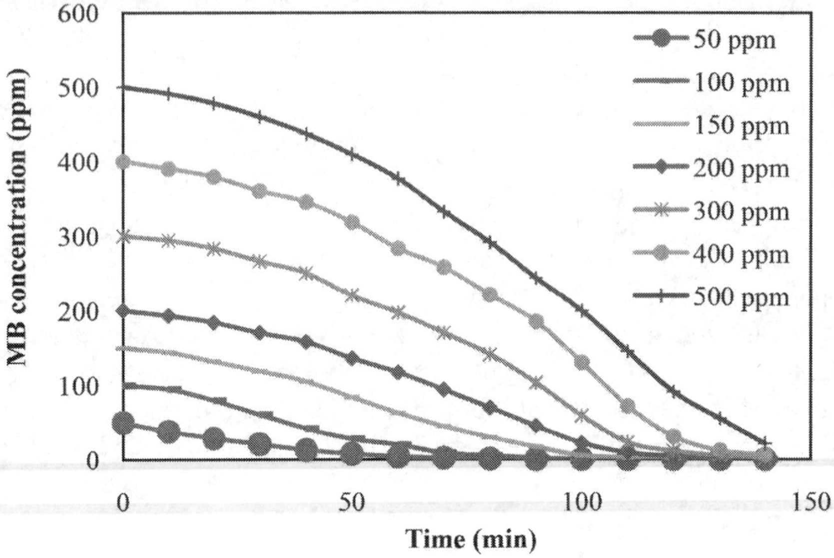

FIGURE 3.5(a) Variation of MB concentration with time.

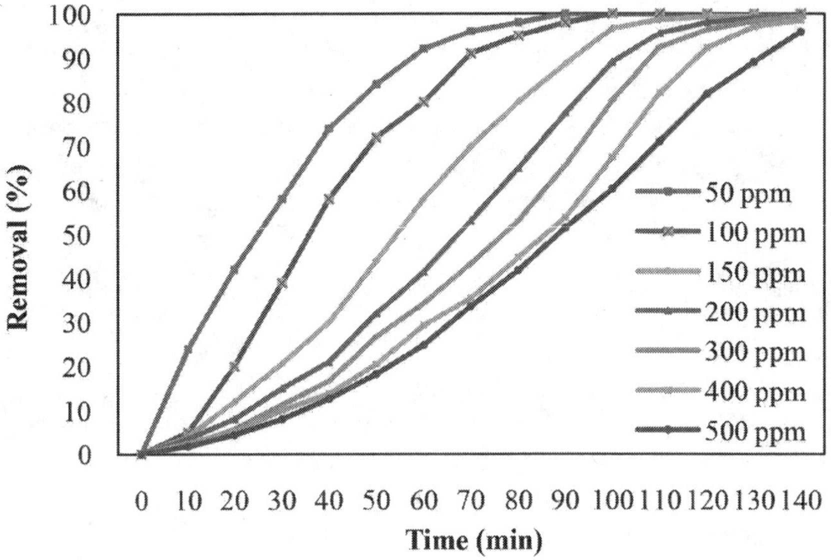

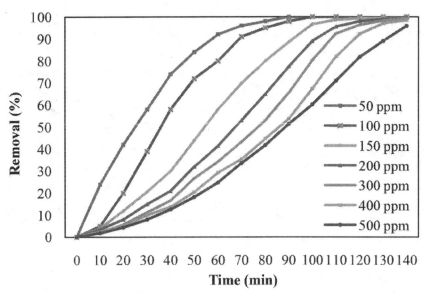

FIGURE 3.5(b) Variation of removal efficiency with time for different MB concentrations.

found to be 0.1 g. Figure 3.6 represents effect of quantity of surfactant on emulsion droplet size.

3.3.2 *EFFECT OF FEED CONCENTRATION*

Effects of feed concentrations of Methylene Blue on the liquid membrane system are shown in Figures 3.7 and 3.8. It is observed from this figure

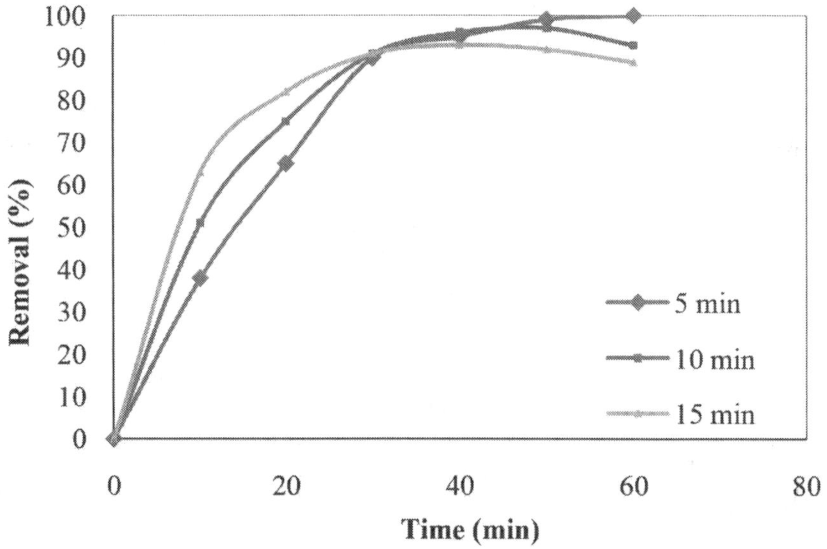

FIGURE 3.6(a) Effect of sonication time (min) on removal of methylene blue at 20 ppm of Methylene Blue and 4 mL Hexane.

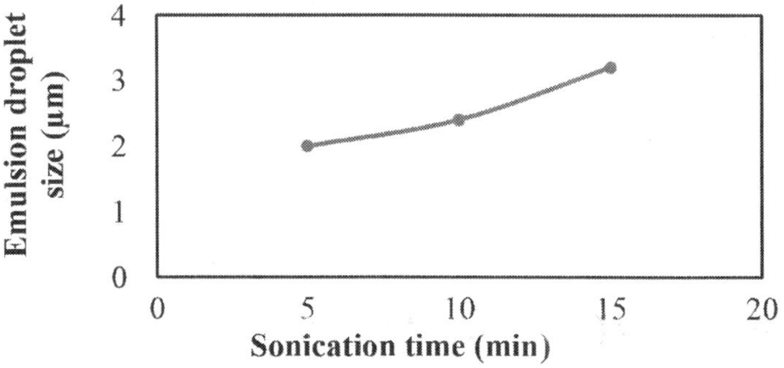

FIGURE 3.6(b) Effect of sonication time on emulsion droplet size.

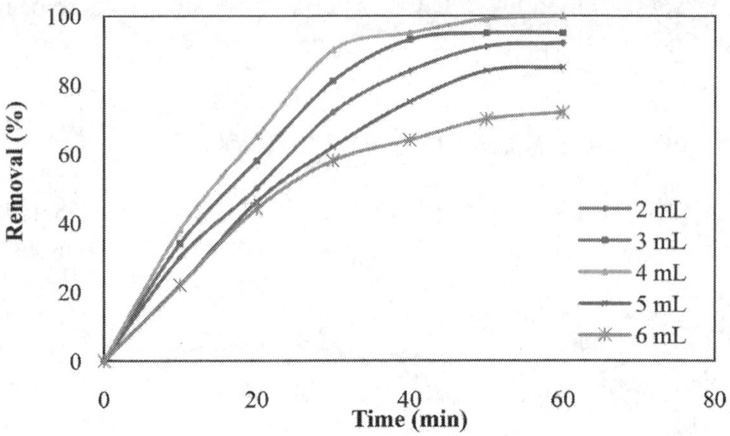

FIGURE 3.7 Effect of hexane quantity on Removal of Methylene Blue at 20 ppm Methylene Blue.

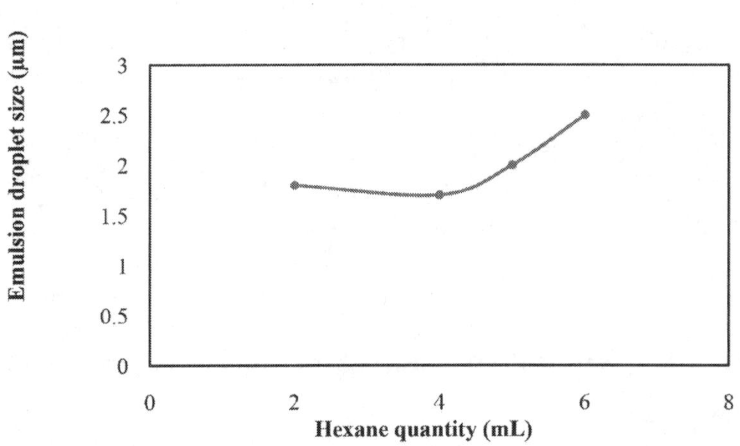

FIGURE 3.8(a) Effect of Hexane quantity on emulsion droplet size.

that dye concentration in the external phase decreases slightly up to about 60 min and then remains falling down till 80 min. This effect is prominent at higher dye concentration. With increase in dye concentration in the external phase, the driving force for dye transport to the internal phase increases. This leads to a sharp decline of dye concentration of the external phase up to about 60 min. The dye concentration in the external phase remains approximately fell down after that. It is observed from Figures 3.7

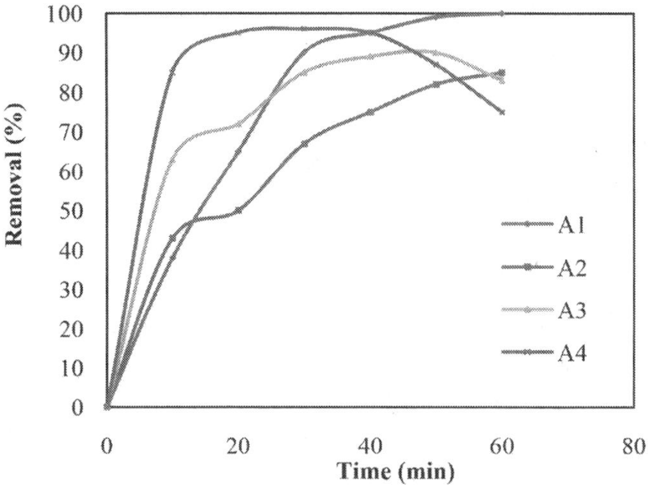

FIGURE 3.8(b) Effect of Emulsion Liquid Membrane: Aqueous liquid volume ratio at 20 ppm Methylene Blue, 4 mL Hexane and 5 min sonication time.

and 3.8 as discussed earlier that there is no leakage of Methylene Blue and hence, dye concentration in external phase remains constant almost at zero level indicating almost 99% extraction. Hence, the percentage removal of Methylene Blue is dependent upon initial concentration.

3.3.3 EFFECT OF SONICATION TIME ON EMULSION DROPLET SIZE AND REMOVAL EFFICIENCY

Sonication time during extraction is an important factor. In order to study the influence of sonication time on emulsion droplet size and removal efficiency, Tween 80 was used as the emulsifier with concentration of 0.1 g. Effect of sonication time on removal of Methylene Blue efficiency is shown in Figure 3.9. It has been observed that at initial period of operation, extraction takes more time. This trend is observed during initial 5 min. With increase in sonication time, the temperature of the emulsion is increased; therefore the interfacial tension and viscosity are expected to decrease considerably. The reduction in emulsion viscosity increases the cavitational activity because of the decrease in cavitational threshold. This effect together with the reduction in interfacial tension improves emulsification. However, increasing the

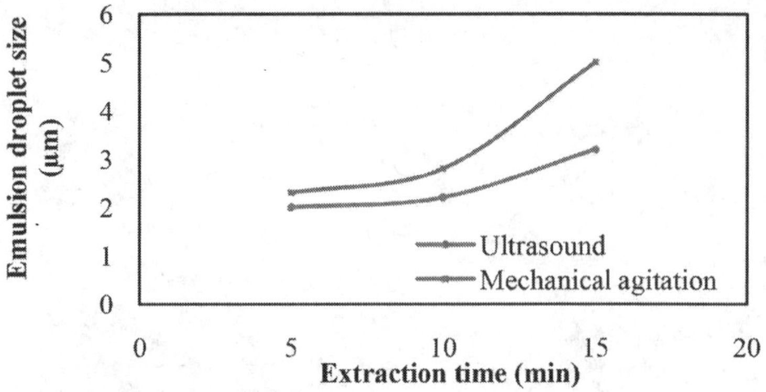

FIGURE 3.9 Effect of emulsification methods on MB in feed phase.

sonication time above a certain limit has an adverse effect on emulsification as it leads to "over processing" which results in re-coalescence of emulsion droplets. The dispersed phase droplet size is gradually decreased with an increase in sonication time up to 15 min. If the sonication time is increased further, the emulsion droplet size is increased but it adversely affected the stability of emulsion globules leading to breakage and the results are shown in Figure 3.10. Therefore, percentage extraction decreases in the long run and maximum removal (92%) occurred at 40 min. Beyond 30 min, percentage removal decreases from time to time. Increasing the sonication

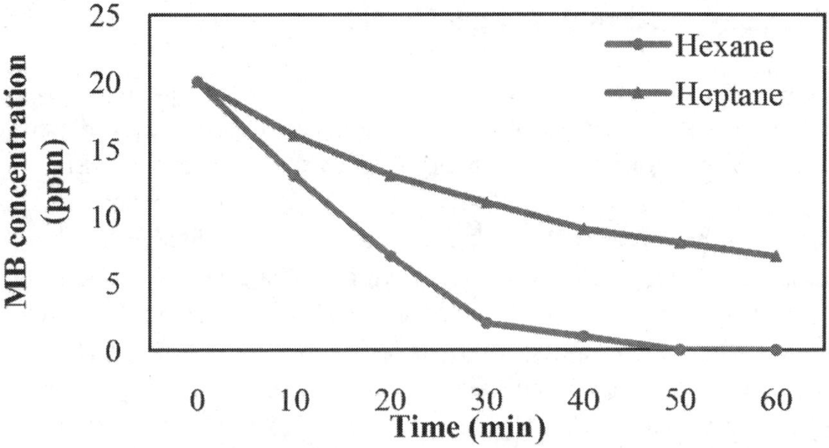

FIGURE 3.10(a) Effect of diluent on Methylene Blue concentration.

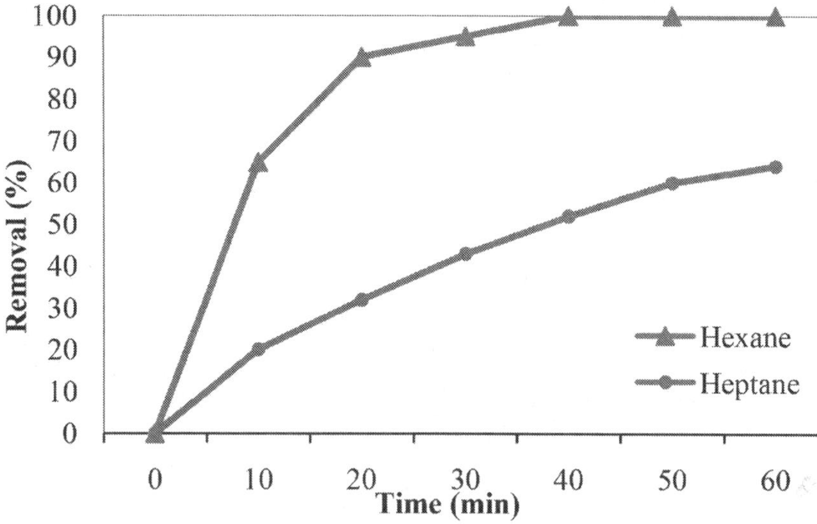

FIGURE 3.10(b) Effect of diluent on Methylene Blue concentration.

time decreases the removal efficiency because the internal droplet size is increased and the mass transfer area is decreased. However, further increase in sonication time decreases the droplet size which leads to a reduced extraction amount.

3.3.4 EFFECT OF HEXANE (OIL PHASE) QUANTITY

Experiments were conducted using an ultrasonic device, sonication time of 5 min, and keeping volume ratio of organic phase to aqueous phase as 50:1. The effect of carrier volume ranging from 2 to 6 mL is shown in Figures 3.11 and 3.12. From this figure, the emulsion stability increases with increasing the carrier quantity, which may be due to the interfacial properties of the carrier. In extraction, volume of carrier is higher than 4 mL, the W/O emulsions become unstable. Additionally, very high content of carrier in the membrane does not result in a benefit due to the increase in viscosity, which leads to larger globules. Increasing quantity of carrier also promotes the permeation swelling, which dilutes the aqueous receiving phase and decreases the efficiency of the process. The best value of the carrier volume was found to be 4 mL.

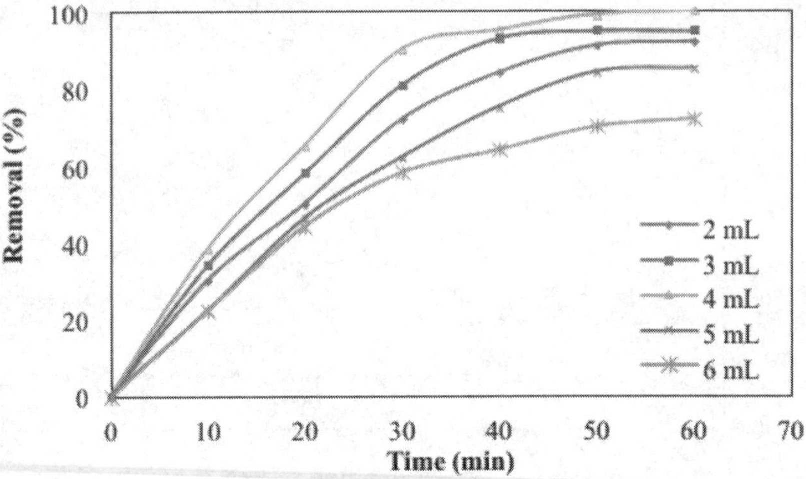

FIGURE 3.11 Effect of Hexane quantity on Removal of Methyle Blue at 20 ppm Methylene Blue.

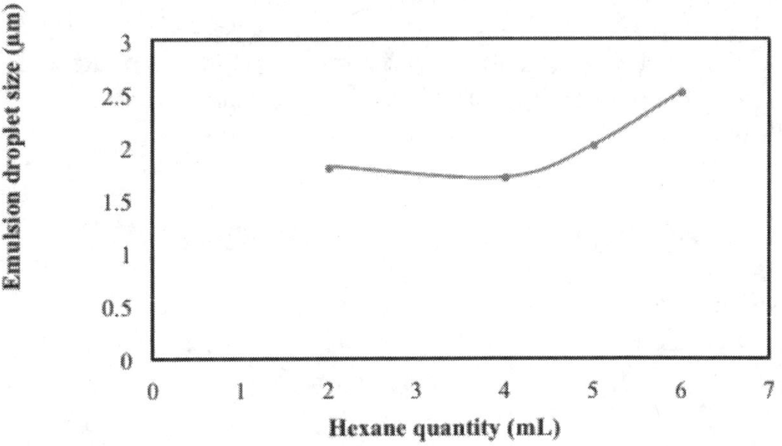

FIGURE 3.12 Effect of Hexane quantity on emulsion droplet size.

3.3.5 EFFECT OF VOLUME RATIO OF THE OIL PHASE TO AQUEOUS PHASE (O/A)

In the removal of organic and inorganic pollutants from the solutions using the liquid emulsion membrane (LEM) technology, the volume ratio of the oil phase to the aqueous phase (O/A) plays an important role. Oil phase provides more resistance to the solute transport but at the same time it offers more stability to emulsion droplet. Figure 3.13 shows that removal of Methylene Blue is achieved best when the volume ratio of the oil phase to aqueous phase is A1. With increase in this ratio, the extraction decreases as concentration of Methylene Blue in the aqueous phase decreases. It is also seen from the figure that there is little emulsion breakage in most of the cases and it is most profound in case of minimum aqueous phase volume. The removal efficiency is also maximum point for the ratio (O/A) of A1 shown in Figure 3.13. The maximum recovery of dye was in the range of 99%. Maximum efficiency for removal of dyes found in this work is about 96–99% for single component system.

Aqueous liquid: ELM = A: ELM
A (Aqueous liquid) = 50 mL
ELM (Emulsion Liquid Membrane) = 1 mL, 2 mL, 3 mL, 4 mL

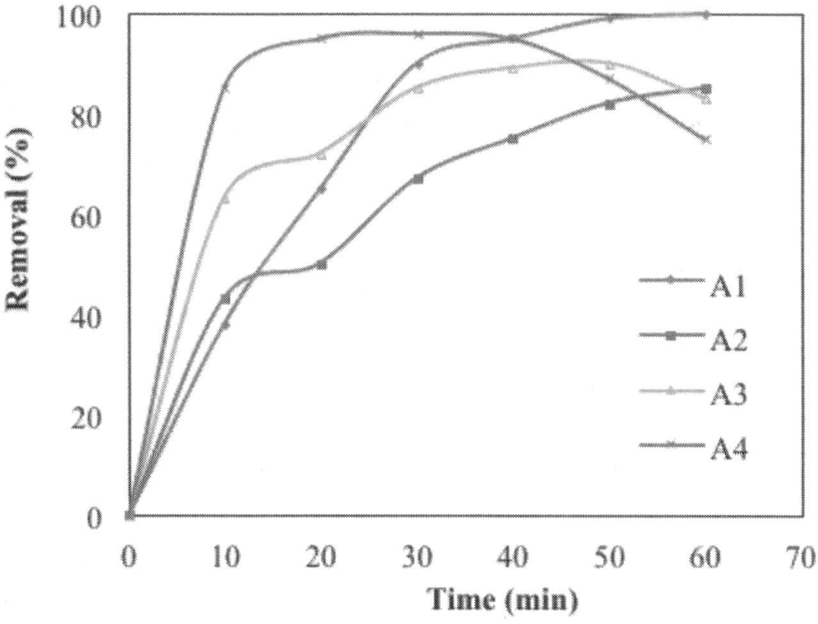

FIGURE 3.13 Effect of Emulsion Liquid Membrane: Aqueous liquid volume ratio at 20 ppm Methylene Blue, 4mL Hexane and 5 minutes sonication time.

3.3.6 DIFFERENCE BETWEEN ULTRASOUND SONICATION DEVICE AND MECHANICAL AGITATION

Ultrasound has become an increasingly popular tool in the modification of metal surfaces, imbuing them with various desired characteristics and functionalities. The exact role played by ultrasound in such processes remains largely speculative and thus requires clarification. After the emulsion was prepared by mixer, it was sampled to analyze the internal droplet size. After that it was sonicated for 5 min and then sampled again. The internal droplet size is shown in Figure 3.14. It should be noted that Tween 80 has been used in the mentioned emulsion composition with a concentration of 0.1 g. Ultrasound waves do not have a reasonable effect on reducing the droplet size. The reason is high viscosity of paraffin which increases the cavitational threshold, resulting in a decreased cavitational activity. However, emulsions prepared by ultrasound waves show smaller droplets and narrower distribution of the droplet size in comparison with the per-emulsions prepared by mixer. The removal efficiency of the two mentioned emulsions can also be compared as seen in Figure 3.14. From the results, it is observed that the extraction is efficient during the first 5 min of process.

3.3.7 EFFECT OF DILUENT

The stability of the W/O emulsions was studied against contact time using two difference diluents (hexane or heptane). Experimental conditions used

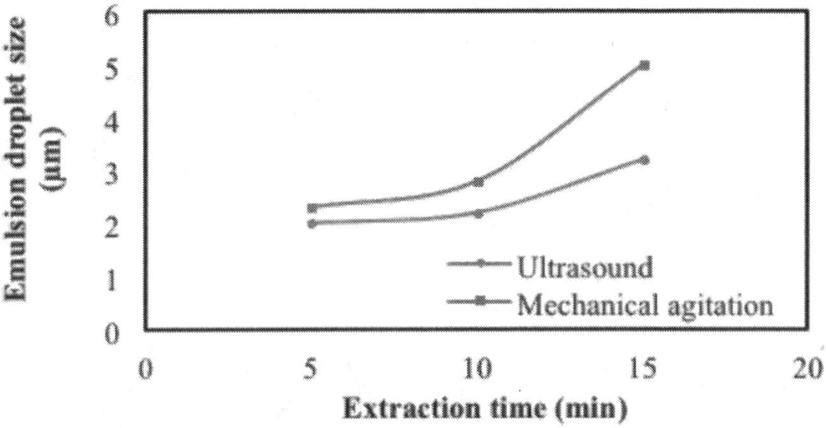

FIGURE 3.14 Effect of emulsification methods on MB in feed phase.

are identical to that used previously. The contact time was varied from 0 to 60 min. Figure 3.15 shows variation of concentration of Methylene Blue with contact time using hexane or heptane as diluent. It is observed that hexane provides to the system a better stability compared to heptane. This could be explained by the fact that heptane is more viscous than hexane and thus, cavitation bubbles are more easily formed using solvents with low viscosity. In the case of hexane, the W/O emulsions breakage remains within the acceptable limits until a contact time of 60 min. Figure 3.16 shows the effect of diluents on methylene blue extraction.

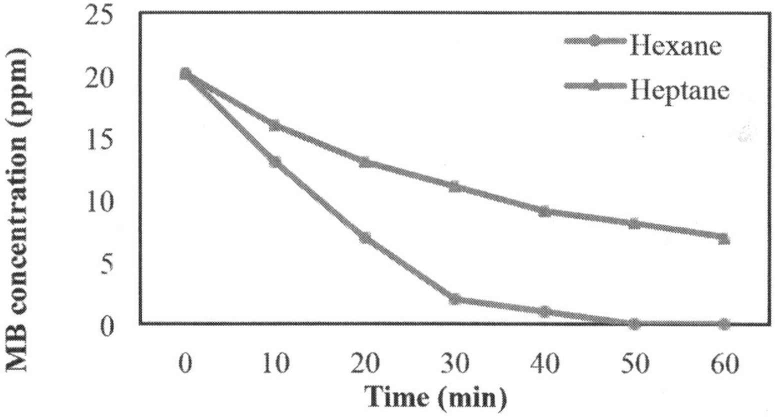

FIGURE 3.15 Effect of diluent on Methylene Blue concentration

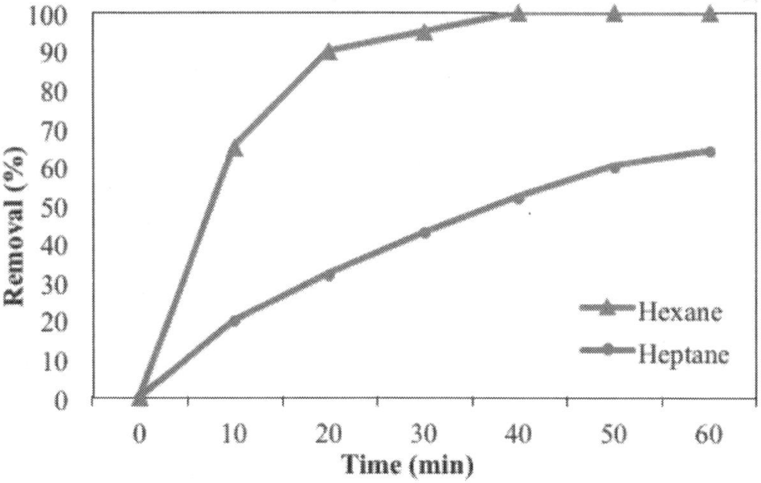

FIGURE 3.16 Effect of diluent on Methylene Blue extraction.

3.4 CONCLUSION

High intensity ultrasound can be used for the production of emulsions. Stable emulsion with uniform droplet size is one of the important tasks in number of applications. It is demonstrated that the ELM technique is very promising in extracting reactive dye from simulated wastewater. In order to achieve the objective several parameters have been studied such as concentration of methylene blue, extraction efficiency, concentration of Tween 80, concentration of diluents and sonication time. The most suitable conditions are at 50 mL distilled water, 4 mL hexane, 0.1 g Tween 80, 1 mL emulsion solution, 50 mL dye solution of 20 ppm concentration and 5 min sonication time. It is observed that maximum dye removal efficiency was achieved at the above mentioned values of the parameters.

NOMENCLATURE

A:ELM	ratio of aqueous liquid is to emulsion liquid membrane
AD	Alzheimer's disease
C_o	initial concentration
C_t	concentration at time, t
ELM	emulsion liquid membrane
ILM	immobilized liquid membrane
MB	methylene blue
O/A	volume ratio of the oil phase to the aqueous phase
O/W	oil in water phase
O/W/O	oil/water/oil systems
t	time
W/O	water in oil phase
W/O/W	water/oil/water systems
μm	micrometer

ACKNOWLEDGMENTS

The authors like to acknowledge the Institute for the facilities provided to carry out the experimental work.

KEYWORDS

- **diluent**
- **emulsion liquid membrane**
- **methylene blue**
- **surfactant**
- **ultrasonication**

REFERENCES

1. Franhenfeld, J. W., & Li, N. N., (1977). Wastewater treatment by liquid ion exchange in liquid membrane system. In: Li, N. N., (Ed.), Recent Developments in Separation Science, vol. 3. CRC Press, pp. 285–292.
2. Kitagawa, T., Nishikawa, Y., Franhenfeld, J. W., & Li, N. N., (1977). Wastewater treatment by liquid membrane process. *Environmental Science and Technology*, *11*, 602–605.
3. Chandan Dasa, Meha Rungta, Gagandeep Arya, Sunando Das Gupta, & Sirshendu De, (2008). Removal of dyes and their mixtures from aqueous solution using liquid emulsion membrane, *Journal of Hazardous Materials*, *159*, 365–371
4. Attef Dâas, & Oualid Hamdaoui, (2010). Extraction of anionic dye from aqueous solutions by emulsion liquid membrane, *Journal of Hazardous Materials*, *178*, 973–981.
5. Meriem Djenouhat, Oualid Hamdaoui, Mahdi Chiha, & Mohamed H. Samar, (2008). Ultrasonication-assisted preparation of water-in-oil emulsions and application to the removal of cationic dyes from water by emulsion liquid membrane Part 2. *Permeation and Stripping Separation and Purification Technology*, *63*, 231–238.
6. Chakravarti, A. K., Chowdhury, S. B, & Mukherjee, D. C., (2000). Liquid membrane multiple emulsion process of separation of copper (π) from wastewaters. *Colloids and Surfaces A*, *166*, 7–25.
7. Datta, S., Bhattacharya, P. K., & Verma, N., (2003). Removal of aniline from aqueous solution in a mixed flow reactor using emulsion liquid membrane, *J. Membr.* Sci, *226*, 185–201.

8. Muthuraman, G. & Palanivelu, K., (2005). Transport of textile anionic dyes using cationic carrier by bulk liquid membrane, *J. Sci. Ind. Res.*, *64*, 529–533.
9. Parida, M. & Pradhan, A. C., (2010). Removal of phenolic compounds from aqueous solutions by adsorption onto manganese nodule leached residue. *J. Hazard. Mater, 173*, 758–764.
10. Kiania, S. & Mousavia, B. S. M., (2013). Ultrasound assisted preparation of water in oil emulsions and their application in arsenic (V) removal from water in an emulsion liquid membrane process, *Ultrasonics Sonochemistry*, *20*(1), 373–377.
11. Gaikwad, S. G. & Pandit, A. B., (2008). Ultrasound emulsification: effect of ultrasonic and physicochemical properties on dispersed phase volume and droplet size. *Ultrason. Sonochem*, *15*, 554–563.
12. Das, C., Meha, R., Gagandeep, A., Das Gupta, S., & Sirshendu, D., (2008). Removal of dyes and their mixtures from aqueous solution using liquid emulsion membrane, *Journal of Hazardous Materials*, *159*, 365–371.
13. Koshio, A., Yudasaka, M., Zhang, M., & Iijima, S., (2001). A simple way to chemically react single-wall carbon nanotubes with organic materials using ultrasonication, *Nano Letters*, *1*(7), 361–363.
14. Zaghbani, N., Amor Hafiane, Mahmoud Dhahbi, (2007). Separation of methylene blue from Aqueous solution by micellar enhanced ultrafiltration, *Separation and Purification Technology*, *55*, 117–124.
15. Ahmad, A. L., Kusumastuti, A., Derek, C. J. C., & Ooi, B. S., (2011). Emulsion liquid membrane for heavy metal removal: an overview on emulsion stabilization and destabilization. *Chem. Eng J.*, *171*, 870–882.
16. Othman, N., Zailani, S. N., & Mili, N., (2011). Recovery of synthetic dye from simulated wastewater using emulsion liquid membrane process containing tri-dodecyl amine as a mobile carrier. *J. Hazard. Mater, 198*, 103–112.
17. Kumbasar, R. A., (2010). Selective extraction of chromium (VI) from multicomponent acidic solutions by emulsion liquid membranes using tributhylphosphate as carrier. *J. Hazard. Mater, 178*, 875–882.

CHAPTER 4

WASTEWATER TREATMENT BY INVERSE FLUIDIZATION TECHNOLOGY

ABANTI SAHOO

Chemical Engineering Department, National Institute of Technology, Rourkela, 769008, Orissa, India

CONTENTS

4.1 INTRODUCTION

Wastes generated by our activities can be classified as solid, liquid and gaseous types which are main sources of environmental pollution. Major junk of liquid wastes is the wastewater which is water no longer needed or suitable for its most recent use and added into an effluent that can be either returned to the water cycle with minimal environmental issues or reused. Wastewater can originate from a combination of domestic, industrial, commercial or agricultural activities, surface runoff or storm water, and from sewer inflow or infiltration. Industrial wastewater is one of the major pollution sources in the pollution of the water environment. With the Industrial Revolution many industries came up. With the rapid development of various industries large quantities of fresh water was used as a raw material. Many kinds of raw materials, intermediate products and wastes are brought into the water when water passes through the industrial process. Thus, this wastewater becomes an essential by-product of almost all industries. Therefore, this by-product wastewater was discharged into water bodies like rivers, lakes and coastal areas resulting serious pollution problems in water environment. As a result the entire eco-system and human life are badly affected. Deteriorating water quality and decrease in vegetation abundance in turn health problems are grave consequences of discharging industrial effluent into the water bodies which have resulted in growing public concern. Again rapid population growth and economic developments lead to the increased water consumption. Limited water resource demands the conservation of water for which people are compelled to treat the wastewater for re-use. Wastewater can be treated in wastewater treatment plants which include physical, chemical and biological treatment processes. Municipal wastewater is treated in sewage treatment plants (which may also be referred to as wastewater treatment plants). Agricultural wastewater may be treated in agricultural wastewater treatment processes, whereas industrial wastewater is treated in industrial wastewater treatment processes.

4.1.1 TYPES AND COMPONENTS OF INDUSTRIAL WASTEWATER

Different industries produce different pollutants with different amounts thereby producing different types of wastewater [1] as mentioned in Table 4.1.

Each industry sector produces its own particular combination of contaminants. Generally industrial wastewater can be divided in two types:

TABLE 4.1 Water Pollutants by the Industrial Sector

Sector	Pollutant
Iron and steel	BOD, COD, oil, metals, acid, phenols, and cyanide
Textile and leather	BOD, solids, sulphates, and chromium
Pulp and paper	BOD, COD, solids and chlorinated organic compounds
Petrochemicals and refineries	BOD, COD, mineral oils, phenols, and chromium
Chemicals	COD, organic chemicals, heavy metals, SS, and cyanide
Non-ferrous metals	Fluorine and SS
Mining	SS, metals, acids, and salts
Micro-electronics	COD and organic chemicals

Inorganic industrial wastewater and Organic industrial wastewater. Inorganic industrial wastewater contains large proportion of suspended matter and inorganic substances in dissolved and undissolved form. These contaminants are separated by sedimentation and often together with chemical flocculation. Organic industrial wastewater contains organic industrial waste flow from those chemical industries which mainly use organic substances for chemical reactions. These organic wastes are eliminated by some special pre-treatment of wastewater followed by biological treatment. Mostly the following industries produce organic wastewater [1].

i. Pharmaceutical and cosmetic industries;
ii. Industries producing organic dye stuff, glue and adhesives;
iii. Soap-Detergent industries;
iv. Pesticides and herbicides industries;
v. Tanneries and leather factories;
vi. Textile industries;
vii. Cellulose and paper industries;
viii. Oil refining industries;
ix. Brewery and fermentation industries;
x. Metal processing industries.

4.1.2 ORGANIC INDUSTRIAL WASTEWATER TREATMENT

Industrial wastewater treatment is understood as the mechanisms and processes which are used to treat waters that have been contaminated by anthropogenic industrial or commercial activities prior to its release into the environment or its re-use. There are various processes available that can be used to treat the wastewaters depending on the type and extent of pollutants. Examples of some industrial wastewater [2] are given in the following subsections.

4.1.2.1 Pharmaceutical Industries

Quality of wastewater produced from pharmaceutical industries is bad. Concentration of COD in such water is in the range of 5000–15000 mg/L and concentration of BOD_5 is relatively low. The value of BOD_5/COD is less than 30% which implies poor biodegradable nature. Such wastewater has bad color and high (or low) pH value. Therefore, strong pretreatment followed by biological treatment process with long reaction time is required for such wastewater.

4.1.2.2 Tannery Plants

In tannery plants large amount of wastewater is generated in different processes such as soaking and washing, liming, rinsing, plumping and bating, chrome tanning, bark tanning and washing and drumming. The quality of wastewater depends only to a slight degree on the types of halides, the mechanical and chemical methods used in tannings. Such wastewater is acidic and contains high chloride content (5 g Cl/L). Concentration of COD is found to be about 1500–2500 mg/L. High amounts of settable substances (10–20 g/L) and emulsified fats are also found to be present in the wastewater that tends to form foam. Dichromate content is found to be up to 2000 mg/L. Therefore, such wastewater needs to be discharged after good treatment.

4.1.2.3 Brewery Industries

This type of wastewater contains high concentrations of suspended solids and detergents. High concentration of COD (1500–5000 mg/L) and BOD_5

(1000–3000 mg/L) are also found to be caused by soluble and insoluble organics. P_{total} (5–30 mg/L), P_{PO4} (2–5 mg/L) and settable solids (3–30 mg/L) are also found in such wastewater which are biodegradable. Therefore, after removal of suspended solids anaerobic and aerobic biological treatment processes can be applied to remove organic contaminants.

4.1.2.4 Iron and Steel Industry

Cooling waters are contaminated with products especially ammonia and cyanide. Contamination of waste streams includes gasification products, such as benzene, naphthalene, anthracene, cyanide, ammonia, phenols, cresols together with a range of more complex organic compounds known collectively as polycyclic aromatic hydrocarbons (PAH). Conversion of iron or steel into sheet, wire or rods requires hot and cold mechanical transformation stages frequently employing water as a lubricant and coolant by which hydraulic oils, tallow and particulate solids are included in contaminants. Wastewaters include acidic rinse waters together with waste acids from the final treatment process of steel. Although many plants operate acid recovery plants (particularly those using hydrochloric acid), where the mineral acid is boiled away from the iron salts, there remains a large volume of highly acid ferrous sulfate or ferrous chloride that to be disposed.

4.1.2.5 Mines and Quarries

The principal wastewaters associated with mines and quarries are slurries of rock particles in water. Coal washing can produce wastewater contaminated by fine particulate hematite and surfactants along with oils and hydraulic oils as common contaminants. Following crushing and extraction of the desirable materials, undesirable materials may become contaminated in the wastewater. For metal mines, this can include unwanted metals such as zinc and other materials such as arsenic. Extraction of high value metals such as gold and silver may generate slimes containing very fine particles in where physical removal of contaminants becomes particularly difficult.

4.1.2.6 Food Industry

Wastewater generated from agricultural and food operations are biodegradable and nontoxic, but that has high concentrations of biochemical oxygen demand (BOD) and suspended solids (SS). The constituents of food and agriculture wastewater are often complex to predict due to the differences in BOD and pH in effluents from vegetable, fruit, and meat products and due to the seasonal nature of food processing and post harvesting. Vegetable washing generates waters with high loads of particulate matter and some dissolved organics. It may also contain surfactants. Animal slaughter and processing produces very strong organic waste from body fluids, such as blood, and gut contents. This wastewater is frequently contaminated by significant levels of antibiotics and growth hormones from the animals and by a variety of pesticides used to control external parasites. Insecticide residues in fleeces are a particular problem in treating waters generated in wool processing. Processing food for sale produces wastes generated from cooking which are often rich in plant organic material and may also contain salt, flavorings, coloring material and acids or alkali. Very significant quantities of oil or fats may also be present.

4.1.2.7 Nuclear Industry

The waste production from the nuclear and radio-chemicals industry is dealt with radioactive wastes.

4.1.3 TREATMENT OF INDUSTRIAL WASTEWATER

Many industries have a need to treat water to obtain very high quality water for demanding purposes. Water treatment produces organic and mineral sludges from filtration and sedimentation. Ion exchange using natural or synthetic resins removes calcium, magnesium and carbonate ions from water, replacing them with hydrogen and hydroxyl ions. Regeneration of ion exchange columns with strong acids and alkalis produces a wastewater rich in hardness ions which are readily precipitated out, especially when in admixture with other wastewaters. Different types of contamination of wastewater require a variety of strategies to remove the contaminants.

4.1.4 SOLIDS REMOVAL

Most solids can be removed using simple sedimentation techniques with the solids recovered as slurry or sludge. Very fine solids and solids with densities close to the density of water pose special problems. In such case filtration or ultra-filtration may be required although; flocculation may be carried out by using alum salts or the polyelectrolytes.

4.1.5 OILS AND GREASE REMOVAL

Many oils can be recovered from open water surfaces by skimming devices. If floating grease forms into solid clumps or mats, a spray bar, aerator or mechanical apparatus can be used to facilitate removal. Dissolving or emulsifying oil using surfactants or solvents usually exacerbates the problem rather than solving it, producing wastewater that is more difficult to treat.

The wastewaters from oil refineries, petrochemical plants, chemical plants, and natural gas processing plants commonly contain gross amounts of oil and suspended solids. Those industries use a device known as an API oil-water separator which works to separate the oil and suspended solids from their wastewater effluents based on the specific gravity difference. The suspended solids settles to the bottom of the separator as a sediment layer, the oil rises to top of the separator and the cleansed wastewater is the middle layer between the oil layer and the solids.

Typically, the oil layer is skimmed off and subsequently re-processed or disposed of and the bottom sediment layer is removed by a chain and flight scraper (or similar device) and a sludge pump. The water layer is sent to further treatment consisting usually of an Electro flotation module for additional removal of any residual oil and then to some type of biological treatment unit for removal of undesirable dissolved chemical compounds. Removal of emulsified oil from water is observed to be possible by inverse fluidization of hydrophobic aero gels [3].

4.2 REMOVAL OF BIODEGRADABLE ORGANICS

Biodegradable organics cannot be removed by means of any instrument. By using suitable process these organics can be removed but complete removal depends upon many process parameters. Thus, any biodegradable organic material of plant or animal origin is usually possible to treat using extended conventional wastewater treatment processes such as activated sludge or trickling filter. Problems can arise if the wastewater is excessively diluted with washing water or is highly concentrated such as neat blood or milk. The presence of cleaning agents, disinfectants, pesticides, or antibiotics can have detrimental impacts on treatment processes. There are different processes [4] available for treatment of biodegradable organic materials. Some of these processes are described below in brief.

4.2.1 ACTIVATED SLUDGE PROCESS

Activated sludge is a biochemical process for treating sewage and industrial wastewater that uses air (or oxygen) and microorganisms to biologically oxidize organic pollutants, producing a waste sludge (or floc) containing the oxidized material. In general, an activated sludge process includes:

- An aeration tank where air (or oxygen) is injected and thoroughly mixed into the wastewater.
- A settling tank (usually referred to as a "clarifier" or "settler") to allow the waste sludge to settle. Part of the waste sludge is recycled to the aeration tank and the remaining waste sludge is removed for further treatment and ultimate disposal.
- A general process for most of the Industrial wastewater where the following technologies are used:
    1. ASP: Activated Sludge process.
    2. SAFF system of Submerged aerobic fixed film system.
    3. MBBR: Moving bed bioreactor (Anox invented this now is considered generic technology).
    4. MBR: Membrane bioreactor.
    5. DAF clarifiers.

6. TBR: Turbo bioreactor Technology (A patented technology of Wockoliver).
7. Filtration technologies.

More information about above can be found on various commercial manufactures like WOIL.

4.2.2 TRICKLING FILTER PROCESS

A trickling filter consists of a bed of rocks, gravel, slag, peat moss, or plastic media over which wastewater flows downward and contacts a layer (or film) of microbial slime covering the bed media. Aerobic conditions are maintained by forced air flowing through the bed or by natural convection of air. The process involves adsorption of organic compounds in the wastewater by the microbial slime layer, diffusion of air into the slime layer to provide the oxygen required for the biochemical oxidation of the organic compounds. The end products include carbon dioxide gas, water and other products of the oxidation. As the slime layer thickens, it becomes difficult for the air to penetrate the layer and an inner anaerobic layer is formed. The components of a complete trickling filter system are: fundamental components:

- A bed of filter medium upon which a layer of microbial slime is promoted and developed.
- An enclosure or a container which houses the bed of filter medium.
- A system for distributing the flow of wastewater over the filter medium.
- A system for removing and disposing of any sludge from the treated effluent.

The treatment of sewage or other wastewater with trickling filters is among the oldest and most well characterized treatment technologies.

4.2.3 REATMENT OF OTHER ORGANICS

Synthetic organic materials including solvents, paints, pharmaceuticals, pesticides, coking products and so forth can be very difficult to treat. Treatment methods are often specific to the material being treated. Methods include advanced oxidation processing, distillation, adsorption, verifications, incineration, chemical immobilization or landfill disposal. Some materials such as some detergents may be capable of biological degradation and in such cases a modified form of wastewater treatment can be used.

4.2.4 TREATMENT OF ACIDS AND ALKALIS

Acids and alkalis can usually be neutralized under controlled conditions. Neutralization frequently produces a precipitate that will require treatment as a solid residue that may also be toxic. In some cases, gasses may be evolved requiring treatment for the gas stream. Some other forms of treatment are usually required following neutralization. Waste streams rich in hardness ions as from de-ionization processes can readily lose the hardness ions in a buildup of precipitated calcium and magnesium salts. This precipitation process can cause severe furring of pipes and can, in extreme cases, because the blockage of disposal pipes. Treatment is by concentration of de-ionized wastewaters and disposal to landfill or by careful pH management of the released wastewater.

4.2.5 TREATMENT OF TOXIC MATERIALS

Toxic materials including many organic materials, metals (such as zinc, silver, cadmium, thallium, etc.) acids, alkalis, non-metallic elements (such as arsenic or selenium) are generally resistant to biological processes unless very dilute. Metals can often be precipitated out by changing the pH or by treatment with other chemicals. Many, however, are resistant to treatment or mitigation and may require concentration followed by land filling or recycling. Dissolved organics can be incinerated within the wastewater by Advanced Oxidation Process.

4.2.6 INVERSE FLUIDIZATION PROCESS

The inverse fluidization process has gained significant importance during the last decade in the field of environmental, biochemical engineering, and oil–water separation. It is also used in the developments in polymer processing, mineral recovery, food processing, biomedical engineering and wastewater treatment (anaerobic digestion of distillery effluent, anaerobic digestion of wine distillery wastewater) [5–7]. This chapter discusses about inverse fluidization in details.

4.3 WHY THE INVERSE FLUIDIZATION

The advantage of inverse fluidization compared to normal fluidization in biochemical engineering is that the bio-film thickness can be controlled in a very narrow range. Some more advantages are that solids can be fluidized at low liquid velocity, the energy expenditure is low and also the solid's attrition is minimum. The other advantages are the high mass transfer rates, minimum carryover of coated microorganisms due to less attrition of solids than normal fluidization and ease of re-fluidization in case of power failure. A binary mixture of different densities of particles can be sorted out in inverse fluidization, as the lower density particles float on the top surface of the liquid.

The inverse fluidized bed technology is considered to be very effective for biological treatment of wastewater in comparison with the normal fluidization. First of all anaerobic or aerobic treatment of wastewaters can be carried out in the inverse fluidization system where suitable gas is passed upward through the bed. Thus, inverse fluidized bed reactor (IFBR) is observed to be more efficient than normal fluidized bed reactor as the control of biofilm thickness is achieved within a very narrow range.

4.3.1 INVERSE FLUIDIZATION TECHNOLOGY

In the gas–liquid–solid system of fluidization, liquid is usually the continuous phase and the process is mainly concerned with the co-current up-flow of the gas and liquid. In this type of fluidization, the particles with density higher than that of the liquid are fluidized by the upward flow of gas and liquid counter to the net gravitational force of the particles. When the density of the particles is smaller than that of the liquid, fluidization can be achieved with downward flow of the liquid counter to the net upward buoyancy force on the particles. In this type of process, the gas flow is upward, counter to the liquid flow. This type of fluidization is termed inverse fluidization [8–10]. Downward or inverse fluidization can be achieved when the density of the particles is less than that of the liquid and the liquid is the continuous phase. This technique is mainly used in biochemical engineering operations, for example, for fermentation and wastewater treatment. Many researchers have conducted many experiments to study the hydrodynamics of inverse gas–liquid–solid fluidized beds using very light particles [11]. A good design is required for the proper functioning of IFBR for which studies on bed dynamics is essential before treating any wastewater.

4.3.2 THEORY AND MECHANISM

Many studies relating to this type of fluidization are reported in the literature, by different researchers on three-phase fluidized beds using solid particles lighter than the liquid phase. The three-phase fluidized beds can be operated with downward flow of the liquid counter to the net upward buoyancy force on the particles. The gas flow is upward, counter to the liquid flow and bed expansion can be supported by the (downward) liquid phase and/or the (upward) gas bubbles [8, 12, 13]. If the density of the particles remains smaller than but close to that of the liquid, the fluidization can be achieved only with an upward gas flow. These multiphase systems are often named as inverse fluidization systems. They have applications in for instance, biotechnology and catalytic chemical processes [14]. At low liquid or gas velocities, the particles form a buoyant packed bed at the top of the column supported by the mesh. As the liquid or gas velocity is increased, bottom layer of the particles start to fluidize and the rest remains in packed condition. With further increase in the velocity, more and more particles at the bottom of the packed bed are fluidized and the bed height increases.

4.3.3 HYDRODYNAMIC BEHAVIORS OF IFB

At one particular velocity, the entire bed is in fluidized condition. The velocity corresponding to this condition is termed as 'minimum fluidization velocity.' Though the entire bed is fluidized, the concentration of solids is not uniform along the axis of the bed. High concentration of solids is observed near the liquid distributor. With further increase in the velocity, the solid holdup becomes uniform throughout the bed. This velocity is termed uniform fluidization velocity [15, 16]. Most of the researchers [15–18] have used water-solid systems during experimentation. Some of these works have shown empirical correlations to predict the minimum inverse fluidization velocity. Experimental studies on the bed expansion characteristics [19] are correlated using bed expansion data for inverse fluidization in terms of either the Richardson or Zaki equation [8, 17–19]. Some have developed new empirical correlations [20]. Sanchez et al. [21] have reported layer inversion from the experimental observations where binary system consisting of high density and low density spherical particle mixtures had been used in the inverse fluidization process. The schematic diagram of inverse fluidized system shown in Figure 4.1 [17, 18]. The column is filled with solid particles first and then topped up with water to the column height. Water and air are used as the liquid and gas phases, all the experiments are performed with no throughput of liquid. Airflow is

FIGURE 4.1(A) Picture of laboratory unit for inverse fluidization.

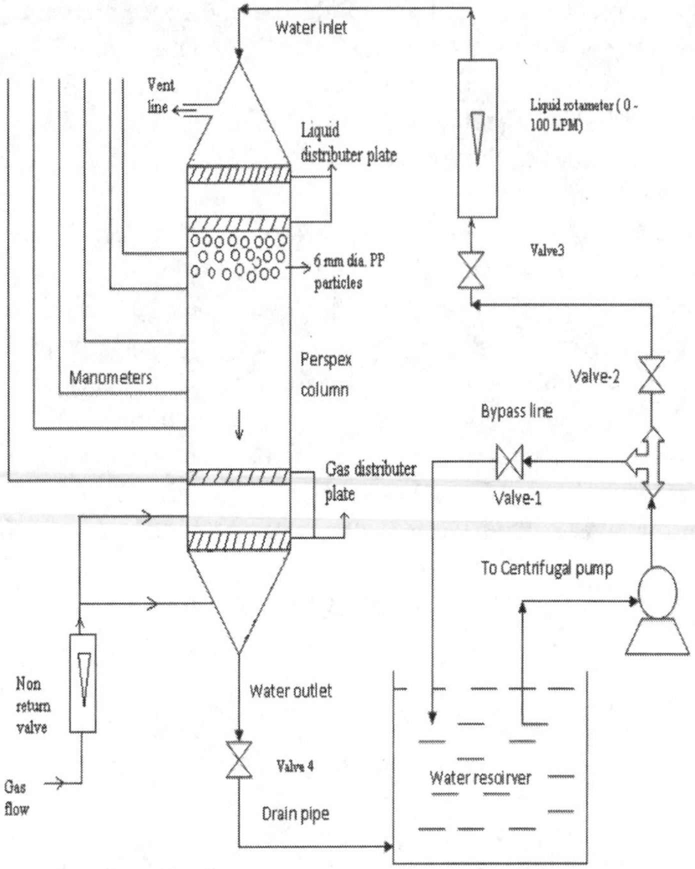

FIGURE 4.1 (b) Picture of laboratory unit for inverse fluidization.

measured by a rotameter before it is allowed to pass through the bed. Three gas spargers, a perforated plate and two flexible membrane spargers are used. The flexible membrane spargers are elastic rubber sheets with punctures which expand with increase in the through put of the gas.

Many researchers have observed the distinct hydrodynamic behaviors of IFB in terms of the pressure drop profiles and the axial solid concentration profiles against the superficial gas velocity U_g. The schematic representations of Inverse fluidization unit with increasing U_g are shown in Figure 4.1(B) [13]. Picture of laboratory unit for inverse fluidization is shown in Figure 4.1(A).

The following observations with respect to bed hydrodynamics (viz. bed voidage, bed pressure drop and bed expansion) are made by various researchers through experimentation.

4.3.3.1 With Respect to Bed Voidage

(a) First case represents the case with no gas flow rate where the bed of particles remained stationary in the system at the top of the column (case 1 in Figure 4.2).

(b) As the gas is injected, the lower portion of the bed begins to be fluidized. With the increase in gas velocity the remaining packed portion of bed moves progressively until the entire bed is fluidized Case-2 in Figure 4.2). Once the packed condition of bed is 'broken,' a minimum superficial gas velocity is required to keep the particles in motion: the gas flow rate can be reduced to a minimum value corresponding to the beginning of the re-aggregation of the particles and characterized by a specific velocity U_{gl}. This velocity is known as the minimum fluidization velocity.

(c) Just beyond U_{gl}, the bed is in the state of minimum fluidization, marking the onset of the expanded- bed regime. As the gas velocity is increased beyond U_{gl}, the bed begins to expand which is shown in

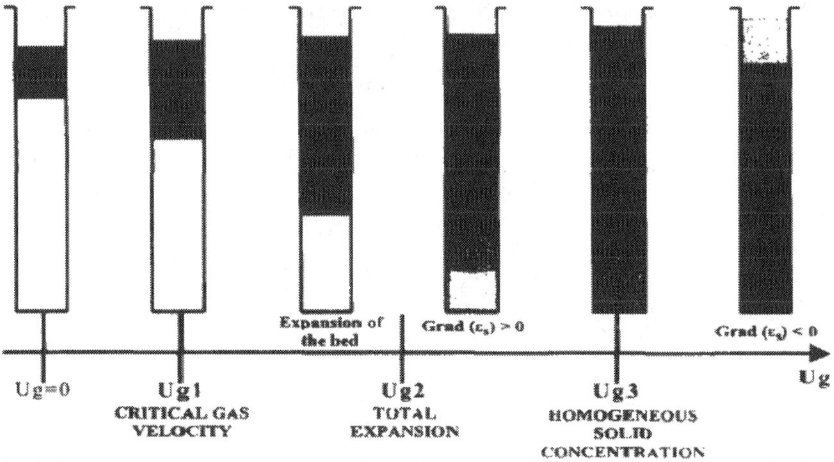

FIGUER 4.2 Schematic representation of Inverse fluidized bed at different velocities of air.

Case-3 in Figure 4.2 and velocity is denoted as U_{g2}. The bed in this condition is composed of two parts: a three-phase fluidized system, on top of a two-phase gas-liquid system. If it is little less well defined as in case-3 of Figure 4.2, the height of the bed is well defined with the membrane spargers within the system. With the perforated plate, the interface of the fluidized bed is more ambiguous and fluctuating.

(d) With further increase in gas velocity, up to a given value, namely U_{g2}, the solid is distributed along the total height of the column. However, particles are not yet uniformly distributed in the column. The solid concentration is higher at the top of the column and progressively decreases axially downward (Case-4 in Figure 4.2).

(e) As the gas velocity is increased beyond U_{g2}, this axial solid concentration gradient decreases and disappears at a given value of the gas velocity, namely, U_{g3}. The solid concentration is then maintained uniform throughout the bed and in such case, ε_g is less than 0. This is shown in Case-5 in Figure 4.2.

(f) When the gas velocity is more than U_{g3}, the axial solid concentration tends to reappear, but inversed, the concentration of solids being higher at the bottom than at the top of the column. In the range of very high gas velocities, the solids would settle down. In such case (Case-6), a three-phase gas–liquid–solid bed is observed at the bottom of the reactor, with a two-phase gas liquid system on top. However, this settling phenomenon is often hindered by the great liquid circulation which drags the solids upward [22]. In such case, ε_g exceeds 0.

Many researchers carried out experiments to study the pressure drop profiles [8, 12, 16, 22–24] in Inverse Fluidized Bed. Effect of liquid velocity on bed voidage is studied by the author who reveals that bed voidage increases with increase in liquid flow. By increasing the static bed height, for example, with more bed weight the bed voidage is observed to decrease (Figure 4.3) which is due to the presence of more particles in the bed under similar conditions.

4.3.3.2 With Respect to Bed Pressure Drop

Attempt is made by the author to study the effects of different system parameters namely, static bed height (Hs), Bed height during experi-

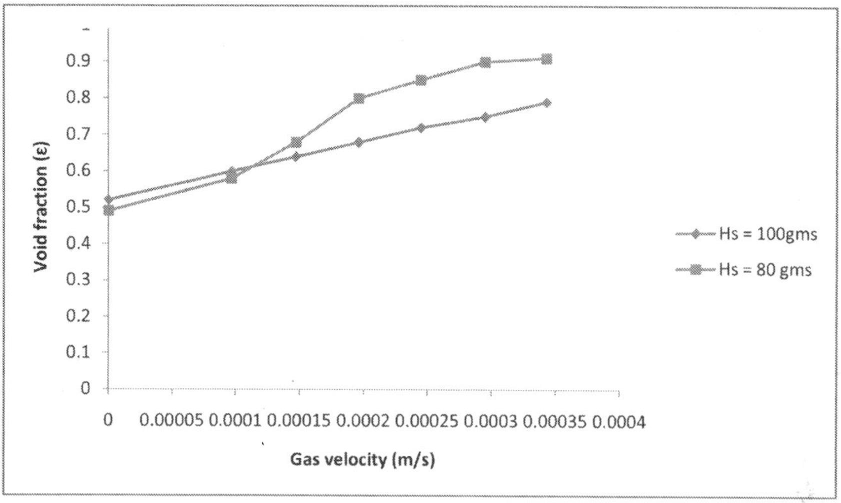

FIGURE 4.3 Effect of liquid velocity on bed voidage for different static bed heights.

mentation (Z), Orifice size of distributor (Do), air flow rate (Fa) and water flow rate (Fw) on bed pressure drop. Attempt is also made to study the pressure profile at different sections of the bed measured through different manometers against liquid flow rate. Bed pressure drop is observed to increase with increase in each of these system parameters (Figures 4.5–4.8). The following correlation (Eq. 1) describes the effect of parameters on the bed pressure drop. The correlation plot is shown in Figure 4.4.

$$\Delta P = 276.19 \times \left[(H_s)^{0.3774}(D_o)^{0.2095}(F_w)^{0.1946}(F_a)^{0.2033}(Z)^{0.1298}\right] \quad (1)$$

4.3.3.3 With Respect to Bed Expansion

Bed expansion characteristic of inverse fluidization is important in designing the IFBR. The experimentally observed data for the bed expansion in the liquid-solid system can be correlated both empirically and semi-empirically for the gas–liquid–solid system in which the gas and liquid flows are counter current. A correlation for the bed expansion is proposed for the inverse gas–liquid–solid fluidization.

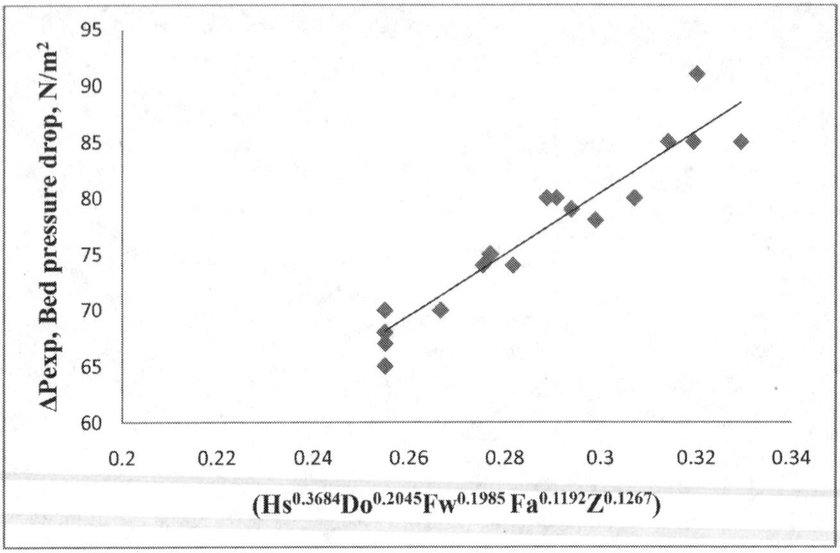

FIGURE 4.4 Correlation plot for bed pressure drop against system parameters.

Different models have been developed for the bed expansion with the superficial liquid velocity. Among all these correlations, the model proposed by Richardson and Zaki [19] has widely been used because of its simplicity and good agreement with the experimental data. The Richardson and Zaki model is mentioned in the following equation:

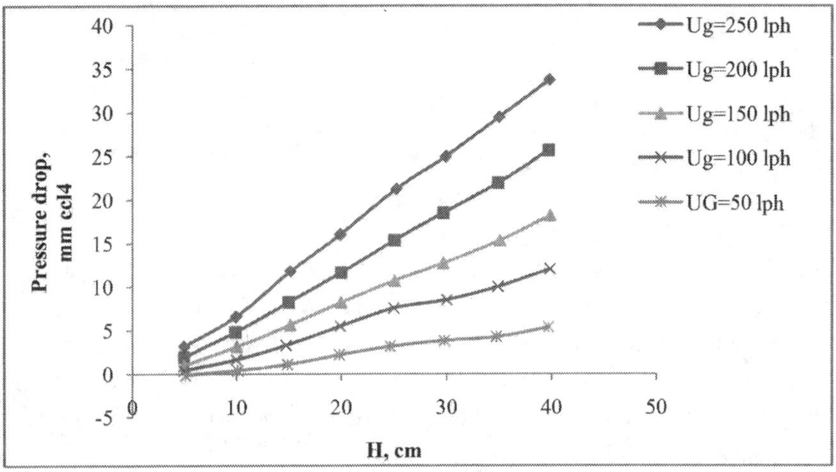

FIGURE 4.5 Pressure variation against bed height for different gas flow rates.

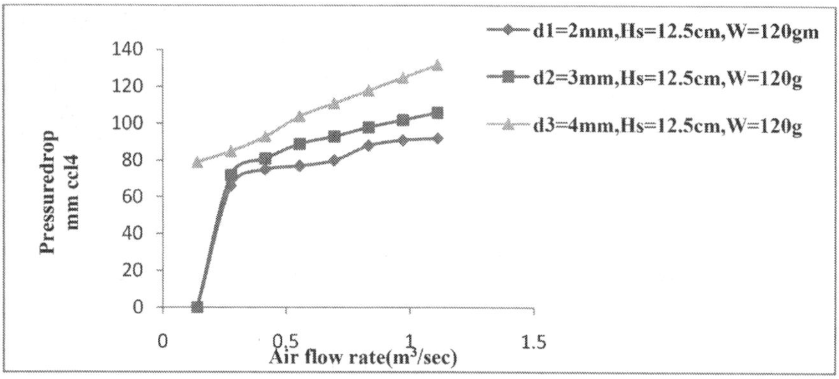

FIGURE 4.6 Comparison of bed pressure drop for different distributor openings.

$$U_t / U_t = \varepsilon^n \qquad (2)$$

where n is the Richardson–Zaki index and is varied with the different range of Reynolds numbers. Karamaner and Nikolov [17] experimentally confirmed that the expressions for the predictions of the exponent, n in the Richardson and Zaki model [19] is effective to describe bed expansion of liquid-solid inverse fluidization.

Using pond water and steel plant wastewater effect of static bed height on bed expansion is investigated (Figure 4.9). Figure 4.9(a) reveals that

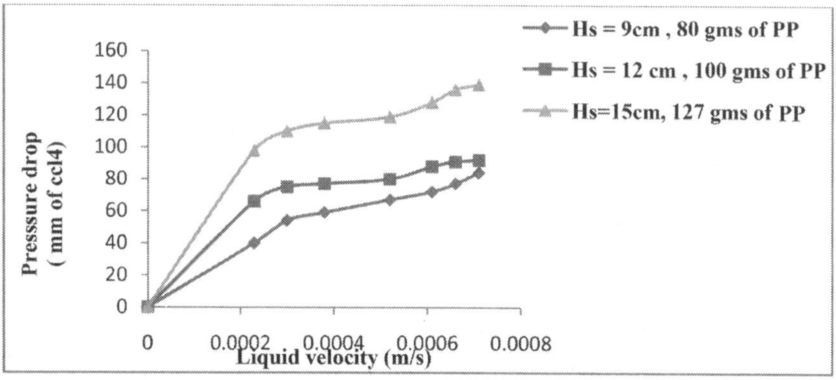

FIGURE 4.7 Bed pressure drop against liquid velocity for different static bed heights.

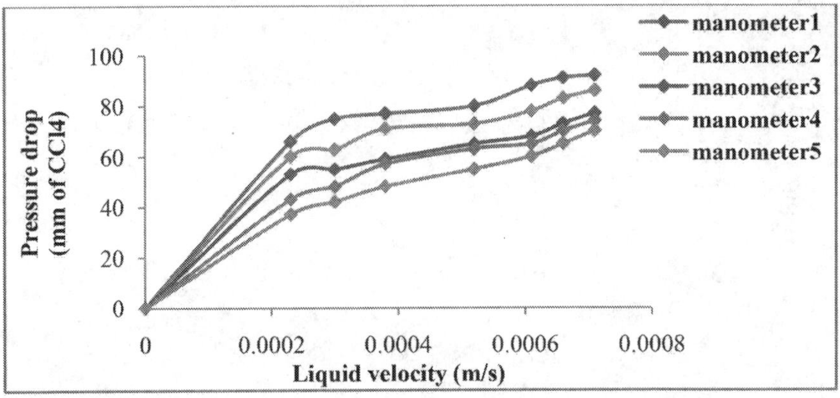

FIGURE 4.8 Comparison of axial pressure drop profile against liquid velocity measured through different manometers.

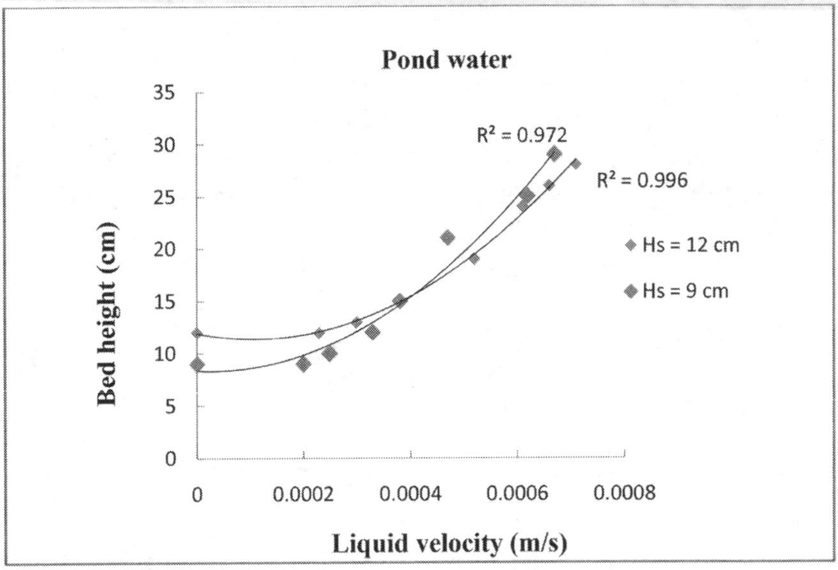

FIGURE 4.9(a) Effect of liquid velocity on bed expansion for different static bed heights.

pond water is relatively clean for which bed expands more for less materials (i.e., Hs = 9 cm) than for Hs = 12 cm. But it is observed from Figure 4.9(b) that static bed height has no impact on bed expansion. But bed expansion increases with the increase in liquid flow rate and more materials move up to more level.

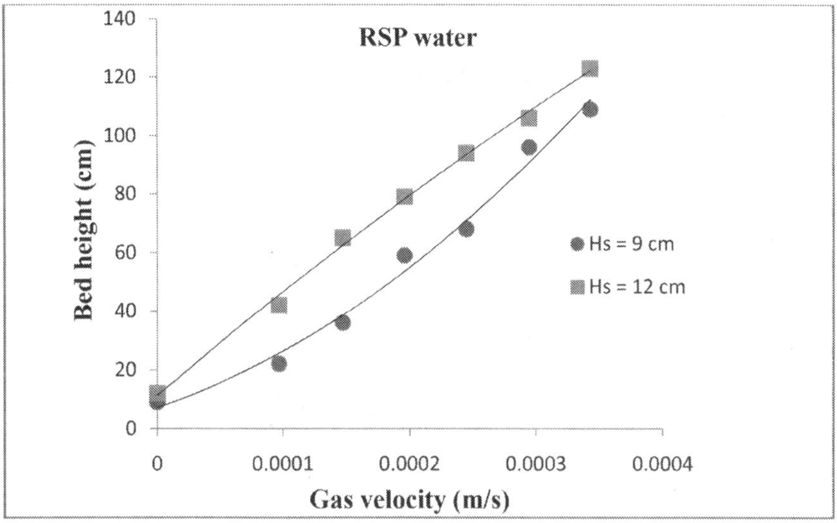

FIGURE 4.9(b) Effect of gas velocity on bed expansion for different static bed heights.

4.3.3.4 With Respect to Gas Holdup

Gas hold up, ε_g is also an important characteristic of inverse fluidization and has been investigated by many researchers. Fan et al. [8] observed that in the inverse bubbling fluidizing bed regime the liquid flow rate has negligible effect on ε_g. However, the superficial gas velocity is found to have little effect on gas hold up during inverse fluidization as per the following empirical correlation.

$$\varepsilon_g = 0.322\varepsilon^{1.35}\left(\frac{U_{go}}{U_{lo}}\right)^{0.18} \tag{3}$$

Zhang et al. [25] have shown some effects of gas hold-up on the hydrodynamics of IFB. They found that increased gas hold-up in the upper region of the bed eventually becomes sufficient to reduce the effective fluid (gas–liquid emulsion) density immediately below the particles to a point where the bed expands downwards into a fluidized state.

Han et al. [26] investigated effects of the superficial gas velocity and particle loading on the gas holdup in the riser. It was concluded that for any particle loading, the gas holdup in the riser increases with increasing superficial gas velocities. For a given particle loading, the increase of the gas holdup in the riser becomes less steep with the increase in the superficial gas velocity at high gas velocity. This may be explained by the fact that gas bubbles tend to coalesce with the increase in the superficial gas velocity.

Experimentally the gas holdup at any point in the riser is measured using the following equations.

$$H_1 \rho_1 g = H_0 + (\varepsilon_l \rho_l + \varepsilon_g \rho_g) g + H_2 \rho_l \tag{4}$$

Since, $\rho_l \rangle \rho_g$ and $\varepsilon_g + \varepsilon_l = 1$ the following equation are also obtained.

$$\varepsilon_g = \frac{\Delta H}{H_0} \tag{5}$$

where, ΔH represents the level difference in the two limbs of the U-tube manometer and H_0 the distance between two measured points.

The gas holdup is observed to be almost uniform, within the range of liquid and gas velocities studied, for example, value of velocity range to be investigated. It has also been observed that the gas holdup is relatively small varying between 0.5% and 2% under the experimental conditions [15].

4.3.3.5 With Respect to Solid Holdup

Buffieare et al. [5] studied the solid hold-up for a gas–liquid–solid system using a laboratory scale inverse turbulent bed reactor and the following relation was found to be justified.

$$\varepsilon_s = \frac{M}{AH_\rho} \tag{6}$$

where, M is the mass of solid taken in the bed (kg), A the cross-sectional area of the column (m^2), H the bed height (m), and ρ is the solid density (kgm^{-3}).

Krishna et al. [15] studied the axial variation of solid holdup in IFB for different liquid velocities at a constant gas velocity. It was observed that at low liquid velocity, the bed is in packed state and the axial solid holdup is constant around 0.6. At higher liquid velocity the solid holdup gradually decreased from the top to the bottom of the bed. Beyond the bed, the solid holdup drops down to that in the freeboard region. It is also observed that the solid holdup along the bed decreases with increase in liquid velocity at constant gas velocity.

4.3.3.6 With Respect to Liquid Holdup

Some researchers studied the variation of average phase holdups with liquid velocity at constant gas *velocity* [15, 26–28]. It is observed that at constant gas velocity, the liquid holdup increases and solid hold up decreases with increasing liquid velocity due to the expansion of the bed under fluidized bed condition. However, the gas hold up is observed to be almost constant with the increase in liquid velocity. Effect of bed height on phase hold ups is studied by the author. It is observed that when the static bed height is changed gas holdup is higher for higher static bed height under conditions of constant water and air flow rate. It is also observed that the gas holdup is relatively small, varying within 6–12%. It is found that more solids are distributed at the top portion of the column than at the bottom portion. At constant gas and water flow rate, the liquid holdup is observed to decrease with increasing static bed heights due to the compactness of the bed with more solid materials. Comparisons of all three-phase holdups against different static bed heights are shown in Figure 4.10. It is observed that with increasing static bed height the liquid holdup decreases but solid and gas holdups increase.

4.3.3.7 With Respect to Minimum Fluidization Velocity

The minimum fluidization velocity, u_{mf} is important in the hydrodynamic studies as it is involved in the design of the system. It is defined as the lowest superficial velocity at which the downward weight of the particles and the drag force due to downward flow of the liquid just counters the

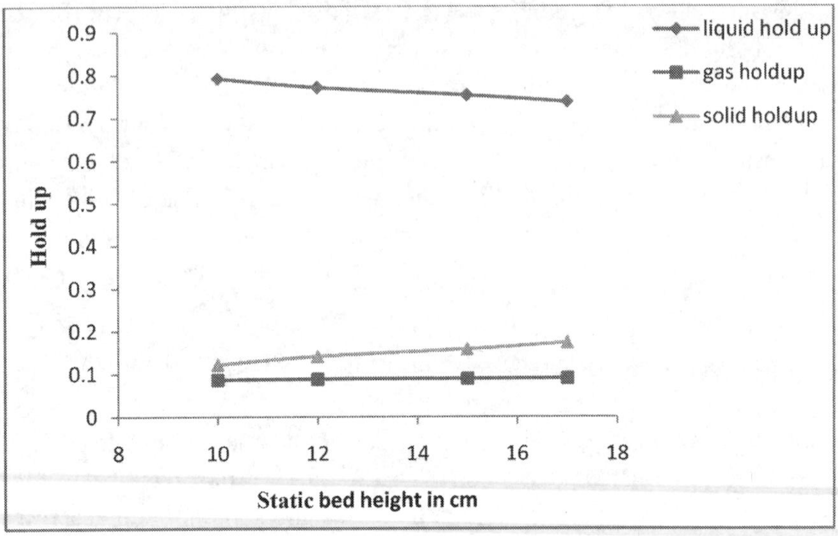

FIGURE 4.10 Phase holdup profiles against static bed height.

upward buoyancy force of the solid particles, for example, the net upward force is equal to the net downward force. Actual bed pressure drop (ΔP) is measured for a particular liquid velocity, U_f by subtracting the solid-free pressure drop, ΔP_o, due to liquid flow at identical liquid velocity, U_f from the observed pressure drop, ΔP_a with solids present. Hence, it can be written as

$$\Delta P = \Delta P_a - \Delta P_0 \tag{7}$$

The plot of $(\Delta \Pi/\Delta Z)_{bed}$ against U_f is used for the determination of minimum fluidization velocity (Figure 4.7) for the inverse fluidization system. The plots of P versus U_f for different sections are shown in Figure 4.8. From these figures it is observed that for the single component system there is a sharp transition between fixed and fluidized bed where, the transition is very gradual in nature for the binary system. The minimum inverse fluidization velocity, u_m is obtained from the point of intersection of the fixed and fluidized bed lines. Figure 4.7 shows the plot of pressure drop, ΔP versus the liquid velocity U_f, for various bed weights using single sized bed materials. From this figure, it is also observed that the pressure drop increases as the bed weight increases but the minimum inverse fluidization velocity is almost constant and is independent of the bed weight. For each gas velocity the minimum liquid fluidization velocity corresponds to the velocity of liquid at which the pressure gradient within the bed is minimum [15]. Hence, the minimum liquid fluidization velocity is obtained from a plot of pressure gradient vs. liquid velocity at a constant gas velocity as shown in Figure 4.7. As the gas velocity is increased, the liquid velocity required for maintaining the bed under minimum fluidization conditions is reduced as observed by several other authors [7, 15, 16, 24, 27].

4.4 APPLICATIONS IN WASTEWATER TREATMENT

4.4.1 ANAEROBIC DIGESTION OF WASTEWATER

Garcia et al. [29] applied inverse fluidization technology for the anaerobic digestion of red wine distillery wastewater. The ground perlite, an expanded volcanic rock was employed as carrier. It was observed that the system achieved 85% TOC removal. It was also found that the main advantage of this system is low energy requirement (because of the low fluidization velocity requirement). There is no need of a settling device for this system as the solids with high-biomass content (whose specific gravity exceeds 1000 kg m^{-3}) accumulate at the bottom of the reactor and can easily be recovered.

The anaerobic fluidized bed reactor utilizes small, fluidized media particles to induce extensive cell immobilization thereby achieving a high reactor biomass hold-up and a long mean cell residence time [30, 31].

The organic loading rate (OLR) and the carbon removal yield (Y) were calculated by Garcia et al. [30] as per the following expressions.

$$OLR = (Q_{in})(C_{in}) / V_{\exp} \quad m^{-3}d^{-1} \tag{8}$$

$$\gamma = (C_{in} - C_{out}) / C_{in} \tag{9}$$

IFB reactor can be considered to be a good option for anaerobic treatment of high strength wastewater, particularly for the treatment of wine distillery wastewater. The system can attain high organic loading rate (OLR) with good chemical oxygen demand (COD) removal rates. Thus, a good stability to the variations in OLR and HRT (hydraulic retention time) can be exhibited. The effect of additives on liquid coalescing properties and thereby the effect on hydrodynamic properties of a three phase inverse fluidized bed were also checked by the researcher.

Sowmeyan and Swaminathan [32, 33] evaluated the feasibility of an inverse fluidized bed reactor for the anaerobic digestion of distillery effluent with a carrier material (perlite having specific surface area of 7.010 m^2/g and specific gravity of 295 kg/m^3) that allows low energy requirements for fluidization process thereby providing also a good surface for biomass attachment and development. This work was intended to test a new kind of anaerobic digestion with attached biomass, for example, the inverse turbulent bed. The application to the treatment of distillery wastewater presented satisfactory results compared to other reactors. The physical characteristics of carrier material (perlite) without and with biomass growth on the carrier material and the biogas production during an apparent steady state period in an inverse anaerobic fluidized bed reactor (IAFBR) used for treating high strength organic wastewater were also studied by these authors by varying different hydraulic retention time and organic loads.

4.4.2 AEROBIC DIGESTION OF WASTEWATER

Allia et al. [34] investigated biological treatment in two different systems namely, aerated closed system and three phase fluidized bed reactors for removal of hydrocarbons from refinery wastewaters. This typical bioreactor allowed a good removal rate of pollutants. According to them the elimination of hydrocarbons in a three-phase fluidized bed bioreactor is lower than that in an aerated closed bioreactor. It was observed that the elimination in an aerated closed reactor is around 100%, while that in a three phase fluidized bed reactor is within 87.77–95%.

Sokol et al. [35, 36] investigated the biological wastewater treatment in an inverse fluidized reactor (IFBR) in which polypropylene particles of density 910 kg/m^3 were fluidized by an upward flow of air. Measurements of chemical oxygen demand (COD) against residence time *t* were carried out for various ratios of settled bed volume to reactor volume (V_b/V_R) and air velocities U_g. The largest COD removal was attained when the reactor was operated at $(V_b/V_R)_m$= 0.55 and an air velocity, U_{gm}= 0.024 m/s. Under these conditions, the value of COD was practically at steady state for around more than 30 hours of time duration. Thus, these values of $(V_b/V_R)_m$, U_{gm} and tcan be considered as the optimal operating parameters for a bioreactor when used in treatment of industrial wastewaters. As per their observation a decrease in COD from 36,650 to 1950 mg/L (i.e., a 95% COD reduction) is achieved when the reactor is optimally controlled at $(V_b/V_R)_m$= 0.55, U_{gm}= 0.024 m/s and t = 30 h. The pH was controlled in the range 6.5–7.0 and the temperature was maintained at 28–30°C. The biomass loading is successfully controlled in an IFBR with support particles whose particle density was smaller than that of liquid. The steady-state biomass loading depends on the ratio (V_b/V_R) and an air velocity, U_g. The steady-state biomass loading is attained after approximately 2-weeks operation (during culture time after switching from the batch to the continuous operation). During the culture period, after changing the ratio of (V_b/V_R) at a particular U_g, the steady-state growth rate of cells on the particles is achieved after about 6-days of operation. For a set ratio (V_b/V_R), the biomass loading depends on *Ug*. With the change in Ugat a set (V_b/V_R) ratio, the new steady-state biomass loading occurs after the culturing for about 2 days.

Polypropylene particles of density 910 kg/m^3 are selected by the author and are allowed to be fluidized by an upward counter current flow of gas and downward liquid flow. Measurements of chemical oxygen demand (COD) versus residence time, *t* are performed for various ratios of settled bed volume to bioreactor volume, (V_b/V_R) and air velocities *u* to determine the optimal operating parameters for a reactor, for example, the values of the parameters (V_b/V_R), *u* and *t* for which the largest reduction in COD occurs [29, 36].

The aerobic treatment of wastewaters is investigated in the inverse fluidized bed biofilm reactor (IFBBR) in which polypropylene particles of density 910 kg/m^3 were fluidized by an upward co-current flow of gas and liquid by Sokol and Korpal [35]. Measurements of chemical oxygen demand (COD) versus residence time *t* were performedfor various ratios of settled bed volume to bioreactor volume (V_b/V_R) and air velocities (u) by the researchers. The optimum operating parameters for a reactor (i.e., the values of V_b/V_R ratio, u and t for which the largest reduction in COD occurs) are also determined by them.

4.4.3 TREATMENT OF OIL-IN-WATER EMULSION

Wang et al. [37] fluidized different size ranges of surface-treated hydrophobic silica aerogels (Nanogel) by a downward flow of an oil-in-water emulsion in an inverse fluidization mode.

The oil removal efficiency for a dilute oil in water emulsion (1000 ppm COD or lower) was stabilized and the capacity of the inverse fluidized for Nanogel granules were also studied by Josh et al. [38]. The concentration of oil was also monitored by analyzing the chemical oxygen demand (COD) at several time intervals. Some variables that were changed to compare the oil removal efficiency are the fluid superficial velocity, particle size, particle type, amount of particles, initial bed height and concentration of oil upstream of the fluidized bed. The variables that were monitored during the oil removal are the bed height and COD levels. Oil removal by an inverse fluidized bed was considered acceptable only if the COD levels remained below 100 mg/L or 100 ppm. The fluidized bed height and the time at which the 100 mg/L COD level are reached are considered as proper. It is thus concluded that the longer the time required to reach the 100 mg/L COD level, the better is the oil removal efficiency.

Downward or inverse fluidization can be achieved when the density of the particles is less than that of the liquid and the liquid is the continuous phase. This technique is mainly used in biochemical engineering operations, for example, for fermentation and wastewater treatment. Many researchers have conducted many experiments to study the hydrodynamics of inverse fluidization using light particles [8, 11, 12, 16]. The use of low density particles in a three-phase IFB allows the control of biomass loading in the bioreactor. Steady biomass loading is easily attained in a bioreactor with lighter particles. The biomass loading is also successfully controlled in an inverse fluidized bed bio-reactor where the particles are fluidized by down flow of the liquid. In this study, the aerobic treatment of phenolic wastewaters (effluent from Steel plant) is carried out using an inverse fluidized bed bioreactor containing polypropylene particles of density 910 kg/m^3. In particular, the settled bed volume to bioreactor volume (Vb/Vr) ratioand the gas flow rate at which the largest COD reduction is obtained are determined. The efficiencies of the bioreactor in treatment of wastewater without and with the addition of mineral salts are also studied.

In the gas–liquid–solid system of fluidization liquid is the continuous phase and the process is mainly concerned with the co-current up-flow of the gas and liquid. In this type of fluidization, the particles with density higher than that of the liquid are fluidized by the upward flow of gas and liquid counter to the net gravitational force of the particles. When the density of the particles is less than that of the liquid, fluidization can be achieved with downward flow of the liquid counter to the net upward buoyancy force on the particles. In this type of process the gas flow is upward, counter to the liquid flow. This type of fluidization is termed inverse fluidization.

In a bioreactor, fluidization of lighter bed materials can be carried out either by co-current or by counter current flow of gas and liquid. In both the cases inverse fluidization occurs. Usually for co-current flow both the gas and liquid flow upward through the bed and counter current flow is achieved by a downward flow of liquid with upward flow of gas [7, 22]. In both cases the bed expands downwards into the less dense mixture of gas and liquid thereby providing large surface area for bio-film formation on each particle. Thus, the biomass loading on the particles can be increased and biomass growth takes place. A fluidized bed biological treatment technology owes its high-rate success to much higher surface area and biomass concentration than those that can be achieved in the conventional treatment processes. That is why the importance of a three-phase IFB bioreactor is increasing in comparison with the conventional suspended growth and fixed-film wastewater treatment processes [29, 36]. Treatment of refinery wastewater in a three phase IFB bioreactor with polypropylene particles gives about 70% reduction in total COD. In aerobic wastewater treatment, the air and liquid flow rates are carefully controlled to maintain adequate bed expansion and gas–liquid mass transfer thereby minimizing shear effects and particle elutriation [36].

4.5 EXPERIMENTATION: CASE STUDIES

4.5.1 CHEMICAL OXYGEN DEMAND

In environmental chemistry, the chemical oxygen demand (COD) test is commonly used to measure indirectly the amount of organic compounds in water. It is often used as a rough approximation of the pollution level of concerned wastewater. In the present work COD has been calculated using the following formula:

$$COD = \left[\frac{(A - B) \times n \times 1000 \times 8}{V_o} \right] \quad (10)$$

where, A = mL of ferrous ammonium sulphate used for blank one; B = mL of ferrous ammonium sulphate used for original sample; n = normality of ferrous ammonium sulphate; and V_0 = mL of the sample.

4.5.2 BIOCHEMICAL OXYGEN DEMAND

The biochemical oxygen demand (BOD) is a measure of the oxygen utilized by micro-organisms under specified experimental condition during biological oxidation of organic matter contained in the liquid waste. BOD at 25°C temperature of the sample has been calculated using the following formula.

$$BOD = \left[\frac{(D_0 - D_5) \times V}{V_0} \right] \quad (11)$$

where, D_0 is the initial D.O. of diluted sample; D_5 is the D.O. of the diluted sample at the end of 5 days; Vis total volume of wastewater and unseeded dilution water; and V_0 is the volume of the sample wastewater.

4.5.3 TOTAL SUSPENDED SOLIDS

Total suspended solids (TSS) is also an another water quality measurement which is usually expressed in mg/L. Thus, total suspended solid in this study has been calculated using the following formula.

$$TSS = \left[\frac{(W_2 - W_1) \times 1000}{V_0} \right] \tag{12}$$

where, W_1 and W_2 are initial and final weight of filter in gram.

4.5.4 METHODS

The experimental set up as shown in Figure 4.1 consists of a cylindrical column (ID = 100 mm, thickness = 3 mm and height = 1240 mm) of Perspex material. There is one conical head at the top with a distributor for uniform liquid distribution. Another conical bottom is provided at bottom section for allowing uniform gas distribution. The distributors at the top and bottom of the reactor also prevent the flow of bed materials out of the reactor. The pressure taps with the manometers are provided alternately on either side of the wall of the column with a spacing of 100 mm gap. Carbon tetrachloride is used as the manometric fluid. The liquid discharge section connects a pipe to the reservoir to transfer the liquid into the tank so that it can be re-circulated if required. A control valve is also provided in the discharge line to adjust the flow rate of fluid.

The column is loaded with solid particles of a particular size and density for the required bed height from the top of the column. The industrial effluent is used as the liquid medium and oxygen/air is supplied from the bottom as the gas. Then the system is allowed to fluidize till steady state is attained by adjustment of inlet and discharge flow rate. The pressure drops across the test section are noted through different manometers. The biomass sloughed off from the particles is separated from the effluent in a vessel and removed from the system. pH value of fluid inside the reactor is controlled by using pH controller connected to the bio-reactor. Different system parameters are varied to study their effects on the COD reduction during wastewater treatment. COD, BOD and TSS are also calculated experimentally as per the literature [39] which are expressed in Eqs. (10)–(12) and the methodologies are described in the following sections.

4.5.5 *FEED AND MICROORGANISMS*

The wastewater from the selected industry is used as the growing medium. The chemical properties of the wastewater before and after the treatment are found out. The wastewater is enriched with mineral salts by adding the following chemicals (mg/L): $(NH4)_2SO_4$ – 500; KH_2PO_4 – 200; $MgCl_2$ – 30; NaCl – 30; $CaCl_2$ – 20; and $FeCl_3$ – 7 as recommended by Sokol and Migiro [40]. The selected microbe is used as the inoculum.

4.5.5.1 Micro-organism Culture

It is observed from literature that the largest values of oxygen hold-up (ε_g) and thereby the largest interfacial area is obtained when a bioreactor with polypropylene particles is controlled at Vb/Vr ratio in the range 0.50–0.60. Therefore, experiments are carried out for the Vb/Vr ratios equal to 0.50, 0.55 and 0.60 separately. Different species of microorganisms are selected depending upon the nature or composition of the wastewater collected from different industries.

The particles and the growth medium are introduced into the bioreactor to get the V_b/V_r ratio of 0.50, 0.55 and 0.6 separately. A batch culture is initiated by introducing about 1 liter of the inoculum into the bioreactor to start the growth of the microorganisms on the particles. Then the culture is incubated for approximately 48 h to encourage cell growth and for the adhesion of freely suspended biomass on the support [21]. The oxygen is supplied at a flow rate of 0.02 m^3/s which is sufficient for biomass growth and the culturing is continued until constant biomass loading is achieved in the bioreactor. The occurrence of the steady state biomass loading is established by increasing the oxygen flow rate. If the constant biomass loading is observed to be attained in approximately three-four days of culturing, it can be understood as attainment of steady state. It is said that the steady state occurs when the weight of biomass in two consecutive samples differs by less than 5%. Mass of cells grown on the support is measured by weighing. For measuring the mass, the biomass (cells grown on support particles) is scraped out using a knife and dried at 105°C for 1 hour and then weighed. The microbial growth kinetics in wastewater added with microbial culture is studied by varying the optical density of the sample with time. It is observed that steady state growth of microbial culture for wastewater occurs at approximately 62 h of time which has been shown in Figure 4.11. When the steady state of biomass loading is achieved, a liquid sample is withdrawn from the bioreactor. COD measurement is carried out as per the procedure [30] and is calculated as per Eq. (10). It is observed that COD value remains constant as has already been concluded by many researchers that once the constant biomass loading occurs in the bioreactor, the COD value is practically at steady state. Thus, the effect of the oxygen flow rate on the culturing is investigated by varying the oxygen flow rate till the steady state biomass loading is achieved. Again V_b/V_r ratio is varied for steady state biomass loading. Finally the COD is measured for optimum conditions. COD measurements are also carried out against time with and without the addition of mineral salts to the raw sample [36]. The effect of the amount of nutrient salts on biomass growth is also studied and has been shown in Figure 4.12. The comparison plot for COD removal from wastewater with and without enrichment by mineral salt is given in this figure. It is observed from this figure that in treatment of 'raw' wastewater, COD value decreased from 380 to 235 mg/L, which is, only about 40% COD removal. In the treatment of wastewater enriched with nutrient salts, a decrease in COD value is observed from 380 to 120 mg/L which is, approximately 68% COD

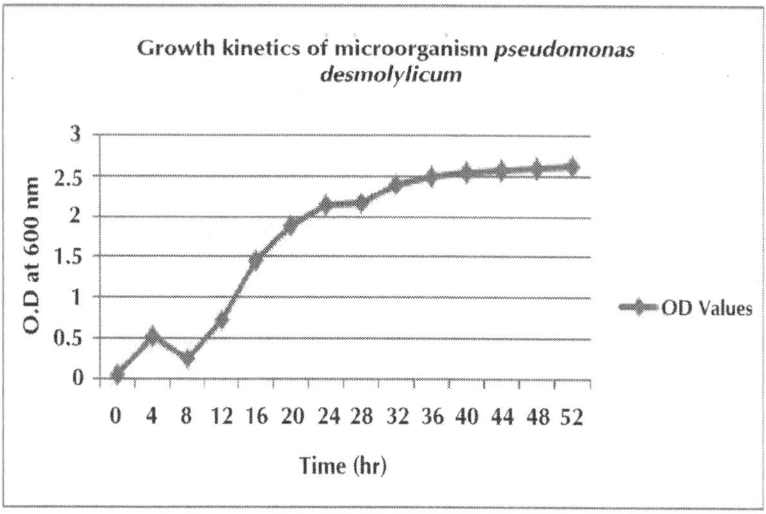

FIGURE 4.11 Growth kinetics of microorganism for dairy farm waste water.

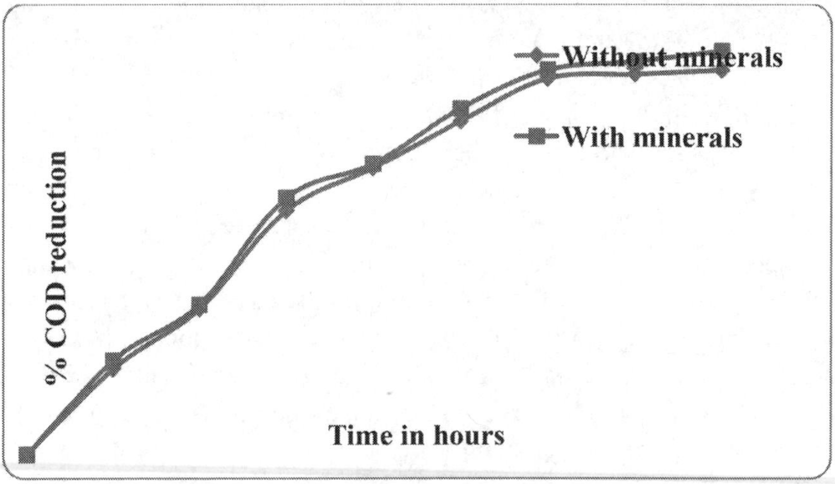

FIGURE 4.12 Effect of addition of salt on %COD reduction.

reduction. No improvement in treatment efficiency is observed with further addition of mineral salts. Thus, the recommended amount of mineral salt addition to wastewater before treatment was found to be sufficient for biomass growth.

The wastewater enriched with nutrient salts is introduced into the bioreactor at a dilution rate of 0.020 h^{-1}. It is observed that the V_b/V_r ratios used for the bioreactor are smaller than the critical V_b/V_r ratio because of which movement of the whole bed was possible. Similarly, when the oxygen flow velocity used in the experiments is maintained smaller than the critical oxygen velocity, the entire bed settles at the bottom of the bioreactor. Under minimum fluidization condition, the bioreactor is operated at a V_b/V_r ratio smaller than the V_b/V_r ratio corresponding to steady biomass growth. Wastewater without addition of any mineral salt is also treated at different conditions of V_b/V_r ratios and different oxygen velocities.

From Figure 4.11, it is observed that steady COD value is attained at duration of 62 h. The bioreactor is observed to be optimally controlled at V_b/V_r= 0.55 and gas flow rate= 40.0 liters/hour. The constant biomass loading is found to be obtained on approximately 6 days of culturing.

4.5.6 WASTEWATER FROM STEEL INDUSTRY

4.5.6.1 Microorganism: Pseudomonas aeruginosae

Wastewater from steel industry contains mainly phenol along with ammonia and cyanides. Phenols (both monohydric and polyhydric) and derivatives of phenols are organic pollutants; potentially toxic to human, aquatic and plant life. Thus, attempt is made to treat the phenolic wastewater. A three-phase inverse (gas–liquid–solid) fluidized bed bioreactor (IFB) has been successfully applied for aerobic/anaerobic biological treatment of industrial and municipal wastewaters [8, 11, 15]. Three phase IFB has an inherent advantage over the two-phase inverse (liquid–solid) fluidized bed bioreactor due to the improved supply of oxygen to microorganisms and the elimination of the need for pre-aeration of the liquid [12, 13, 17]. Effect of different system parameters namely, static bed height (Hs), gas flow rate (Fg), water flow rate (Fw), time of contact of cultured microbe with the wastewater (t) and V_b/V_r ratio on COD reduction was Analyzed. The following correlations were developed on the basis of dimensional analysis and fractional factorial design approach. The correlation plot is shown in Figure 4.13.

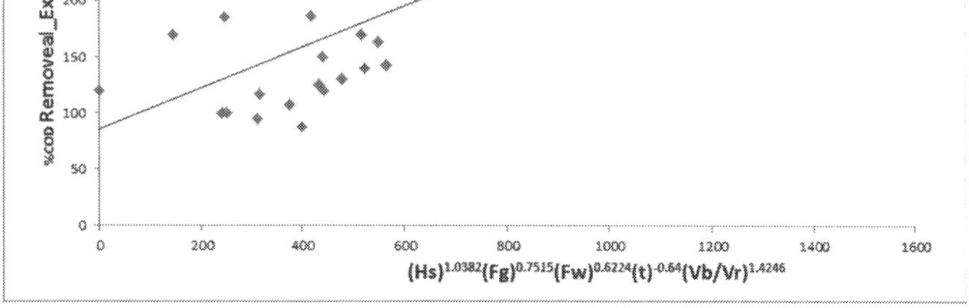

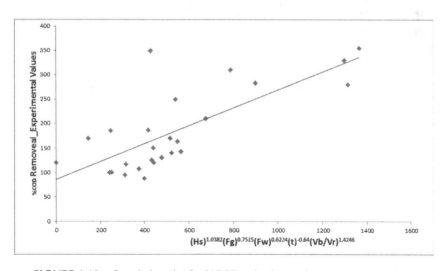

FIGURE 4.13 Correlation plot for %COD reduction against system parameters.

4.5.6.2 By Dimensional Analysis

$$\%COD_{reduction} = 0.7297 \times \left[(H_s)^{1.038} (F_g)^{0.752} (F_w)^{0.622} (t)^{-0.64} \left(\frac{V_b}{V_r} \right)^{1.425} \right]^{0.862} \quad (13)$$

4.5.6.3 By Statistical Analysis (2^{5-2} Fractional Factorial Design)

$$\%COD_{reduction} = 235 + 6.25 * H_s - 16.25 * F_g - 5 * F_W - 85 * t + 22.5\left(\frac{V_b}{V_r} \right) - 8.75 * H_s.F_W + 47.5 * F_g.F_W \quad (14)$$

4.5.7 WASTEWATER FROM SUGAR INDUSTRY

Wastewater from sugarcane industry generally contains organic materials such as carbohydrates and proteins. Sugarcane industry has significant wastewater production. Unfortunately, due to the lack of know-how and financial support, most of sugarcane industries in developing countries discharge their wastewater without adequate treatment.

Therefore, it is required to examine the biodegradability of organic components of wastewater from sugarcane industry wastewater and to examine the possibility of using upflow aerobic/anaerobic sludge blanket reactor to treat wastewater from sugarcane industry by evaluating the start-up process and the operation performance. Thus, the microbe, *Staphylococcus aureus* was selected and cultured (Figure 4.14). Same procedure was adopted for treatment of wastewater from sugar industry.

4.5.8 WASTEWATER FROM PAPER AND PULP INDUSTRY

Wastewater from Paper and pulp industry contains chlorinated phenolic compounds and adsorbable organic halides (AOX). Pulp and paper industry is considered to be a significant contributor of pollutant dis-

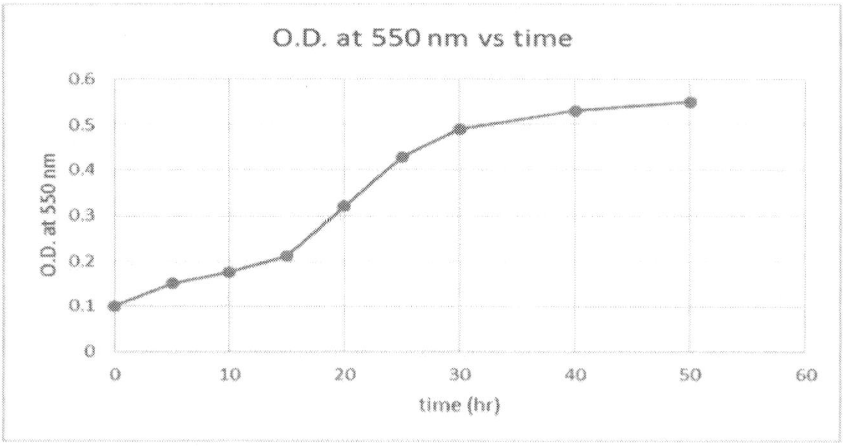

FIGURE 4.14 Growth Kinetics of microorganism – *Staphylococcus aureus*.

charges to the environment. Raw wastewater from pulp and paper mills are very polluting and its chemical oxygen demand can be as high as 11,000 mg/L for which it needs to be treated before discharging to reduce any possible impacts on the aquatic environment. Therefore, microbe, Pseudomonas Desmolyticum was selected for treating wastewater in the Inverse Fluidized Bed. Experimental setup is shown in Figure 4.1. The same procedure is followed for the treatment of wastewater. Characteristics of a typical paper mill wastewater is given in Table 4.2. The effect of different system parameters on COD reduction are studied and shown in Figure 4.15.

TABLE 4.2 Characteristics of Paper Mill Industry Wastewater

Parameter	Values (mg/L)
pH	4.4–5.3
Chemical oxygen demand	11200–12500
Biological oxygen demand	5300–6200
Total solid	8253–9451
Total organic carbon	312
Total suspended solid	2450
Total dissolved solid	5020–5500

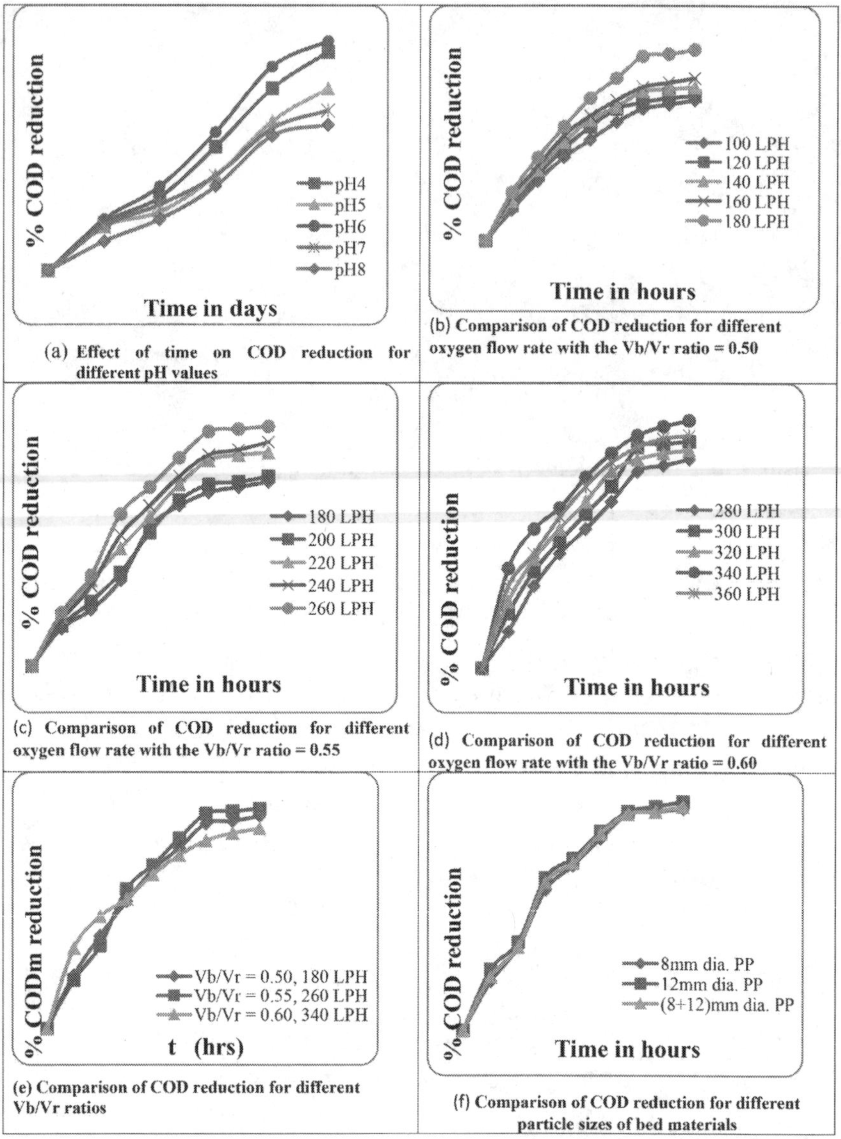

FIGURE 4.15 Effects of different system parameters on %COD reduction for paper mill wastewater

4.6 RESULTS AND DISCUSSION

It is observed that the minimum fluidization velocity and solid hold-up for the IFB increase with the increase in initial static bed height (Figures 4.8 and 4.10). It is also observed that the different system parameters affect the hydrodynamic behavior of the bed thereby affecting the IFB process. The main objective of this study was to concern the biomass growth onto the solid carrier which would be useful for treatment of the industrial effluents. But the hydrodynamic condition of the IFB affects the biomass growth. Therefore, it is essential to design the IFB properly so that it can be used as an effective bioreactor thereby controlling the BOD/COD requirement for the wastewater treatment.

The developed correlation for the pressure drop can be used safely for the design of the bio-reactor to be used for treatment of industrial wastewater treatment. These studies reveals that BOD, COD, TSS and TOC of wastewater can be controlled effectively by IFB. pH of effluents can also be maintained at optimum level. Therefore, inverse fluidized bed can be used as an effective bioreactor to treat all types of industrial effluents biologically. It can also be said that inverse fluidization technique is a better alternative successful for treatment of wastewater from paper mill

KEYWORDS

- bed dynamics
- BOD
- COD
- microorganism effect
- statistical analysis
- V_b/V_r ratio
- waste water treatment

REFERENCES

1. Hanchang Shi, Point Sources of Pollution: Local effects and its Control, Vol. 1, Industrial Wastewater Types, Amounts and Effects. (2002), 191–203.
2. http://www.iwawaterwiki.org/xwiki/bin/view/Articles/IndustrialWastewaterTreatment.
3. Ding Wang, Trent Silbaugh, Robert Pfeffer, & Lin, Y. S. (2010). Removal of emulsified oil from water by inverse fluidization of hydrophobic aerogels, *Powder Technology, 298*–309.
4. http://www.wikiwand.com/en/Industrial_wastewater_treatment#/cite_note-Metcalf-1
5. Buffieare, P., Bergeon, J., & Moletta, R., (2000). The inverse turbulent bed: a novel bioreactor for anaerobic treatment. *Water Res. 34*(2), 673–677.
6. Rajasimman, M., &Karthikeyan, C., (2007). Aerobic digestion of starch wastewater in a fluidized bed bioreactor with low-density biomass support. *Journal of Hazardous Materials, 143*(1–2), 82–86.
7. Arnaiz, C., Buffiere, P., Elmaleh, S., Lebrato, J. & Moletta, R. (2003). Anaerobic digestion of dairy wastewater by inverse fluidization: The inverse fluidized bed and the inverse turbulent bed reactors, *Environmental Technology, 24*(11), 1431–1443.
8. Fan, L., Katsuhiko, M., & Chern, S., (1982). Hydrodynamic characteristics of inverse fluidization in liquid-solid and gas–liquid–solid systems, *The Chemical Engineering Journal, 24*, 143–150.
9. Benedict & Moletta, (1999). Some hydrodynamic characteristics of inverse three-phase fluidized bed reactor. *Chemical Engineering Science, 54*, 1233–1242.
10. Chen, X., & Zheng, P., (2010). Bed expansion behavior and sensitivity analysis for super-high rate anaerobic bioreactor. *Univ. Sci. B (Biomed & Biotechnol), 11*(2), 79–86.
11. Chern, H. S., Muroyama, K., & Fan, L. S., (1982). Hydrodynamics of constrained inverse fluidization and semi-fluidization in a gas–liquid–solid system, *Chemical Engineering Science, 38*(8), 1167–1174.
12. Krishnaiah, K., Guru, S., & Sekar, V., (1993). Hydrodynamic studies on inverse gas—liquid—solid fluidization, *The Chemical Engineering Journal, 51*, 109–112.
13. Comte, M. P., Bastoul, D., Hebrard, G., Roustan, M., & Lazarova, V., (1997). Hydrodynamics of a three-phase fluidized bed—the inverse turbulent bed, *Chemical Engineering Science, 52*(21–22), 3971–3977.
14. Krishna, S. V., Bandaru, S. R., Murthy, D. V. S., & Krishnaiah, K., (2003). Some hydrodynamic aspects of 3-phase inverse fluidized bed, *China Particology, 5*, 351–356.
15. Krishna, S., Murthy D., & Krishnaiah, K., (2007). Some hydrodynamic aspects of 3-phase inverse fluidized bed. *China Particology, 5*, 351–356.
16. Ibrahim, Y., Briens, A., Margaritis, C., & Bergongnou, M. A. (1996). Hydrodynamic Characteristics of a Three-Phase Inverse Fluidized-Bed Column, *AIChE Journal, 42,* 1889–1900.
17. Karamaner, D. G.; Nikolov, L. N. (1992). Bed Expansion of Liquid−Solid Inverse Fluidization. *AIChE J. 38*, 1916.

18. Renganathan T., & Krishnaiah, K., (2003). Prediction of minimum fluidization velocity in two and three phase inverse fluidized beds, *The Canadian Journal of Chemical Engineering Journal, 81*, 853–860.
19. Richardson, J. &Zaki, F., (1954). Sedimentation and fluidization, *Trans I Chem.* Part E, 32–35.
20. Briens, C. L., Ibrahim, Y. A. A., Margaritis, A., & Bergougnou, M. A., (1999). Effect of coalescence inhibitors on the performance of three-phase inverse fluidized-bed columns, Chemical Engineering Science, *54*, 4975–4980.
21. Sanchez, O., Michaud, S., Escudi, R., Delgenes, J., & Bernet, N., (2005). Liquid mixing and gas–liquid mass transfer in a three-phase inverse turbulent bed reactor, *Chemical Engineering Journal, 114*, 1–7.
22. Das, B., Ganguly, U., & Das, S., (2010). Inverse fluidization using non-Newtonian liquids,*Chemical Engineering and Processing, 49*(11), 1169–1175.
23. Craig T., Barker, & Noel De Nervers, (1982). Small-medium chamber volume boundary in the formation of bubbles, *Chemical Engineering Science, 37*(10), 1572–1575.
24. Ulaganathan, N., & Krishnaiah, K., (1996). Hydrodynamic Characteristics of Two-Phase Inverse Fluidized Bed, *Bioprocess Engineering, 15*, 159–164.
25. Zhang, J. P., Epstein, N., & Grace, J. R. (1988). Minimum fluidization velocities for gas-liquid-solid 3-phase systems, *Powder Technology, 100*, 113–118.
26. Han, S. J., Tan, R. B. H., & Loh, K. C. (2000). Hydrodynamic behavior in a new gas-liquid-solid inverse fluidization airlift bioreactor, *Trans I Chem. E, 78*, Part C.
27. Renganathan, T. & Krishnaiah, K., (2004).Liquid phase mixing in 2-phase liquid–solid inverse fluidized bed, *Chemical Engineering Journal, 60*, 213–218.
28. Renganathan, T. & Krishnaiah, K., (2005). Voidage characteristics and prediction of bed expansion in liquid–solid inverse fluidized bed, *Chemical Engineering Science, 98*, 2545–2555.
29. Garcia, D., Buffiere, P., Molettal, R., & Elmaleh, S., (1998).Anaerobic digestion of wine distillery wastewater in down-flow fluidized bed, *Wat. Res., 32*(12), 3593–3600.
30. Shieh, W., & Hsu, Y., (1996). Biomass loss from an anaerobic fluidized bed reactor, *Water Res. 30*, 1253–1257.
31. Son, S. M., Kang, S. H., Kim, U. Y., & Kang, Y., (2007). Bubble Properties in Three-Phase Inverse Fluidized Beds with Viscous Liquid Medium, *Chemical Engineering and Processing, 46,* 736–741.
32. Sowmeyan, R., & Swaminathan, G., (2008). Evaluation of inverse anaerobic fluidized bed reactor for treating high strength organic wastewater; *Bioresource Technology 99*, 3877–3880.
33. Sowmeyan, R., & Swaminathan, G., Performance of inverse anaerobic fluidized bed reactor for treating high strength organic wastewater during start-up phase, *Bioresource Technology, 99*, 6280–6284.
34. Allia, K., Tahar, N., Toumi, L., & Salem, Z., (2006). Biological treatment of water contaminated by hydrocarbons in three-phase gas-liquid-solid fluidized bed, Global NEST Journal, *8*(1), 9–15.
35. Sokol, W., & Korpal, W., (2006). Aerobic treatment of wastewaters in the inverse fluidized bed reactor. *Chemical Engineering Journal, 118,* 199–205.

36. Sokol, W., Ambaw, A., & Woldeyes, B., (2009). Biological Wastewater Treatment in the Inverse Fluidised Bed Reactor, *Chemical Engineering Journal*, *150*(1), 63–68.
37. Wang, D., Silbaugh, T., Pfeffer, R., & Lin, Y. S., (2010). Removal of emulsified oil from water by inverse fluidization of hydrophobic aerogels, *Powder Technology*, *203*, 298–309.
38. Jose, A. Q., Patel, G., & Peffer, R. (2009). Removal of Oil from Water by Inverse Fluidization of Aerogels, *Ind. Eng. Chem. Res.*, *48*(1), 191–201.
39. Sahoo, A., & Lima, T., (2014). Effect of Parameters on Treatment of Industrial Effluents Using Inverse Fluidized Bed Bioreactor: Statistical Analysis, *Particulate Science and Technology, 32*(2), 151–157.
40. Sokół, W., & Migiro, C. L. C., (1992). Metabolic responses of microorganisms growing on inhibitory substrates in nonsteady state culture, *J. Chem. Technol. Biotechnol.*, *54*, 223–229.

CHAPTER 5

REMOVAL OF PB^{2+} FROM WATER USING SILICA NANO SPHERES SYNTHESIZED ON $CACO_3$ AS A TEMPLATE: ADSORPTION KINETICS

MILTON MANYANGADZE,[1] J. GOVHA,[1] T. BALA NARSAIAH,[2] CH. SHILPA CHAKRA,[3] and P. AKHILA SWANTHANTHRA[4]

[1]*Lecturer, Harare Institute of Technology, Chemical and Process Systems Engineering Department, Harare, Zimbabwe, Mobile: +263-775-691-485, E-mail: manyangadzem@gmail.com*

[2]*Associate Professor, Centre for Chemical Sciences and Technology, Institute of Science and Technology, JNTUH, Telangana, India*

[3]*Assistant Professor, Centre for Nano Science and Technology, Institute of Science and Technology, JNTUH, Telangana, India*

[4]*Assistant Professor, Dept. of Chemical Engineering, SV University, Tirupati, A.P., India*

CONTENTS

ABSTRACT

This chapter intends to establish the kinetics of the adsorption of one of the toxic heavy metals, Pb^{2+} by employing nanoparticles of silica (SiO_2) as an adsorbent in wastewater treatment. Synthesis of SiO_2 particles was done using a three-stage process in which $CaCO_3$ nanoparticles were prepared first as a template. This was followed by treatment of the formed oxide particles with Sodium Silicate ($NaSiO_3$) to give the oxide composite. Finally, the carbonate template was destroyed using HCl. Characterization of the resultant nano particles was done using analytical techniques which include XRD, SEM, and TGA. The kinetics for the adsorption of Pb^{2+} was studied using the pseudo-first order, pseudo-second order, Elovich and Intraparticle diffusion models. Obtained results showed that the adsorption process followed the pseudo-second order kinetics.

5.1 INTRODUCTION

In recent years, a surge in pollutant levels in both ground and surface waters has been reported posing a serious threat to the availability of fresh water the world over. A number of factors have been associated with these escalating pollution levels which include a continuous growth in global population, rapid industrialization and industrial growth as well as persistent droughts [1–3].

The growth of the chemical process industry in recent years, has directly or otherwise contributed to the increase in the use of heavy metals and these have subsequently found their way into fresh water bodies through industrial effluents [1, 3, 4]. Common examples include, metal plating, paints and dyes, tanneries, ceramic and glass, battery manufacturing and mineral processing industries, just to name a few [2, 5–10]. Because of their toxicity, bio-accumulation tendency and non-biodegradability [4, 8], heavy metals have been reported dangerous to both human and aquatic life causing several diseases and health disorders. Research has reported chronic poisoning with such metals as Arsenic and Chromium for example. The metals have been noted to be carcinogenic [7, 11].

Prolonged exposure to Arsenic, through taking contaminated water, has been noted to cause cancers of the skin, lungs, bladder, the kidneys apart from other skin problems like pigmentation changes and thickening [3]. Badruddoza et al. [7] have shown that Chromium is not only a carcinogen, but also a mutagen as well as teratogen. Lead is another heavy metal that has been labeled highly toxic. For example, according to Sing et al. [3], it has been linked to such problems as damage of vital internal body systems such as the nervous and reproductive systems as well as the kidneys. The metal has been associated with problems of high blood pressure and anemia. Lead has also been reported to cause brain damage apart from mental disturbances and retardation as well as serious reduction in hemoglobin production [2, 8, 10, 12].

All these problems, associated with heavy metals, have led to stricter and more stringent environmental regulations on the release of heavy metals into surface water streams. According to Rostamian et al. [5] there is need to come up with effective and highly efficient mitigation methods. The methods should be such that they are cheap both in terms of operation and installation. They should be environmentally friendly, with minimal or zero secondary pollution and lastly they should be able to handle large volumes of effluent and even very low pollutant concentrations.

Various traditional methods have been and are still being used as pollution mitigants for the adsorption of the various heavy metals from wastewater. Examples include: membrane separation, ion exchange, ultrafiltration, filtration, sedimentation, electrodialysis, chemical reduction, electrochemical treatments, biological treatment, reverse osmosis, flocculation/coagulation as well as adsorption among others [1, 2, 5, 11–14]. Most of these methods however have been noted to have high capital and operational costs, unacceptable sensitivities to operational conditions, high-energy consumption as well as high sludge generation [8].

A lot of literature however has revealed adsorption as a better option with immense potential in heavy metal removal. The technology has been noted the most promising and widely used, mainly due to its superior efficiency. It has also been reported to be economically competitive and that it is easily applied [15, 16]. Adsorption has generally been shown to be simple, efficient and sludge free. The operation has been noted to be cheap, easy to handle with potential and/or capacity for regeneration [6–8, 17]. It is from this background therefore, that adsorption has been considered a better alternative compared to the other technologies.

On the same note, several materials have been used as adsorbents (both organic and inorganic). Good examples include: clay minerals, trivalent and tetravalent metal phosphates, biosorbents, synthetic resins, kaoline, zeolites, fly ash, peat, agricultural residues as well as activated carbon [9, 18, 19]. Of these materials, activated carbon has emerged the most commonly used adsorbent [6, 9, 20]. This has also been confirmed by Shin et al. [19] who noted that the carbon-based materials have to this day attracted a lot of interest due to a number of factors. They have been known to be inert to their environment, mechanically stable and possess a highly porous structure that is accompanied by specific surface chemical properties.

In spite of all these pros, activated carbon for example, has had one major drawback among others, the cost. The adsorbent's high cost has limited its use especially in highly contaminated wastewaters [9, 21]. Apart from the cost, activated carbon and the rest of the commercially available sorbents have been noted to have other disadvantages like low loading capacities, relatively small metal ion binding constants and small selectivity [18]. In view of these setbacks, extensive research is underway to come up with a low cost adsorbent with efficiency well above that of activated carbon and other commercially available adsorbents. Research is underway to try and utilize agricultural as well as industrial waste as low cost adsorbents [8, 9, 12, 20]. Tan et al. [8] however have pointed out that in most cases the low cost adsorbents are associated with low adsorption efficiencies, suggesting that other material properties like say adsorption capacity should also be considered along with the cost aspect [9]. An adsorbent should in essence not only satisfy low costs, but should also ensure competitive surface properties that translates to high adsorption capacities, material surface areas and available active sites. Basically therefore, an ideal adsorbent is expected to exhibit among other desired characteristics, a strong affinity for the desired sorbate, a large surface area, as well as a large number of active or binding sites. The adsorbent should also be thermally and chemically inert to resist harsh environmental conditions. An ideal adsorbent should be easily regenerated and separated from the treated water for reuse [16, 17].

From our previous work, it has been highlighted that high material porosity is one of the properties considered for an effective adsorbent [22]. Work by Guoliang et al. [16] confirmed this as they noted that a lot of research has focused on porous adsorbents which have shown remarkable improvement in heavy metal removal efficiency. This somewhat high efficiency is explained by the fact that porous materials exhibit both external and internal surface areas all of which are available for the adsorption process.

Silica or silicon dioxide (SiO_2) is one good example of porous materials and a lot of research has been done on this material. In all its forms, the material has received considerable attention due to its uniquely large surface area, well-defined shape and pore size [23]. Also according to Guoliang et al. [16] mesoporous materials with the already mentioned unique features have been extensively studied for their varied applications in the areas of adsorption, catalysis, sensors, semiconductor materials in electronics as well as in separation processes. Silica nanoparticles in particular have attracted a lot of attention in view of their excellent optical, electrical and thermal properties [24]. The nanoparticles have also been applied in various fields such as cosmetics, catalysis, paints and drug delivery [24, 25].

The study of silica nanoparticles and their derivatives on the removal of heavy metals has also been carried out extensively. It should however be highlighted that the current studies on silica nanoparticles and their application as heavy metal adsorbents are mainly focusing on their modified forms, either functionalized or as composites. For example, Bois et al. [23] studied the adsorption of heavy metal ions using silica functionalized with three different functional groups: Aminopropyl, [Amino-ethylamino] propy, as well as [(2-amino-ethylamino)-ethylamino] propyl. Magnetic Silica, functionalized with a thiol-group was also studied for the adsorption of Hg^{2+} and Pb^{2+} ions [16]. Aguado et al. [6] and Heidari et al. [10] studied the removal of metal ions Ni^{2+}, Cd^{2+}, Pb^{2+}, Cu^{2+} and Zn^{2+} using amino-functionalized mesoporous Silica. On nanocomposites, Cheng et al. [26] for example looked at the synthesis and characterization of alumina nanoparticles coated with a uniform silica shell. Li et al. [11] investigated the removal of Cr^{6+} from wastewater using silica coated Iron nanocomposites. Finally, Lim et al. [27] worked on the synthesis of polypyrrole (PPy)/silica nanocomposites for the removal of Hg^{2+}, Pb^{2+} and Ag^{+}.

This current work looks at synthesis and characterization of hollow Silica nanoparticles on Calcium Carbonate ($CaCO_3$) as a template for adsorption of Pb^{2+} from wastewater. It should be noted that very little work has been done on silica as nano adsorbents using this synthesis route. In cases where it was done, the application was not as an adsorbent. Nekhamanurak et al. [28] for instance studied the effect of the $CaCO_3@SiO_2$ composite on the mechanical properties and fracture behavior of poly (lactic acid) nanocomposite. Li et al. [29] also worked on the preparation of hollow Silica on $CaCO_3$ template for applications in drug release control. Lastly, Won et al. [30] investigated biological applications of $CaCO_3@$ $CdSe/ZnS/SiO_2$ microcomposites.

5.2 EXPERIMENTAL WORK

5.2.1 *MATERIALS AND METHODS*

5.2.1.1 Materials

Synthesis of nano the silica hollow spheres (NSHS) involved, anhydrous Calcium Chloride (AR $CaCl_2.2H_2O$, 99%), Potassium carbonate (AR, K_2CO_3, 99.9%), Sodium hydroxide (AR NaOH, 99.9%), Hydrochloric acid (AR, HCl, 36%), and Deionized Water (Millipore).

5.2.1.2 Methods

The synthesis route for silica nanoparticles was in three stages apart from the preparation of sodium silicate. The three stage involved the preparation of calcium chloride, which acted as a template, preparation of $CaCO_3@$ SiO_2 and then finally NSHS itself.

5.2.1.2.1 Preparation of Calcium Carbonate Nanoparticles

The synthesis process followed the method by Won et al. [30], with some changes. In the procedure, 11 g of anhydrous Calcium Chloride was prepared in 200 mL of deionized water. Also, 13.82 g of Potassium Carbonate was separately prepared in 200 mL to give equimolar solutions of the two salts. The prepared potassium carbonate solution was added to the Calcium Chloride solution to give a white precipitate. The mixture was then stirred for 20 min before separation by filtration. The solid product was washed with deionized water and was left to dry in air.

5.2.1.2.2 Preparation of Sodium Silicate

Method was adopted from Manyangadze et al. [22]. In this procedure, 32 g of Sodium Hydroxide were weighed and separately 24 g of Silica Gel were also weighed. The Sodium Hydroxide was dissolved in 120 mL of deionized water under heating at 230°C. To the dissolved Sodium Hydroxide, Silica Gel was then added in small portions so that it was dissolved in hot Sodium Hydroxide. The mixture was heated between additions until all the 24 g of the gel were dissolved to give a Sodium Silicate solution. The hot solution was left to cool before transferring into a 2000 mL volumetric flask which was filled to mark using deionized water.

5.2.1.2.3 Preparation of the Nanocomposite

Work by Manyangadze et al. [22] and Chen et al. [31] was adopted for this synthesis procedure. To prepare $CaCO_3$@SiO_2 nanoparticles, 2.0 g of the prepared Calcium Carbonate nanoparticles were weighed into a conical flask. The material was dispersed in 100 mL of deionized water with stirring at 700 rpm. To the suspension, 100 mL of 0.4 M Sodium Silicate was added drop wise. The mixture was then heated to 80°C on a hot plate with stirring. This was followed by adjusting the pH of the mixture to between 9.0 and 10.0 using 1.0M Hydrochloric Acid which was added drop wise with continuous stirring. On attaining the required pH, the suspension was further stirred for 40 min at 800–900 rpm. The formed nanoparticles were then separated via filtration and then washed thoroughly with deionized water. They were then dried at 100°C for two hours in an air oven. The dried nanoparticles were pulverized in an agate mortar before calcination. The calcination process was done at 700°C for 4 h after which the nanoparticles were now ready for characterization.

5.2.1.2.4 Preparation of Nano Silica Hollow Spheres (NSHS)

Pure silica nanoparticles were prepared by destroying the calcium carbonate template from the previous method in hydrochloric acid. Thus, the procedure is also based on the work by Chen et al. [31]. For the synthesis of silica nano hollow spheres, 4 g of the prepared calcium carbonate nanoparticles were weighed into a 500 mL beaker and 200 mL of deionized water were added. The formed suspension was stirred at 700 rpm. To the suspension, 150 mL of 0.4 M sodium silicate was added drop wise and the mixture was then heated to 80°C with stirring. This was followed by a further drop wise addition of 75 mL of 2.0 M hydrochloric acid and stirring was further continued for 40 min at between 800 and 900 rpm after which the mixture was left overnight, forming a gel. The formed gel was separated by filtration and was washed several times using deionized water before drying in an air-oven at 100°C for 5 h. The dried solid was then pulverized in an agate mortar before the calcination process. Calcination was carried out at 450°C for 3 h in a muffle furnace.

5.2.2 CHARACTERIZATION

The crystal structure of the prepared nanoparticles was determined using the X-ray diffraction (XRD) technique. Also from the XRD analysis, the average crystallite size was calculated from the Scherrer equation (Eq. 1). The general morphology of the nanoparticles was given the scanning electron microscope (SEM), which also gave particle size measurements. Thermal analyzes for the determination of such properties as mass changes, thermal capacity, enthalpy and the coefficient of heat expansion was carried out using TG-DTA. This analysis was however mostly confined to mass and enthalpy changes.

$$d_{XRD} = K\lambda/\beta\cos\theta \qquad (1)$$

5.2.3 BATCH ADSORPTION

Batch experiments were in this case carried out. These were done to ascertain how variable parameters on adsorption such as pH, contact time, initial sorbate concentration, adsorbent dose as well as temperature affected the adsorption rate as well as the adsorption capacity. All experimental procedures were adopted from literature [2, 7, 10, 32, 33].

5.2.3.1 Effect of Contact Time

To assess how the contact time affected adsorption rates, 50 mg of the adsorbent were suspended in 50 mL of 200 mg/L concentration of the metal ion under stirring at 450 rpm. Tests were carried out at 20 min intervals up to 120 min at the optimized pH of 5.0. All the other parameters were fixed while the time was varied. Samples were filtered at the elapse of each prescribed run time to separate the solid adsorbent. Concentration of residual heavy metal ion in the filtrate was determined using AAS, and from the attained results, the amount of the sorbate (heavy metal ions) adsorbed, q_t was calculated from Eq. (2). From this experiment, the time it takes to reach equilibrium was noted.

$$q_t = (C_o - C_t)V / W \tag{2}$$

5.2.3.2 Initial Sorbate Concentration

To investigate how the initial concentration of the heavy metal ions affected the rate of adsorption, 50 mg of the adsorbent were again suspended in 50 mL of varied concentrations of the sorbate between 40 and 300 mg/L. Stirring speed was fixed at 450 rpm. The process was carried out at pH of 5.0 and the run time was fixed and equal to the already determined equilibrium time. At the elapse of the prescribed contact or run time, the samples were filtered to separate the solid adsorbent. Just like in the above sections, residual sorbate concentrations at equilibrium were determined using AAS. From the obtained results, the adsorption capacities at equilibrium (q_e) were computed from Eq. (3). Also from this experiment, the optimum initial concentration of the sorbate was noted.

$$q_e = (C_o - C_e)V / W \tag{3}$$

5.2.3.3 Effect of the Amount of Adsorbent

A look at how varying the amount of the adsorbent affected the adsorption capacity was also considered. From the procedure, varying amounts of the adsorbent (0.02–0.10 g) were dispersed in 50 mL of the maximum initial concentration of the sorbate equal to 300 mg/L. As given in the above sections, the stirring speed remained fixed at 450 rpm. The process was also carried at pH 5.0 with the run or contact time equal to the equilibrium time of 2 h. At the elapse of the prescribed contact, the samples were filtered to separate the solid adsorbent and similarly, the residual sorbet concentrations were obtained using AAS. The optimum adsorbent dose was here also noted.

5.2.3.4 Adsorption Kinetics

To determine the adsorption kinetics, 50 mL of the Pb^{2+} ion solution with a concentration of 200 mg/L was prepared in a 100 mL beaker. The solution pH was adjusted to optimum pH 5.0 using 0.1 M nitric acid or sodium hydroxide. This was followed by dispersing 0.05 g of the adsorbent at room temperature. Dispersion was by means of magnetic stirring at 450 rpm.at intervals of 20 min, up to equilibrium, the samples were filtered with the filtrate retained. Residual Pb^{2+} concentrations were determined using AAS. From the attained results, the kinetic model was estimated using Eqs. (4)–(7), for the pseudo-first order and the pseudo-second order kinetic models, respectively, as well as the equilibrium relationship from Eq. (3). The Elovich and Intraparticle diffusion models were also tested for the same.

$$dq/dt = K_1(q_e - q_t) \tag{4}$$

$$\ln(q_e - q_t) = \ln q_e - K_1 t \tag{5}$$

$$dq/dt = K_2(q_e - q_t)^2 \tag{6}$$

$$t/q_t = 1/(K_2 q_t^2) + t/q_e \tag{7}$$

5.3 CHARACTERIZATION

Three characterization techniques were used to assess inherent surface characteristics of the prepared nanomaterials, their morphology, thermal stability and particle size, all of which are typically central to the application of the material in heavy metal removal from aqueous systems. Below are the results from the various analytical techniques.

5.3.1 THERMOGRAVIMETRIC AND DIFFERENTIAL THERMAL ANALYZES (TG/DTA)

Analyzes were done for the prepared nanomaterial to assess especially its thermal stability. The graphs (Figure 5.1) show both mass and energy changes associated with the nanoparticles as they went through thermal treatment.

5.3.1.1 SiO_2 Nanoparticles

As given in Figure 5.1, analysis results for nano silica hollow spheres (NSHS) show a significantly low material loss. The overall material loss is here pegged at approximately 15%. Apparently, a very low material loss is observed around 100°C, amounting to approximately 1% and this endothermic effect is

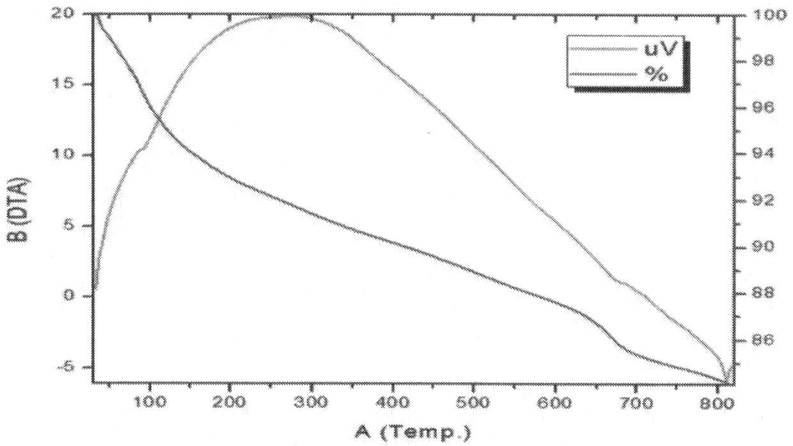

FIGURE 5.1 TG/DTA curves for SiO_2 nanoparticles.

attributed to moisture loss. This however seems to stretch up to 200°C. Other endothermic effects are observed at 650°C with a loss of approximately 2% and around 805°C as shown on both the TG and DTA curves.

5.3.2 *X-RAY DIFFRACTION (XRD) ANALYSIS*

XRD Analyzes were carried out for the prepared nanomaterial to assess especially its surface characteristics as well as particle size and structure. The results here are given in the form of diffraction patterns.

5.3.2.1 NSHS Particles

Figure 5.2 shows the X-ray diffraction (XRD) pattern of the NSHS particles with distinct peaks at 2θ equivalent to 22°, 29.5°, 31.5°, 45°, 56.5°, and 75.5°. Of importance however is the characteristic peak at 2θ = 22°. In their work, in which SiO_2 was synthesized via neutralization reaction of sodium silicate with sulphuric acid, Music et al. [34] located the peak at 21.8°. From the XRD pattern, the amorphous nature of silica is quite evident and using the Scherrer equation (Eq. 1), the average crystal size was also determined and was found to be 39.5 nm.

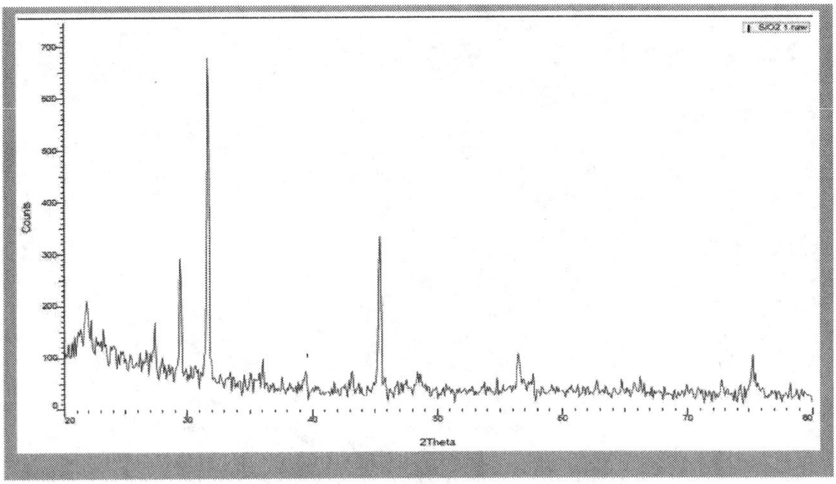

FIGURE 5.2 XRD pattern for NSHS particles.

5.3.3 *SCANNING ELECTRON MICROSCOPY*

Scanning Electron Microscopy (SEM) was done and the important information on the morphology, the texture, size as well as the shape of the nanoparticles are provided.

5.3.3.1 NSHS Particles

Micrograph images of silica nanoparticles, otherwise known as NSHS, are shown in Figure 5.3. From the images, it can be seen that the nanoparticles are mostly spherical. Particle aggregation is also observed. The analysis also reveals the particle size ranging between 100 and 200 nm. Also from the micrographs, distinct particles are not clearly separated; hence the morphology of the material could not be easily ascertained.

5.3.4 *ADSORPTION STUDIES*

For the adsorption studies, the removal of Pb^{2+} ions was investigated using NSHS. The effect of the various adsorption parameters, which include contact time, initial concentration of the sorbate, as well as the amount of adsorbent were looked at.

5.3.4.1 The Contact Time

Table 5.1 presents the results how the contact time affected the adsorption capacity. The results are graphically presented in Figure 5.4. From

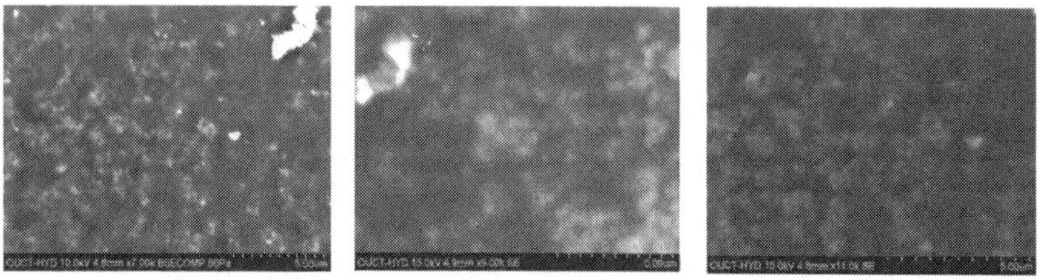

FIGURE 5.3 SEM images for NSHS particles.

TABLE 5.1 Summary of Results Showing How the Contact Time Affected Adsorption Capacity and Removal Efficiency

Time (t)	Adsorption Capacity (q_t)	Removal Efficiency (R %)	ln*t*
20	116.45	58.23	2.996
40	158.61	79.31	3.689
60	190.60	95.30	4.094
80	199.03	99.52	4.382
100	200.00	100.00	4.605
120	200.00	100.00	4.787

the results, rapid adsorption is observed between 20 and 60 min of contact time before slowly increasing up to 80 min. Between 80 and 120 min the adsorption rate becomes constant. The adsorption process thus reached equilibrium in 80 min.

5.3.4.2 Initial Sorbate Concentration

Results on how the initial concentration of the sorbate affected the overall performance of the adsorbent are given in Table 5.2 with a corresponding

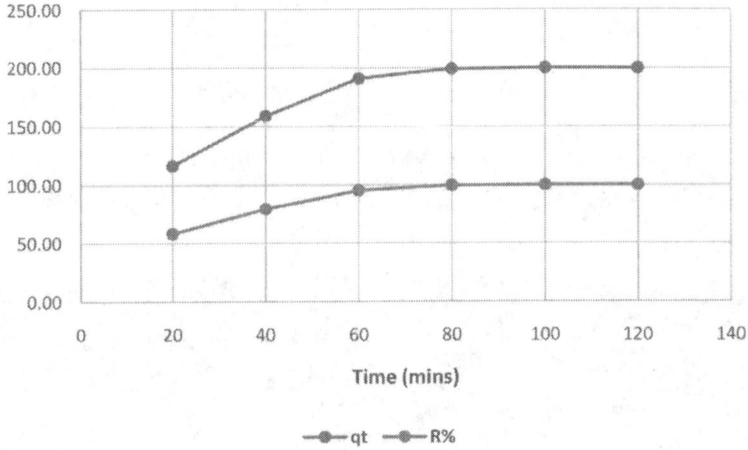

FIGURE 5.4 Contact time effect on adsorption capacity and removal efficiency of NSHS.

TABLE 5.2 Summary of Results Showing Effect of Initial Sorbate Concentration on Adsorption Capacity

C_o	C_e	q_e	C_e/q_e	$\ln q_e$	$\ln C_e$
40	0.00	40.00	0.00	3.69	—
100	0.37	99.63	0.00	4.60	-0.99
200	8.20	191.80	0.04	5.26	2.10
300	37.82	262.18	0.14	5.57	3.63

graphical presentation given in Figure 5.5. From the results, as portrayed in Figure 5.5, a linear relationship between the initial concentration of the sorbate and the corresponding adsorption capacity is observed up to 150 mg/L of the sorbate. Between 150 mg/L and 300 mg/L, the adsorption capacity begins to fall behind, decreasing slowly as the initial concentration increases. This behavior suggests the presence of a threshold limit of the initial sorbate concentration beyond which the adsorbent becomes less effective for a constant adsorbent dose. Keeping on increasing the initial sorbate concentration in this case would have given the threshold value, which lies somewhere beyond 300 mg/L of the sorbate, for an adsorbent dose of 0.05 g.

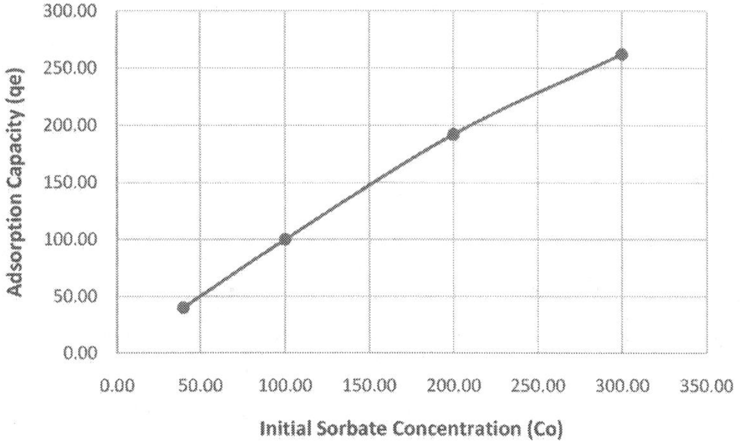

FIGURE 5.5 Initial sorbate concentration effect on the adsorption capacity of NSHS.

5.3.4.3 Amount of Adsorbent Dose

Results showing how increasing the adsorbent amount to a fixed concentration of the sorbate affected adsorption capacity are given in Table 5.3 with the corresponding graphical presentation given in Figure 5.6. From Figure 5.6, two graphs have been plotted, one showing how the adsorption capacity varies as the adsorbent dose is increased while the other shows how the adsorbed amount relates to the increasing adsorbent dose. The adsorption capacity is observed to decrease as the adsorbent dose increases, whilst the amount adsorbed is actually increasing. A decreasing trend is expected since the adsorption capacity is inversely related to the adsorbent dose. An increase in the dose implies a lower

TABLE 5.3 Summary of Results Showing Effect of Adsorbent Dose on Adsorption Capacity

W (g)	C_e (mg/L)	$C_o - C_e$ (mg/L)	q_e (mg/g)
0.02	158.13	141.87	354.67
0.05	16.83	283.17	283.17
0.08	8.55	291.45	182.16
0.10	0.00	300.00	150.00

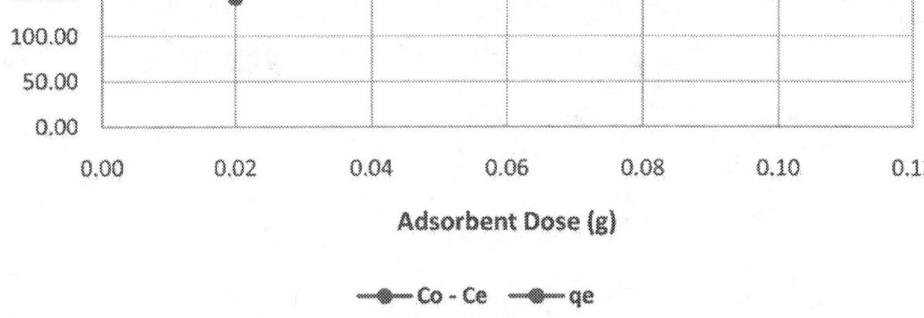

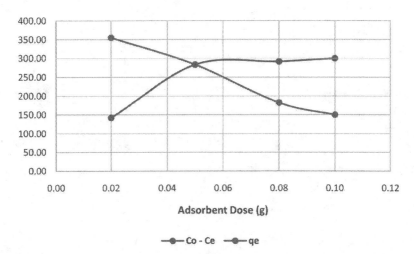

FIGURE 5.6 Effect of the amount of adsorbent on adsorption and adsorption capacity.

adsorption capacity and vice versa. A threshold limit of the adsorbent dose is also observed in Figure 5.6. The adsorption rate or the amount adsorbed is seen to increase as the dose increases up to 0.06 g of the adsorbent. Between 0.06 and 0.10 g, the adsorption rate is shown to slightly decrease. Generally, increasing the adsorbent dose implies a corresponding increase in the available specific surface area as well as the amount of binding or active sites. This explains the initial rapid surge in the adsorption rate. Further increasing the adsorbent dose increases the available binding sites which are however not utilized so that there is no change in the amount adsorbed.

5.3.5 *ADSORPTION KINETICS*

Graphical presentation of the data for the different models for the adsorption kinetics tested for this work is given in Figures 5.7–5.10.

5.3.5.1 Pseudo-First Order Kinetic Model

The graph for the pseudo-first order kinetic model is shown in Figure 5.7. From the scatter, it can be seen that $\ln(q_e - q_t)$does not relate linearly with con-

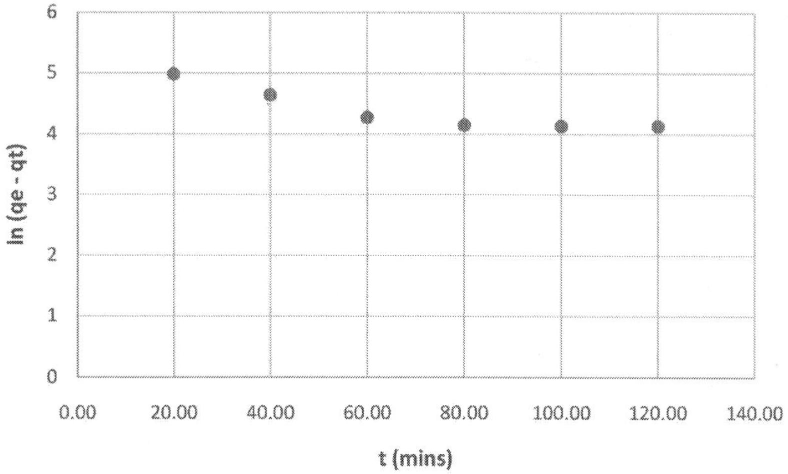

FIGURE 5.7 A plot for the pseudo-first order kinetic model data.

tact time. Thus, to assess whether this relationship best describes the kinetics of the adsorption of Pb^{2+} on NSHS, a regression analysis was performed to draw the line of best fit for the model which gave an r^2 value of 0.8039.

5.3.5.2 Pseudo-Second Order Kinetic Model

The corresponding graph for the pseudo-second order kinetic model is shown in Figure 5.8. The graph shows a close to linear relationship between the contact time and the ratio t/q_t. Similarly, a regression analysis was carried out to draw the line of best fit for the model. The regression analysis gave a perfect fit with correlation coefficient (r^2) of 0.9936.

5.3.5.3 Elovich Model

A plot of the Elovich kinetic model data is given in Figure 5.9. From the scatter, it is evident that there is no linear relationship between the adsorption capacity (q) and ln(t). Thus, to assess whether this relationship best describes the kinetics of the adsorption of Pb^{2+} on NSHS, a regression analysis was also carried out to draw the line of best fit for the model with r^2 value of 0.9271.

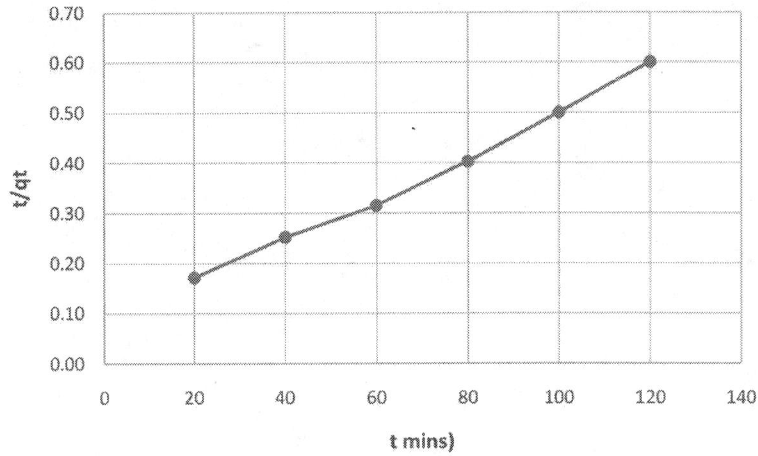

FIGURE 5.8 A plot for the pseudo-second order kinetic model data.

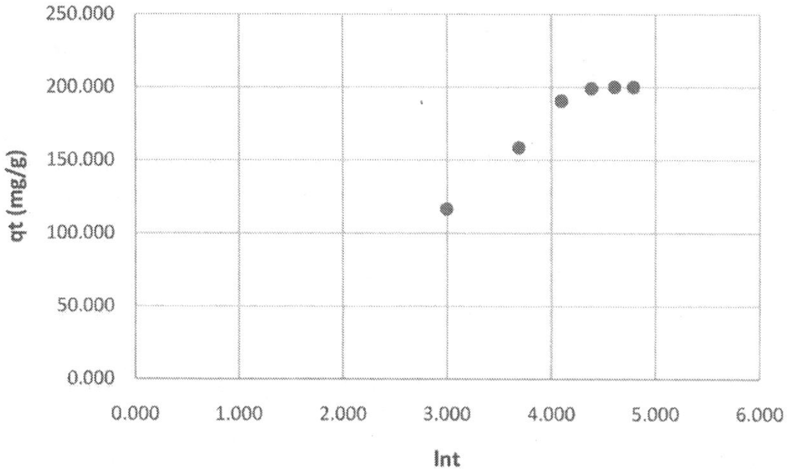

FIGURE 5.9 A plot for the Elovich kinetic model data.

5.3.5.4 Intraparticle Diffusion Model

An assessment of the intraparticle diffusion kinetic model is given in Figure 5.10. From the scatter, it can be seen that the variation of ln*R*with ln(*t*) is not linear. Thus, in order to assess whether this relationship best

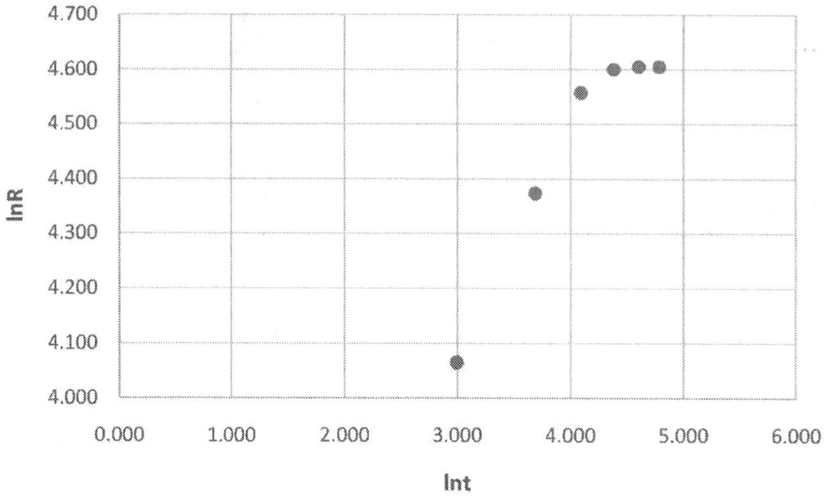

FIGURE 5.10 A plot for the intraparticle diffusion kinetic model data.

describes the kinetics of the adsorption of Pb^{2+} on NSHS, a regression analysis was thus done to come up with the line of best fit for the model to give an r^2 value of 0.9097.

Overall therefore, an assessment of correlation coefficients from the four models (Table 5.4) shows that the pseudo-second order kinetic model, with $r^2 = 0.9936$ best describes the kinetics for the adsorption of Pb^{2+} onto NSHS. This is followed by the Elovich and intraparticle diffusion models with $r^2 = 0.9271$ and $r^2 = 0.9097$, respectively. The pseudo-first order is the least with a corresponding correlation coefficient of 0.8039. Therefore, since the adsorption process is governed by the pseudo-second order kinetic model, it suggests that the mechanism of adsorption follows a chemical reaction, hence chemisorption.

5.4 CONCLUSIONS

From the work carried out, the intended nanomaterial was successfully synthesized based on the three characterization techniques employed in this study. The removal of Pb^{2+} with NSHS was noted to increase as the contact time increased, up to 80 min, which was considered the equilibrium time. Further, the adsorption capacity was also revealed to depend on both the initial concentration of the sorbate as well as the adsorbent dose. A linear relationship was noted up to 150 mg/L for the initial sorbate concentration and 0.06 g for the adsorbent dose so that beyond these values, either there was a corresponding decrease in the adsorption capacity or there was no significant change of the same. Thus, the optimum initial sorbate concentration was 150 mg/L, while the optimum adsorbent dose was 0.06 g in 50 mL of sorbate solution.

From the kinetic study, the adsorption of Pb^{2+} was noted to follow pseudo-second order kinetics, hence a chemisorption adsorption mechanism.

In future, there is need to test the nanomaterial, NSHS on various other heavy metals such as chromium and arsenic. Also in the current study, it

TABLE 5.4 Correlation Coefficients for the Various Kinetic Models Employed

Model	Pseudo-1st Order	Pseudo-2nd Order	Elovich	Intraparticle Diffusion
r^2	0.8039	0.9936	0.9271	0.9097

should be noted that the adsorption process was mainly batch. Thus, an in-depth study using continuous systems is highly recommended. This is because continuous systems give a more realistic resemblance to the industrial set up.

KEYWORDS

- **characterization**
- **Elovich**
- **intraparticle diffusion**
- **kinetics**
- **nanoparticles**

REFERENCES

1. Xu, P., Zeng, G. M., Huang, D. L., Feng, C. L., Hu, S., Zhao, M. H., et al., (2012). Use of iron oxide nanomaterials in wastewater treatment: A review, *Sci Total Environ. 424*, 1–10.
2. Hao, S., Zhong, Y., Pepe, F., & Zhu, W., (2012). Adsorption of Pb^{2+} and Cu^{2+} on anionic surfactant-templated amino-functionalized mesoporoussilicas, *Chem. Eng. J., 189–190*, 160–167.
3. Singh, S. Barick K. C., Bahadur, D., (2013). Functional Oxide nanomaterials and nanocomposites for the removal of heavy metals and dyes, *Nanomaterials and Nanotechnology*, Intech.
4. Ge, F., Li, M. M., Ye, H., & Zhao, B. X., (2012). Effective removal of heavy metal ions Cd^{2+}, Zn^{2+}, Pb^{2+}, Cu^{2+} from aqueous solution by polymer-modified magnetic nanoparticles, *J. Hazard. Mater. 211–212*, 366–372.
5. Rostamian, R., Najafi, M., & Rafati, A. B., (2011). Synthesis and characterization of thiol-functionalized silica nano hollow sphere as a novel adsorbent for removal of poisonous heavy metal ions from water: Kinetics, isotherms and error Analysis, *Chem. Eng. J., 171*, 1004–1011.
6. Aguado, J., Arsuaga, J. M., Arencibia, A., Lindo, M., & Gascon, V., (2009). Aqueous heavy metals removal by adsorption on amine-functionalized silica, *J. Hazard. Mater., 163*, 213–221.
7. Badruddoza, A. Z., Md. Shawon, Z. B. Z., Rahman, Md. T., Hao, K. W., Hidajat, K., & Uddin, M. S., (2013). Ionically modified magnetic nanomaterials for arsenic and chromium removal from water, *Chem. Eng. J.*, 225, 607–615.
8. Tan, Y., Chen, M., & Hao, Y., (2012). High efficient removal of Pb(II) by amino-functionalized Fe_3O_4 magnetic nano-particles, *Chem. Eng. J., 191*, 104–111.

9. Jin, X., Yu, C., Li, Y., Qi, Y., Yang, L., Zhao, G., et al., (2011). Preparation of novel nano-adsorbent based on organic-inorganic hybrid and their adsorption for heavy metals and organic pollutants presented in water environment, *J. Hazard. Mater., 180*, 1672–1680.
10. Heidari, A., Younesi, H., & Mehraban, Z. (2009). Removal of Ni(II), Cd(II), and Pb(II) from a ternary aqueous solution by amino-functionalized mesoporous and nano mesoporous silica, *Chem. Eng. J., 153*, 70–79.
11. Li, Y., Jin, Z., & Li, T., (2012). A novel and simple method to synthesize SiO_{2-}coated Fe nanocomposites with enhanced Cr(VI) removal under various experimental conditions, *Desalination, 288,* 118–125.
12. Rahmani, A., Mousavi, H. Z., & Fazli, M., (2010). Effect of nanostructure alumina on adsorption of heavy metals, *Desalination, 253*, 94–100.
13. Singh, S., Barick, K. C., & Bahadur, D., (2011). Surface engineered magnetic nanoparticles for removal of toxic metal ions and bacterial pathogens, *J. Hazard. Mater. 192*, 1539–1547.
14. Huang, S. H., & Chen, D. H. (2009). Rapid removal of heavy metal cations and anions from aqueous solutions by an amino-functionalized magnetic nano-adsorbent, *J. Hazard. Mater., 163*, 174–179.
15. Zhu, H., Jia, Y., Wu, X., & Wang, H., (2009). Removal of arsenic from water by supported nano zero-valent iron on activated carbon, *J. Hazard. Mater., 172*, 1591–1596
16. Li, G., Zhao, Z, Liu, J., & Jiang, G., (2011). Effective heavy metal removal from aqueous systems by thiol functionalized magnetic mesoporous silica, *J. Hazard. Mater., 192*, 277–283.
17. Li, L., Fan, L., Sun, M., Qiu, H., Li, X., Duan, H., et al., (2013). Adsorbent for chromium removal based on graphene oxide functionalized with magnetic cyclodextr in-chitosan, *Colloids Surf. Biointerfaces, 107,* 76–83.
18. Najafi, M., Yousefi, Y., & Rafati, A. A. (2012). Synthesis, characterization and adsorption studies of several heavy metal ions on amino-functionalized silica nano hollow sphere and silica gel, *Sep. Purif. Technol. 85*, 193–205.
19. Shin, K. Y., Hong, J. Y., & Jang, J., (2011). Heavy metal ion adsorption behavior in nitrogen-doped magnetic carbon nanoparticles: Isotherms and kinetic studies, *J. Hazard. Mater., 190*, 36–44.
20. Huang, Y. H., Hsueh, C. L., Huang, C. P., Su, L. C. & Chen, C. Y., (2007). Adsorption thermodynamic and kinetic studies of Pb(II) removal from water onto a versatile Al_2O_{3-}supported iron oxide, *Sep. Purif. Technol., 55*, 23–29.
21. Afkhami, A., Tehrani, M. S. & Bagheri, H. (2010). Simultaneous removal of heavy-metal ions in wastewater samples using nano-alumina modified with 2, 4-dinitrophenylhydrazine, *J. Hazard. Mater., 181*, 836–844.
22. Manyangadze M., Narsaiah B.T., Kumar P., & Chakra, Ch. P. S. (2014). Synthesis and Characterization of Hollow Silica Nano Particles on Calcium Carbonate for Possible Removal of Lead from Industrial Wastewater, *IJLTEMAS, 3*(10), 3-9.
23. Bois, L., Bonhomme, A., Ribes, A., Pais, B., Raffin, G., & Tessier, F., (2003). Functionalized silica for heavy metal ions adsorption, *Physicochem. Eng. Aspects,221*, 221–230.

24. Sadek, O. M., Reda, S. M., & Al-Bilali, R. K. (2013). Preparation and characterization of silica and clay-silica core-shell nanoparticles using sol-gel method, *Advances in Nanoparticles, 2*, 165–175.
25. Dabbaghian, M. A., Babalou, A. A., Hadi, P., & Jonnatdoust, E., (2010). A parametric study of the synthesis of silica nanoparticles via sol-gel precipitation method, *Int. J. Nanosci. Nanotechnol.*, *6*(2), 104–113.
26. Cheng, Z. P., Yang, Y., Li, F. S., & Pan, Z. H. (2008). Synthesis and characterization of alumina nanoparticles coated with uniform silica shell, *Trans. Non ferrous Met. Soc. China, 18*, 378–382.
27. Lim, C. H., Song, K., & Kim, S. H., (2012). Synthesis of PPy/silica nanocomposites with cratered surfaces and their application in heavy metal extraction, *J. Ind. Eng. Chem., 18*, 24–28.
28. Nekhamanurak, B., Patanathabutr, P., & Hongsriphan, N., (2012). Mechanical properties of hydrophilicity modified $CaCO_3$-poly (Lactic Acid) nanocomposite, *International Journal of Applied Physics and Mathematics, 2*(2), 98-103.
29. Li, Z. Z., Wen, L. X., Shao, L., Chen, J. F., (2004). Fabrication of porous hollow silica nanoparticles and their applications in drug release control, *J. Control. Release, 98*, 245–254.
30. Won, Y. H., Jang, H. S., Chung, D. W., & Stanciu, L., (2010). Multifunctional calcium carbonate microparticles: Synthesis and biological applications. Birck Nanotechnology Centre, Paper 622.
31. Chen, J. F., Ding, H. M., Wang, J. X., & Shao, L., (2004). Preparation and characterization of porous hollow silica nanoparticles for drug delivery application, *Biomaterials, 25,* 723–727.
32. Mahmoud, M. E, Abdelwahab, M. S., & Fathalah, E. M., (2013). Design of novel nano-sorbents based on nano-magnetic iron oxide-bound-nano-silicon oxide-immobilized-triethylenetetramine for implementation in water treatment of heavy metals, *Chem. Eng. J., 223*, 318–327.
33. Phuengprasop, T., Sittiwong, J., & Unob, F., (2011). Removal of heavy metal ions by iron oxide coated sewage sludge, *J. Hazard. Mater., 186,* 502–507.
34. Musić, S., Filipović-Vinceković, N., & Sekovanić, L., (2011). Precipitation of amorphous SiO_2 particles and their properties, *Brazilian Journal of Chemical Engineering, 28*(1), 89–94.

CHAPTER 6

A STUDY ON ADSORPTION KINETICS OF *AZADIRACHTA INDICA* (NEEM) AND *FICUS RELIGIOSA* (PIPAL) FOR REMOVAL OF FLUORIDE FROM DRINKING WATER

M. JAIPAL and P. DINESH SANKAR REDDY

Department of Chemical Engineering, JNTUA College of Engineering, Anantapuramu, 515002, Andhra Pradesh, India, E-mail: pdsreddy@gmail.com

CONTENTS

ABSTRACT

Fluoride in drinking water is a well-known pollutant throughout the world and like any other pollution the fluoride pollution occurs due to both natural and manmade reasons. Though, a minimum concentration of 0.8–1.0 mg/L of fluoride in drinking water helps in the production and maintenance of healthy bones and teeth, consumption of water with a fluoride concentration of more than 1.5 mg/L over a longer period causes fluorosis, and a concentration beyond 3 mg/L may result in neurological problems. In the present work, adsorbents were prepared from leaves of *Azadirachta indica* (Neem) and *Ficus religiosa* (Pipal) to remove excess fluoride from water in order to make it safe for human consumption. Several adsorption experiments were conducted using these natural to assess their suitability as adsorbents in treating the fluoride contaminated water. Further, the effect of various controlling parameters like pH, dosage of adsorbent, contact time, and initial concentration on fluoride removal efficiency were studied to determine optimum parameters for maximum removal efficiency. The optimum values of pH, contact time and adsorbent dosage were found to be 2, 90 min and 8.0 g/L, respectively, for both the adsorbents and it was also observed that *Ficus Religiosa* (Pipal) acts as a better adsorbent compared to *Azadirachta Indica* (Neem).

6.1 INTRODUCTION

Fluorine is present naturally in the earth's crust and it can be found in rocks, coal and clay. Elevated levels of Fluoride in drinking water due to both natural and manmade reasons are a major cause of concern throughout the world. Though, consumption of fluoride with a minimum concentration of 0.8–1.0 mg/L of water is beneficial for the production and preservation of healthy bones and teeth, consuming more than 1.5 mg/L of fluoride for longer periods cause chronic disease called fluorosis, noticeable through dental fluorosis, bone weakness and consumption of fluoride beyond 3 mg/L concentration leads to neurological damage and skeletal fluorosis [1, 2]. In view of the above health effects caused by consumption of water with high concentrations of fluorine, researchers have studied diverse techniques like *adsorption methods, ion exchange methods and precipitation methods* for the removal of excess fluoride content from water.

Adsorption methods by using several adsorbents such as bentonite[2], activated alumina [3, 4], amorphous alumina [5], activated carbon [6], zeolites, charcoals [7, 8], fly ash [9], etc., have been tested in past at different fluoride concentrations. However, these traditional adsorbents were proven to be highly expensive in removal of excess levels of fluoride content from ground water and current research is mainly focused on study of natural adsorbents such as calcite [10], fly ash [11], brick powder [12], fishbone charcoal [13], sunflower plant powder (dry) [16], bagasse ash [17], burnt bone powder [18] which involve lower cost of operation. However, lower efficiency and/or non-applicability of these techniques hamper the use of bio adsorbents for larger scale removal operations.

In the present work, adsorbents prepared from leaves of *Azadirachta indica* and *Ficus religiosa* trees which are abundantly available locally and are having medicinal factors are used for the removal of excess fluoride content from water. The high fiber content of these leaves makes them suitable to use them as biosorbents. In the current work, the effect of different parameters like pH, adsorbent dosage, initial fluoride ion concentration and contact time of adsorbent on adsorption efficiency was studied. A comparative study of these adsorbents on adsorption efficiency was carried out to conclude which adsorbent is more suitable for removal of excess fluoride from drinking water. The data obtained from these adsorption studies was fitted to Langmuir, Freundlich and Lagergren's isotherms.

6.2 MATERIALS AND METHODS

6.2.1 PREPARATION OF ADSORBENT

The adsorbents were prepared from fresh leaves obtained from Neem and Pipal trees. These leaves were dried under sun for 3 to 4 days and crushed manually, followed by sieving with standard screens of mesh no. 52/75. The biomass of leaf powder collected was digested chemically using both acid and alkali treatment. The acid and alkali treatments involve heating the powder with 1N aqueous solution of HNO_3 and 1N aqueous solution of NaOH, respectively, for nearly 20 min and immediate washing of the same with double distilled water until the color was completely removed. Finally, the treated adsorbent was dried in a drier to remove the moisture content until a constant weight is attained.

6.2.2 PREPARATION OF ADSORBATE SOLUTION

The fluoride stock solution was prepared by dissolving 30 mg of Sodium Fluoride (NaF) in one liter of double distilled water and the solution was diluted as required to obtain standard solutions containing 2–30 mg/L of fluoride.

6.2.3 *EXPERIMENTAL STUDIES ON BATCH ADSORPTION*

In the present study, experiments on batch adsorption of excess fluoride were conducted by having the known concentration of adsorbate solution, for example, the fluoride stock solution. In all the experiments, 150 mL of fluoride stock solution is taken in 250 mL glass beaker and to this solution a known amount of prepared adsorbent was added. Then the mixture of stock solution and adsorbent was agitated using a mechanical agitator at a speed of 200 rpm for one hour. After agitation, the contents in the beaker were allowed to settle for few minutes such that the supernatants can be cautiously decanted and filtered through Watmann filter paper (no.42). The filtrate collected was then analyzed for remaining fluoride ion concentration using a UV-Vis spectrophotometer. The absorbance of fluoride was Analyzed at 570 nm of wave length spectrophotometrically and the calibration chart was drawn as shown in Figure 6.1 and was used as a reference for calculating the fluoride concentrations.

The removal percentage of fluoride from synthetic solution was calculated using the following relation:

$$\text{Removal of Flouride (\%)} = [(IFC - FFC)/IFC] \times 100$$

where, *IFC* is initial fluoride concentration and *FFC* is final fluoride concentration.

The amount of Adsorbate (fluoride) adsorbed per unit mass of adsorbents can be expressed using different isotherms [15] as given in the following subsections.

6.2.3.1 Freundlich Isotherm

The linearized Freundlich adsorption isotherm is given as,

$$\log (q_e) = \log (K_f) + (1/n) \log (C_e) \qquad (1)$$

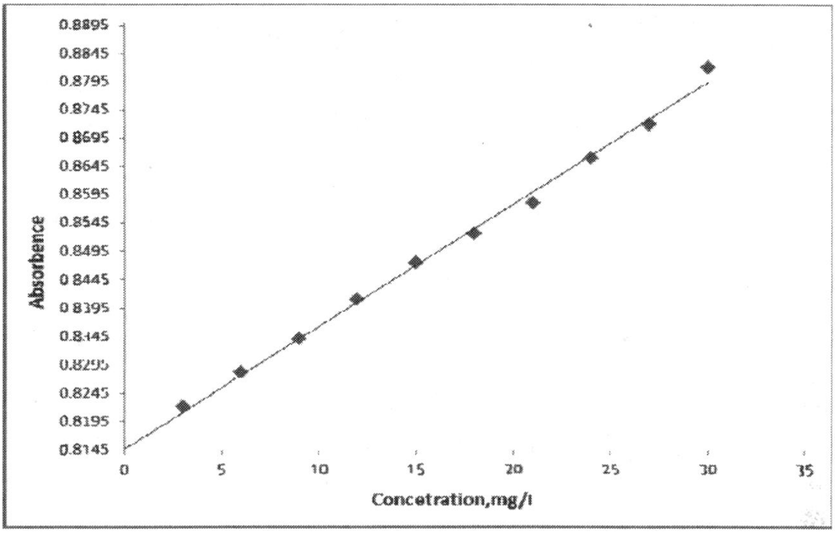

FIGURE 6.1 Calibration chart.

6.2.3.2 Langmuir Isotherm

The linear form of Langmuir isotherm can be expressed as,

$$1/q_e = (1/a) + (1/abC_e) \tag{2}$$

6.2.3.3 Lagergren's Isotherm

Lagergren's equation is given by

$$\log (q_e - q) = \log (q) - [(K \text{ x } t)/2.303] \tag{3}$$

where, q_e is the amount of adsorbate adsorbed per unit weight of adsorbents, mg/g; C_e is the equilibrium adsorbate concentration in solution, mg/L; K_f and $1/n$ are the Freundlich constants; a is number of moles of solute adsorbed per unit weight of adsorbent in forming monolayer on the surface; b is a constant related to energy; q is the amount of fluoride adsorbed at time t (mg/g); K is the rate constant of adsorption (per minute); and t is time (minute).

6.3 RESULTS AND DISCUSSIONS

In this section, the observations of various experiments conducted to know the effect of different parameters like pH, absorbent dosage, initial fluoride concentration and contact time, on adsorption efficiency of Neem and Pipal were discussed. All the experiments were conducted at an ambient temperature (33 ± 2°C).

6.3.1 EFFECT OF PH

In order to study the effect of pH on adsorption efficiency, test samples with initial fluoride ion concentration of 5 mg/L and a bio-sorbent dosage of 8 g/L were taken and the fluoride removal efficiency was studied by adjusting the pH values ranging from acidic to alkaline. The test sample is mixed for 60 min using a magnetic stirrer at different values of pH, for example, at 2, 4, 6, 8 and 10. The desired pH value was maintained with a deviation of ± 0.2 with selective addition of 0.5 N HNO_3 or 0.1 N NaOH. All these different samples are then analyzed for residual fluoride ion concentration using a UV-Vis spectrophotometer. The observations of these experiments with both Neem and Pipal were presented in Table 6.1.

Further, the effect of pH on absorption efficiency of Neem and Pipal was compared in the Figure 6.2(a). The curves in Figure 6.2(a) show that with the increase in the pH of test fluoride solution from 2.0 to 10.0, the fluoride removal efficiency has got reduced from 72% to about 40%. It

TABLE 6.1 Adsorption Efficiency of Neem and Pipal with Changing pH for a Contact Time of 60 min and with Changing Contact Time for a pH Value of 2*

pH	Adsorption efficiency (%)		Contact time (min)	Adsorption efficiency (%)	
	Neem	Pipal		Neem	Pipal
2	72	75	30	65	72
4	62	65	60	73	73
6	55	50	90	80	82
8	40	38	120	81	81
10	40	37	150	80	79

*Initial Fluoride Concentration is 5 mg/L and Adsorbent Dosage is 8 g/L.

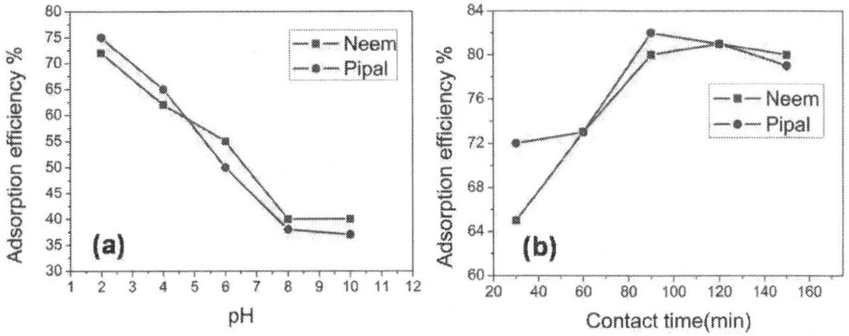

FIGURE 6.2 Plots (a) and (b) show the effect of pH and Contact time, respectively, on adsorption efficiency of both Neem and Pipal adsorbents.

was observed that the removal efficiency decreases with increasing pH for both the adsorbents. Further, it can be observed that the maximum fluoride removal efficiency is achieved at pH value of 2 in both the cases. Since, the presence of large number of H^+ ions at low values of pH will tend to neutralize the OH^- ions on the surface of the biosorbent, the hindrance for the diffusion of fluoride ions gets reduced and the maximum fluoride removal and thereby better adsorption efficiency was observed at low values of pH. Thus, the pH value of 2, at which the maximum adsorption efficiency was achieved is taken to be constant in the further experimental studies.

6.3.2 EFFECT OF CONTACT TIME

The effect of contact time of bio-sorbent on fluoride removal efficiency was studied by varying the time between 30 and 120 min while keeping other parameters like the adsorbent dosage of test mixture (5 g/L), initial fluoride ion concentration (5 mg/L) and pH (2) as constant. Observations were made by conducting experiments at contact times of 30, 60, 90, 120 and 150 min, respectively for both the absorbents and the removal efficiencies are tabulated in Table 6.1. The effect of contact time on removal efficiency for both the bio-sorbents is compared in Figure 6.2(b). The curves in Figure 6.2(b) show that at large contact times the maximum achievable fluoride removal efficiency is about 79% for Pipal and 80% for Neem. It is observed that for both the adsorbents, initially the fluoride removal efficiency increases when the contact time of adsorbent with test solution is increased but at higher contact times, no marginal change in removal efficiency was observed and the value approaches to a constant, indicating attainment of equilibrium. The plot shows that the removal efficiency increases from 65% to 80% in case of Neem and 72% to 79% for Pipal with variation of contact time from 30 min to 150 min, respectively.

6.3.3 EFFECT OF ADSORBENT DOSE

The effect of adsorbent dosage on fluoride removal efficiency was conducted by varying it from 2 g/L to 10 g/L for both Neem and Pipal, while keeping the initial fluoride ion concentration (5 mg/L), contact time (90 min) and pH (2) constant. Experiments were conducted at an adsorbent dosage of 2, 4, 6, 8 and 10 g/L for both absorbents separately. The effect of Adsorbent dosage on adsorption efficiency of both the adsorbents was compared in Figure 6.3(a). From the plot it is observed that with the increase of adsorbent dosage of both the adsorbents the percentage removal of fluoride also gradually increases upto an adsorbent dosage of 8 g/L, but beyond that the removal efficiency remains nearly constant due to the attainment of equilibrium. Results show that 65% of fluoride ions can be removed with Neem adsorbent at a dose of 2 g/L and the same increases to 80% at a dose of 8 g/L. Whereas, with Pipal adsorbent an increase in fluoride removal efficiency from 72% to 81% can be achieved when the dosage of the adsorbent is increased from 2 gL to 10 g/L.

6.3.4 EFFECT OF INITIAL FLUORIDE ION CONCENTRATION

In order to study the effect of initial fluoride ion concentration on removal efficiency, sample test mixtures were prepared with a range of initial fluoride ion concentration, for example, from 3 mg/L to 20 mg/L, while maintaining the adsorbent dosage and pH constant at a value of 8 g/L and 2, respectively. These samples were mixed for 90 min and then the residual fluoride ion concentration is estimated using UV-Vis spectrophotometer. The values of removal efficiency obtained from the experimental studies conducted for both the adsorbents at different F-concentrations are reported in Table 6.2 and are also depicted in Figure 6.3(b). The plot between adsorption efficiency vs. initial fluoride ion concentration show that the efficiency decreases with increasing initial fluoride ion concentration. The removal efficiency decreases drastically from 82% to 31%, and 85% to 35% for Neem and Pipal adsorbents, respectively, when the initial fluoride ion concentration was increased from 2 mg/L to 20 mg/L.

6.3.5 ADSORPTION MODELS FOR NEEM AND PIPAL

In general Adsorption Isotherm is a curve plotted between the quantity of adsorbate and the concentration of the substance in the original solution at constant temperature. There are two widely used models for use in adsorption isotherms: the Langmuir and the Freundlich isotherms, they give qualitative information about the adsorption process and is, therefore, very important and widely used.

6.3.5.1 Freundlich Isotherm

Freundlich Isotherm is the most commonly used adsorption isotherm to fit the data for small concentration ranges and in particular for dilute solutions. The form of the equation indicates that plotting the equilibrium solute concentration (q_e) as ordinate against adsorbate content (C_a) of the

TABLE 6.2 Effect of Adsorbent Dosage for an Initial Fluoride Concentration of 5 mg/L (Contact Time of 60 min) and Initial Fluoride Concentration for an Adsorbent Dosage of 6 g/L (Contact Time of 90 min), on Adsorption Efficiency of Both the Adsorbents*

Adsorbent dosage (g/L)	Adsorption efficiency (%)		Initial fluoride concentration (mg/L)	Adsorption efficiency (%)	
	Neem	Pipal		Neem	Pipal
2	65	72	3	82	85
4	70	80	5	80.2	80
6	72	82	10	68	75
8	80	82.5	15	32	35
10	80	82	20	31	35

*Constant Value of pH = 2.

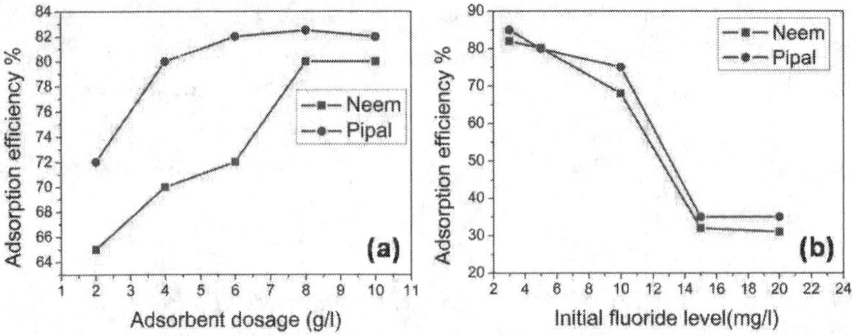

FIGURE 6.3 Plots (a) and (b) show the effect of adsorbent dosage and initial fluoride concentration, respectively, on adsorption efficiency of both Neem and Pipal adsorbents.

solid as abscissa will provide a straight line of slope n and intercept k. Figure 6.4 (a) and 6.4 (b) show the Freundlich isotherms plotted between log q_e and log C_e for Neem and Pipal adsorbents, respectively. The isotherms show that at high solute concentrations, the experimental data slightly deviate from the linear fit and do not follow the equation. The deviation is due to the result of substantial adsorption of the solvent which is not taken into account or inapplicability of the empirical expression.

6.3.5.2 Langmuir Isotherm

The Langmuir isotherm is in general used to study the dependence of surface coverage of adsorbed molecules above the adsorbent surface at a constant temperature. Langmuir isotherms plotted between $1/q_e$ and $1/C_e$

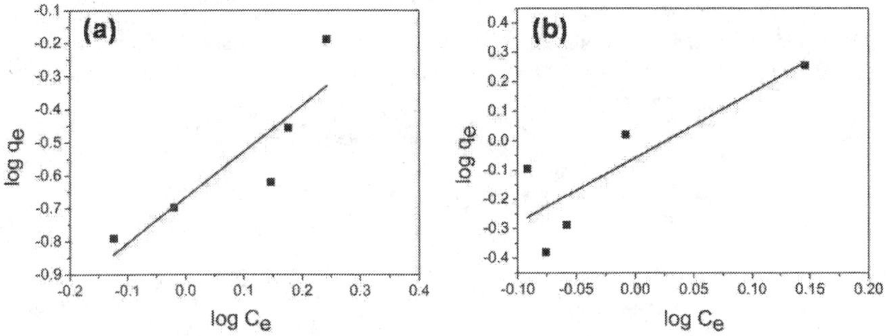

FIGURE 6.4 Plots (a) and (b) show the Freundlich isotherms for Neem and Pipal adsorbents, respectively.

TABLE 6.3 Freundlich Isotherm Values for Neem and Pipal

Neem		Pipal	
log q_e	log C_e	log q_e	log C_e
–0.187	0.243	0.255	0.146
–0.455	0.176	0.021	–0.008
–0.619	0.146	–0.095	–0.0915
–0.698	–0.02	–0.287	–0.0579
–0.79	–0.124	–0.38	–0.0757

are shown below in Figure 6.5(a) and 6.5(b) and the values of q_e and C_e are given in Table 6.4.

Different constants in Freundlich and Langmuir equations obtained from Figures 6.4 and 6.5 are presented in Table 6.5. These constants that are included in the adsorption isotherms have substantial importance when characterizing the adsorption system. The constant K_f and $1/n$ represent adsorption capacity and strength of adsorption of each adsorbent. Since the value of $1/n$ is less than 1 for Neem adsorbent, the bond energy increases with surface density and the strength of the adsorption will also be high for this adsorbent.

6.3.5.3 Lagergren's Isotherm

The kinetics of fluoride sorption by Neem leaves are modeled using the first order rate equation of Lagergren. The plots between Log (q_{e-} q) and time show a linear relationship with negative slope indicating the validity of Lagergren equation. Further, the linear relationship also

TABLE 6.4 Langmuir Isotherm for Neem and Pipal

Neem		Pipal	
$1/q_e$	$1/C_e$	$1/q_e$	$1/C_e$
1.20656	0.78345	0.555	0.714
1.57928	0.83801	0.995	1.020
1.86185	0.86363	1.245	1.142
2.01572	1.0203	1.939	1.146
2.20653	1.13263	2.403	1.190

TABLE 6.5 Kinetic Constants for Biosorbents

Adsorbent	K_f	1/*n*	*a*	*b*	Freundlich isotherm	Langmuir isotherm
Neem	0.223	0.230	1.4084	1.052	$q_e = 0.233Ce^{0.230}$	$(1.481\times Ce)/(1 + 1.052\times C_e)$
Pipal	0.572	2.747	0.8	1.489	$q_e = 0.572Ce^{2.747}$	$(1.191\times Ce)/(1 + 1.481\times Ce)$

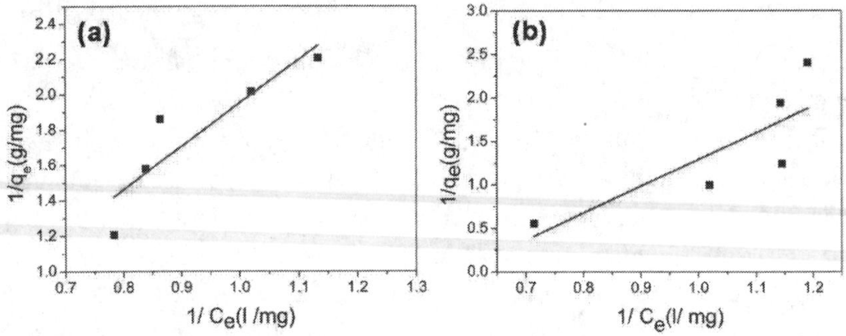

FIGURE 6.5 Plots (a) and (b) show the Langmuir Isothermsfor Neem and Pipal adsorbents, respectively.

confirms the first order nature of the process. The slope values (K) for both the adsorbents calculated from the plots in Figure 6.6 (a) and 6.6 (b) are listed in Table 6.6.

6.4 CONCLUSIONS

For 5 mg/L of initial fluoride ion concentration and at a pH value of 2, Neem biosorbent exhibited a fluoride removal of 80.2% and Pipal biosorbent exhibited a fluoride removal efficiency of 81%. It is also observed that the process of adsorption with treated biosorbents follows

TABLE 6.6 Langergren's Constant for Different Adsorbents

Adsorbent	K
Neem	–1.428
Pipal	–2.14

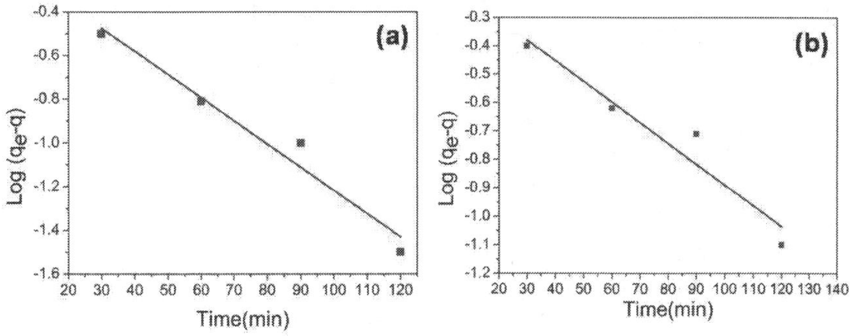

FIGURE 6.6 Figures (a) and (b) show the plots of Log (q_e_q) vs time for Neem and Pipal adsorbents, respectively.

Langmuir isotherm more strongly than Freundlich isotherm. Thus, these adsorbents can be suitably utilized on domestic or community levels to treat the fluoride contamination. Further, it is observed that for a given biosorbent the Fluoride removal efficiency increases with contact time and attains equilibrium within 90–120 min. Also, the fluoride removal efficiency increases with increasing adsorbent dosage at a constant initial solute concentration. Further, a decrease in fluoride removal efficiency is observed with increasing initial fluoride ion concentration at a constant adsorbent dosage and contact time. However, the total capacity of the adsorbent is observed to be increasing with increasing initial fluoride ion concentration. The effective preferable absorbent dosages were found to be 8.0 g/L for both the adsorbents and higher efficiencies of adsorption were achieved at a pH level of 2.

As the adsorbents used in this work are naturally available materials, they are economically viable and ecofriendly. It could be seen that biosorbents prepared from treated leaves of *Ficus Religiosa* (Pipal) are more efficient in removing fluorides from water as compared to leaves of *Azadirachta Indica* (Neem).

KEYWORDS

- **adsorption kinetics**
- **fluoride removal**

- **Langmuir/Freundlich isotherm**
- **neem**
- **Pipal**

REFERENCES

1. Fan, X., Parker, D. J., & Smith, M. D., (2003). Adsorption kinetics of fluoride on low cost materials. *Jour. Water Research, 37*, 4929.
2. Srimurli, M., et al., (1998). A study on removal of fluorides from drinking water by adsorption onto low cost materials. *Jour. Environmental Pollution, 99*, 285.
3. Choi, W. W. & Chen, K. Y., (1979). The removal of fluoride from waters by adsorption. *Journal of AWWA, 71*(10), 562.
4. Wu, Y. C., & Nitya, A. (1979). Water de-fluoridation with activated alumina. *Journal of Environmental Engineering,* 105, 357-367.
5. Li, Y. H., et al., (2001). Adsorption of fluoride from water by amorphous alumina supported on carbon nanotubes. *Journal of Chem. Phys. Lett., 350,* 412.
6. Jhonston, W. S. & McKee, R. H., (1934). Removal of fluoride from drinking water. *Journal of Industrial Engineering Chemistry, 26*(8), 849.
7. Bhargava, D. S. & Killedar, D. J., (1992). Fluoride adsorption on fishbone charcoal through a moving media adsorption column. *Water Research, 26*, 781.
8. Yadav, A. K., Kaushik, C. P., Haritash, A. K. Kansal, Rani, N. (2006). De-fluoridation of Groundwater using Brick as an Adsorbent. *Journal of Hazardous B,* 128, 289–293
9. Jamode, A. V., Sapkal, V. S., Jamode, V. S., & Deshmukh, S. K. (2004). Adsorption kinetics of de-fluoridation using low cost adsorbents. *Journal of Adsorption Science & Technology, 22*(1b), 65.
10. Yang, M. (1999). Fluoride removal in a fixed bed packed with granular calcite. *Water Research, 33*(16), 3395.
11. Piekos, R. & Paslawska, S., (1999). Fluoride uptake characteristics of fly ash. *International Society for Fluoride Research, 32*, 14.
12. Jamode, A. V., Sapkal, V. S., & Jamode V. S. (2004). Defluoridation of Water using Inexpensive Adsorbents. *Indian Institute of Science, 84*, 163-171.
13. Bhargava, D. S. & Killedar, D. J., (1993). Effect of stirring rate and temperature on fluoride removal by fishbone charcoal. *Indian Journal of Environmental Health, 35*(2), 81.
14. Bulusu, K. R. & Nawlakhe, W. G., (1998). Defluoridation of water with activated alumina: Batch operations. *Indian Journal of Environmental Health, 30*(3), 262.
15. Okeola, F. O., & Odebunmi E. O., (2010). Freundlich and Langmuir isotherms parameters for adsorption of methylene blue by activated carbon derived from agrowastes, *Advances in Natural and Applied Sciences, 4*(3), 281–288, ISSN: 1995–0748.

16. Jing, H., Song, H., Liang, Z., Fuxing, G., Yuh-Shan, H. (2010). Equilibrium and thermodynamic parameters of adsorption of methylene blue onto rectorite, by PSP, *Fresenius Environmental Bulletin, 19*(11a), 2651-2656.
17. Jamode, A. V., Sapkal, V. S. & Jamode, V. S.(2004). Defluoridation of water using inexpensive adsorbents, *J. Indian Inst. Sci.*, 84, 163–171.
18. Shaobin Wang, et al., (2006). Environmental-benign utilization of fly ash as low-cost adsorbents. *Journal of Hazardous Materials*, *136*(3), 482–501.
19. Christoffersen, M., Christoffersen, R., Larsen, R., & Møller, I. J., (1991). Regeneration by surface coating of bone char used for defluoridation of water, *Water Research*, *25*(2), 227–229.

CHAPTER 7

ADSORPTION STUDIES ON REMOVAL OF CADMIUM (II) FROM AQUEOUS SOLUTION

P. AKHILA SWANTHANTHRA[1], S. NAWAZ BAHAMANI[2], BHAVANA[3], and P. RAJESH KUMAR[4]

[1]*Assistant Professor, Department of Chemical Engineering, S.V. University, Tirupati, A.P., India, Mobile: +91-9492549980, E-mail: drakhilap@gmail.com*

[2,3]*Student, M.Tech, Department of Chemical Engineering, S.V. University, Tirupati, A.P., India*

[4]*Research Scholar, Department of Chemical Engineering, NIT Warangal, Telangana, India*

CONTENTS

ABSTRACT

As the water scarcity is increasing globally, water related problems are expected to grow worse in the coming decades. So, there is a need to identify efficient techniques for treating water at low cost with less energy. The effluents from small-scale industries, like electroplating, smelting, battery manufacture, tanneries, etc. are contaminating the nearby water source by increasing the heavy metal concentrations mainly heavy metals like Chromium, Copper, Zinc and Cadmium, etc. These cause acute poisoning in humans with high **blood pressure**, kidney damage and destruction of testicular tissue, red blood cells, stomach cramps, skin irritations, vomiting and respiratory disorders. Therefore, this wastewater should be treated using cost effective methods. In recent years, the adsorption of metal ions using low cost adsorbents is very promising method. The present study involves the investigation on removal of Cadmium using low cost adsorbents like *Maize corn leaves* powder and *Syzygiumcumini leaves* powder (popularly known as *jamun*).

Different parameters such as contact time, pH, initial concentration of cadmium, temperature and adsorbent dosage on the adsorption of Cadmium were studied. The effect of initial metal concentrations of the solutions on the removal efficiency was examined. It was found that the adsorption data were fitted well by Langmuir and Freundlich isotherms and the adsorption capacity was estimated for both adsorbents.

7.1 INTRODUCTION

Cadmium is highly toxic to human, plants, animals and it is not degradable. The cadmium enters into the environment through wastes from industrial processes such as, smelting, electroplating, alloy manufacturing, pigments, plastics, cadmium-nickel batteries, fertilizers, pesticides and textile operations, etc. Generally, the cadmium contaminated wastewaters, effluents are generated either direct production of Cd or through secondary sources.

Adsorption is one of the effective methods to remove heavy metals from aqueous solutions. Adsorbents can be considered as cheap or low cost if it is abundant in nature, requires little processing. Solid wastes are in expensive as they have no or low economic value. The aim of the present work is to study the performance of *Maize corn leaves* and *Syzygium-cumini leaves* on Cadmium removal from aqueous solutions by varying Cadmium concentration, pH and adsorbent dosage.

7.2 MATERIALS AND METHODS

The chemicals used in present study are of analytical grade and they are Cadmium chloride mono hydrated ($CdCl_2.H_2O$) Hydrochloric acid, sulphuric acid, buffer tablets of pH 2–8 and distilled water.

7.2.1 ADSORBENT PREPARATION AND ITS ACTIVATION

The *jamun* leaves were collected from Sri Venkateswara University Women's Hostel located in S.V. University Campus, Tirupati, Andhra Pradesh. Initially, the leaves were washed and dried to remove the color of adsorbent. The dried leaves are grinded into powder. Then activation of the adsorbent was done by treating the powder with concentrated sulphuric acid (0.1N) for 24 h and again washed with distilled water to remove the free acid. Then the powder was again dried under sunlight.

The Maize corn leaves were collected from a vendor at local market in Tirupati, dried under sunlight an dash was produced after burning it. The ash was washed with water to remove impurities. Then the ash was activated by treating with concentrated sulphuric acid (0.1N) for 24 h and again washed with distilled water to remove free acid. Finally, the powder was dried under sunlight.

7.2.2 EXPERIMENTAL PROCEDURE

The aqueous solution of Cadmium was prepared by dissolving 1.7935 g of Cadmium dichloride monohydrate in a liter of double distilled water to prepare 1000 mg/L (ppm) of Cadmium stock solution. Dilution of this 1000 ppm solution is prepared to the required concentrations. Using this stock solution, different initial concentrations of 50, 100, 200, 300, 400, and 500 ppm were prepared. The batch experiments were conducted in a 250 mL capacity of glass flasks with initially prepared stock solutions by shaking with 100 mL of the aqueous cadmium solutions for effect of contact time at 32°C on a rotary shaker while maintaining the initial concentration 100 mg/L, pH 6(since from literature) and adsorbent dosage 10 g/L. Adsorption isotherm study is carried out with different initial concentrations 50, 100, 200, 300, 400 and 500 mg/L while maintaining the adsorbent dosage of 10 g/L, pH 6, temperature 32°C and contact time is 3 h. The effects of adsorption dosage were 2, 4, 6, 8 and 10 mg/L studied at 32°C with an initial concentration of cadmium metal ions at 100 mg/L, pH 6 and contact time is 3 h. To determine the optimum pH on adsorption with varying pH values 3, 4, 5, 6, 7, and 8 while maintaining the adsorbent dosage of 10 g/L, temperature 32°C, initial concentration 100 mg/L and contact time is 3 h.

The filtrate was used for determination of metal with the help of Atomic Absorption Spectrophotometer (AAS).

7.3 RESULTS AND DISCUSSION

7.3.1 EFFECT OF CONTACT TIME

The removal of cadmium by using maize corn leaves ash at different initial concentrations (50–500 mg/L) at fixed dosage 10 g/L and contact time of 5 h has been studied. It is observed from the results that with increase in concentration, the % removal of cadmium decreases.

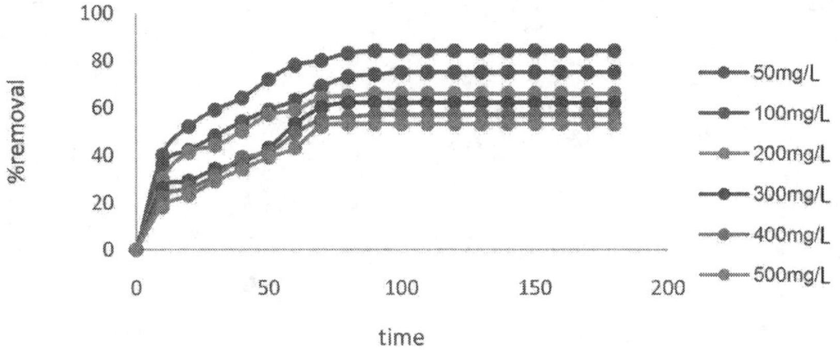

FIGURE 7.1(a) The effect of contact time on initial concentration for activated maize corn leaves ash.

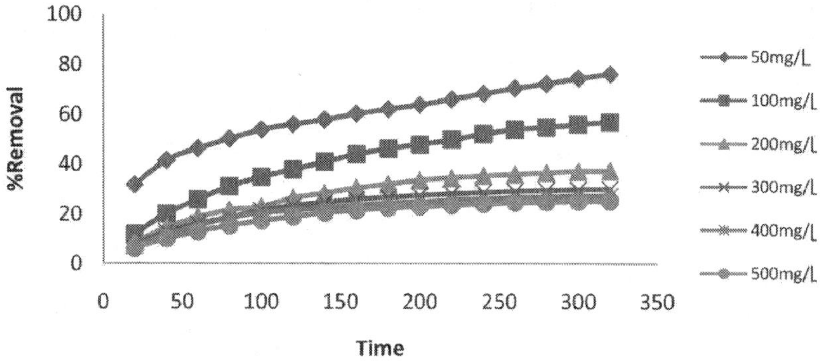

FIGURE 7.1(b) The effect of contact time on initial concentration for activated jamun leaves ash.

7.3.1.1 For Activated Maize Corn Leaves Ash

Figure 7.1(a) shows the effect of contact time on initial concentration for activated maize corn leaves ash.

7.3.1.2 For Activated Jamun Leaves Ash

Figure 7.1(b) shows the effect of contact time on initial concentration for activated jamun leaves ash.

7.3.2 EFFECT OF PH

The uptake of Cd^{+2} ions as a function of hydrogen ion concentration was examined over a pH range 2–8. It was observed that Cd^{+2} was strongly adsorbed at low pH and declined at higher pH. Since, large number of H^{+} ions present at low pH which in turn neutralize the negatively charged adsorbent surface, thereby reducing hindrance to the diffusion of cadmium ions. The decrease in % removal at high pH may be due to large number of OH-ions causing increased hindrance to diffusion of cadmium ions. From the experimental studies it was observed that the maximum %removal of Cd^{+2} at pH value of 6.

7.3.2.1 FOR ACTIVATED MAIZE CORN LEAVES ASH

7.3.2.2 For Activated Jamun Leaves Ash

Figure 7.2 shows the effect of pH on % removal for activated (a) maize corn leaves ash, and (b) jamun leaves ash.

7.3.3 EFFECT OF DOSAGE OF ADSORBENT

7.3.3.1 For Activated Maize Corn Leaves Ash

The percentage removal of Cadmium increases from 77% to 89% by increasing in dosage of adsorbent from 2–10 g/L, for initial concentration of 100 mg/L in the solution. This is because of increase in surface area and adsorption sites available for adsorption (Figure 7.3).

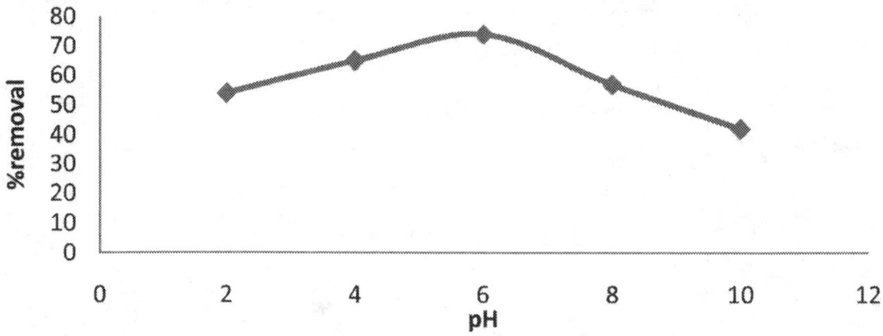

FIGURE 7.2(a) The effect of pH on % removal for activated maize corn leaves ash.

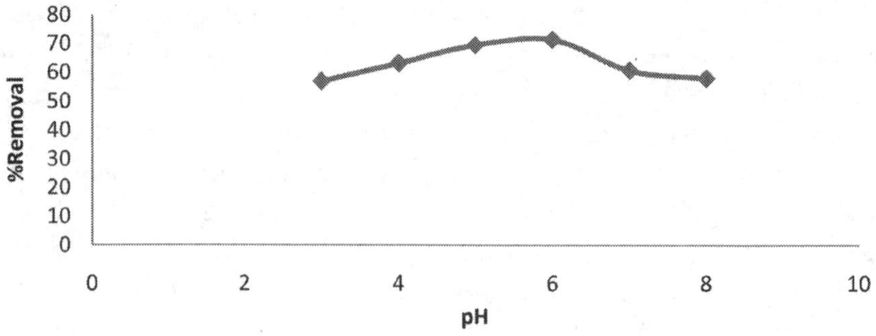

FIGURE 7.2(b) The effect of pH on % removal for activated jamun leaves ash.

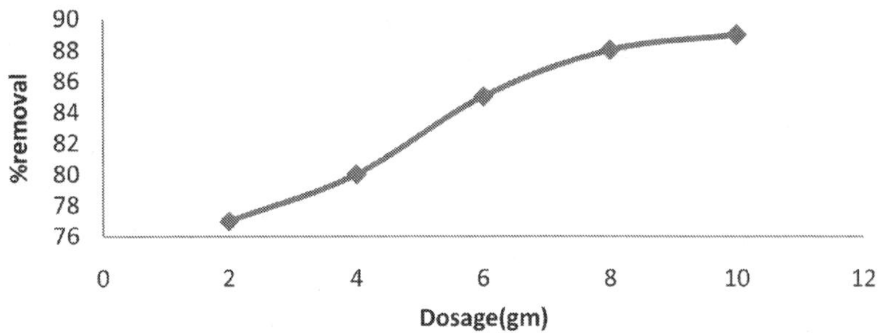

FIGURE 7.3(a) Effect of adsorbent dosage on % removal for activated (a) maize corn leaves ash, and (b) jamun leaves ash.

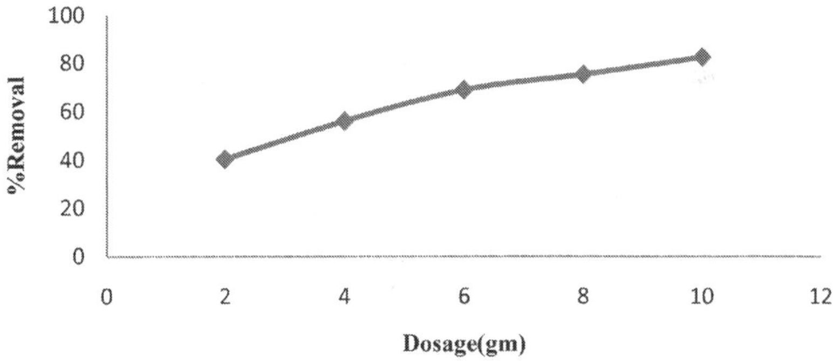

FIGURE 7.3(b) Effect of adsorbent dosage on % removalfor activated jamunleaves ash

7.3.3.2 For Activated Jamun Leaves Ash

The percentage removal of cadmium increases from 40.3% to 82.45% by increasing dosage of adsorbent from 2–10 g/L, for initial concentration of 100 mg/L in the solution. This happened due to increase in surface area and adsorption sites available for adsorption.

7.3.3.3 Effect of Adsorbent Dosage on Adsorption Capacity

It was observed that the adsorption capacity decreases due to the sites remaining in adsorbent surface unsaturated during the adsorption process. It is graphically represented as shown in Figure 7.4.

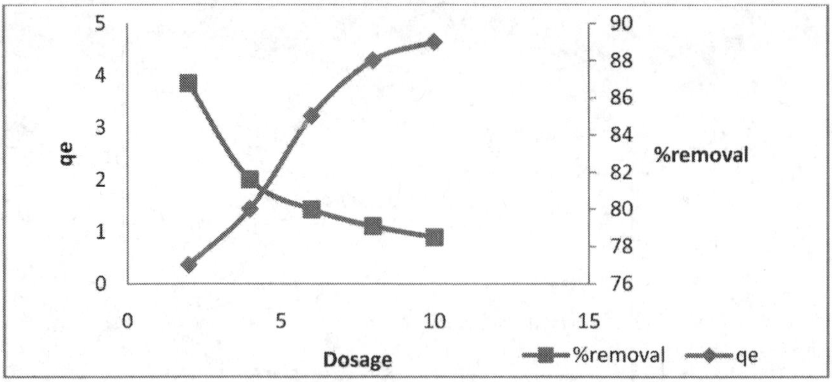

FIGURE 4(a) Effect of adsorbent dosage on adsorption capacityby maizecorn leaves ash

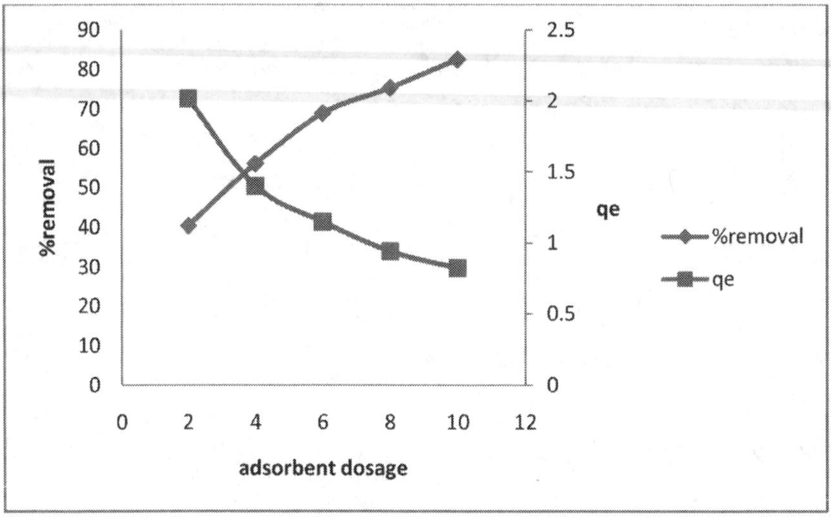

FIGURE 4(b) Effect of adsorbent dosage on adsorption capacityby jamun leaves ash

7.3.3.4 Effect of Initial Concentration on Adsorption Capacity

Figure 7.5 shows the effect of initial concentration on % removal and adsorption capacity by (a) maize corn leaves ash, and (b) jamun leaves ash.

7.3.4 COMPARISON OF ADSORBENTS

Comparison for activated maize corn leaves ash and activated jamun leaves ash regarding agitation time, from the below graph we can say that

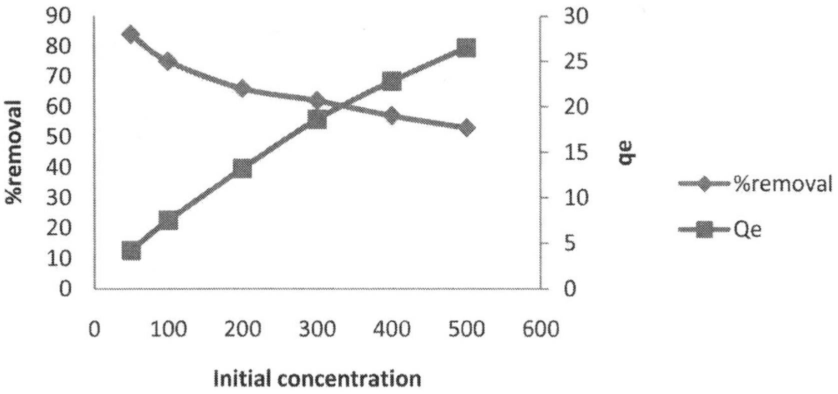

FIGURE 7.5(a) Effect of initial concentration on % removal and adsorption capacity by maize corn leaves ash,

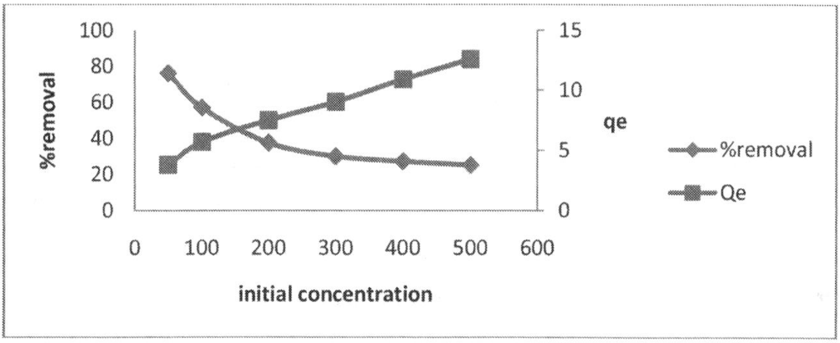

FIGURE 7.5(b) Effect of initial concentration on % removal and adsorption capacity by jamun leaves ash.

activated maize corn leaves ash is better than activated jamun leaves ash. this is so because for the same agitation time we can see the more amounts of adsorption rate for activated maize corn leaves ash than that of the activated jamun leaves (Figure 7.6).

7.4 ADSORPTION ISOTHERMS

The study of the adsorption isotherms is very important in determination of the capacity of adsorbents. The Langmuir equation is applied to the case where adsorption is on completely homogeneous surfaces

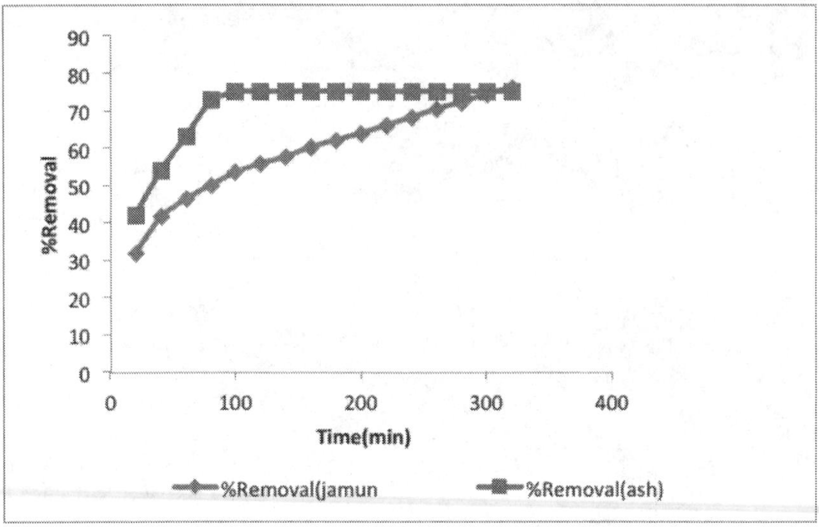

FIGURE 7.6 Effect of agitation time versus % removal for activated maize corn leaves ash and activated jamun leaves ash

where the interactions between adsorbed molecules are negligible. The Freundlich equation used to describe the adsorption in aqueous systems (Figure 7.7).

The Langmuir isotherm is given by the equation:

$$C_e/q_e = 1/(\mathrm{bQm}) + (1/\mathrm{Qm})\, C_e.$$

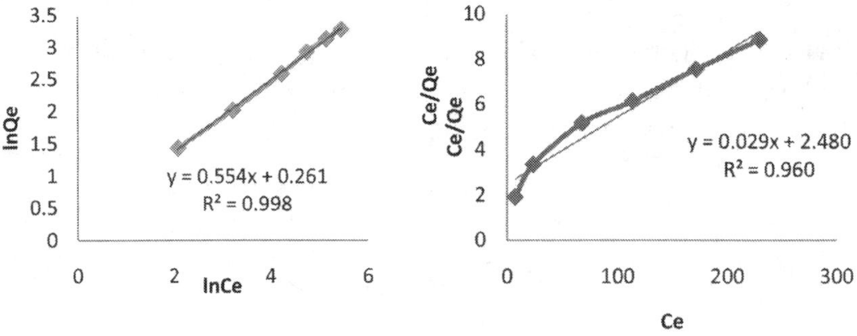

FIGURE 7.7 Freundlich isotherm and Langmuir isotherms for activated maize corn leaves ash.

The Freundlich isotherm is given by the equation:

$$\ln q_e = \ln K_f + (1/n) \ln C_e.$$

7.4.1 FOR ACTIVATED JAMUN LEAVES ASH

Freundlich constants, K_f and n are obtained by plotting the graph between $\ln q_e$ vs. $\ln C_e$. The values of K_f and n are 1.298 and 1.8028, respectively. It was observed that the regression correlation factor is close to unity for both models and it was 0.9603 for Langmuir and 0.9987 for Freundlich isotherms for activated maize corn leaves ash:

7.4.2 FOR ACTIVATED JAMUN LEAVES ASH

Freundlich constants, K_f and n are obtained by plotting the graph between $\ln q_e$ vs. $\ln C_e$. The values of K_f and n are 1.6247 and 3.011, respectively. It was observed that the regression correlation factor is close to unity for both models and it was 0.9473 for Langmuir and 0.9825 for Freundlich isotherms (Figure 7.8).

7.5 CONCLUSIONS

The following conclusions can be made from the results:

1. Adsorbent prepared from maize corn leaves ash and activated jamun leaves ash can be used for cadmium removal from aqueous solutions.

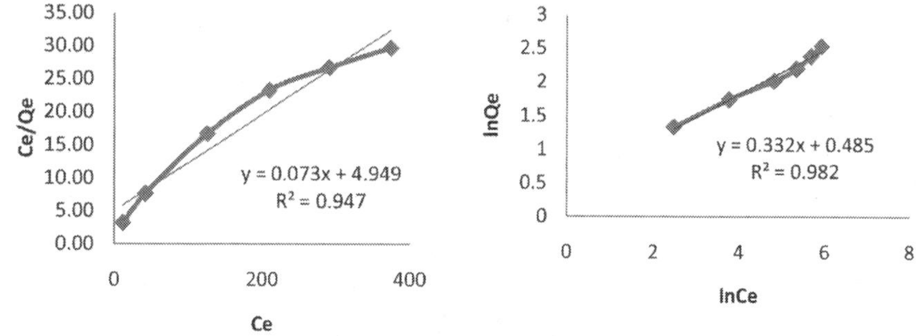

FIGURE 7.8 Langmuir isotherm and Freundlich isotherm for activated jamun leaves ash.

2. The maximum percentage removal is observed at pH value of 6.
3. Cadmium Removal increases with increase of adsorbent dosage at pH 6 at an adsorbent dosage 10 g/L.
4. The adsorption isotherms were studied.
5. Comparison studies maize corn leaves ash (89%) and jamun leaves (82%) revealed that maize corn leaves ash is better adsorbents than jamun leaves ash.
6. Freundlich isotherm better fitted the experimental data since the regression correlation coefficients for Freundlich isotherm was higher than Langmuir isotherm.

KEYWORDS

- **adsorption**
- **adsorption capacity**
- **cadmium**
- **isotherms**
- **maize corn leaves powder**
- **Syzygium cumini leaves powder**

REFERENCES

1. Rao, K. S., Anand, S., & Venkateswarlu, P., (2010b). *Psidiumguvajaval* leaf powder: A potential low cost biosorbent for the removal of cadmium (II) from wastewater. *Adsorption Science and Technology*, *28*(2), 163–178.
2. Pino, G. H., de Mesquita, L. M. S., Torem, M. M. L., & Pinto, G. A. S., (2006). Biosorption of cadmium by green coconut shell powder, *Miner. Eng.*, *19*, 380–387.
3. Qaiser, S., Saleemi, A. R., & Ahmad, M. M., (2007). Heavy metal uptake by agro based waste materials. *Environ. Biotechnol.*, *10*, 409–416.
4. Panda, G. C., Das, S. K., Chatterjee, S., Maity, P. B., Bandopadhyay, T. S., & Guha, A. K., (2006). Adsorption of cadmium on husk of *Lathyrussativus*: Physico-chemical study. *Colloids and Surfaces B: Biointerfaces*, *50*, 49–54.
5. Esteves, A. J. P., Valdman, E., & Leite, S. G. F., (2000). Repeated removal of cadmium and zinc from an industrial effluent by waste biomass *Sargassum* sp. *Biotechnol. Lett.*, *22*(6), 499–502.

6. Ahluwalia, S.S., & Goyal, D., (2005). Removal of heavy metals by waste tea leaves from aqueous solution. *Engineering in Life Sciences*, *5*(2), 158–162.
7. Azab, M. S., & Peterson, P. J., (1989). The removal of cadmium from water by the use of biological sorbents. *Water Sci. Technol.*, *21*, 1705–1706.
8. Babel, S., & Kurniawan, T. A., (2003). Low-cost adsorbents for heavy metals uptake from contaminated water: a review. *J. Hazard. Mater.*, *97*, 219–243.
9. Qaiser, S., Saleemi, A. R., & Ahmad, M. M., (2007). Heavy metal uptake by agro based waste materials. *Environ. Biotechnol, 10*, 409–416.
10. Nitin W. Ingole, Vidya N. Patil (2010). Cadmium removal from aqueous solution by modified low cost adsorbent(s): A state of the art, *International Journal of Engineering, Science and Technology*, *2*(7), 81–103.

CHAPTER 8

HYDRODYANAMIC CAVITATION FOR DISTILLERY WASTEWATER TREATMENT: A REVIEW

DIPAK K. CHANDRE,[1] CHANDRAKANT R. HOLKAR,[2] ANANDA J. JADHAV,[2] DIPAK V. PINJARI,[2] and ANIRUDDHA B. PANDIT

[1]*Department of Chemical Engineering, SVIT, Chincholi, Sinnar, Nashik – 422103, India, Tel.: +91-22-33612032; Fax: +91-22-33611020; E-mail: dv.pinjari@ictmumbai.edu.in, dpinjari@gmail.com*

[2]*Department of Chemical Engineering, Institute of Chemical Technology (ICT), Mumbai – 400019, India*

CONTENTS

8.1 INTRODUCTION

8.1.1 ISSUES IN DISTILLERY WASTEWATER

The distillery effluent stream, also known as spent wash is generated during alcohol production as unwanted residual liquid waste. Pollution caused by distillery effluent stream is most critical environmental issues worldwide. Distilleries industries are one of most polluting industries worldwide and according to the Central Pollution Control Board (CPCB), Govt. of India, these industries are rated as one of the 17 most polluting industries [1].

Molasses based distillery wastewater is dark brown in color because it contain melanoidins polymers, having very high molecular weights and always exits in colloidal form [3], occurs from Maillard reaction between amino acids and carbonyl groups in molasses [2]. Molasses based distillery wastewater contains high-level dissolved organic as well as inorganic matter along with nutrients. They can be very aggressive to the environment if improperly managed. The molasses based wastewater also characterized by moderately acidic pH (3.8–4.4) [4]. When an effluent is released directly in water bodies then it can cause oxygen depletion and creates the problems and/or released in soil then reduction of soil alkalinity and manganese. The melanoidin pigments are toxic to microorganism available in soil and water [2].

Molasses based distillery industries are among the most polluting industries in India which generates large volumes of high strength of wastewater, its volume is approximately 12 to 15 times that of production of alcohol volume [5]. The effluents of distillery industries contains high strength of Biochemical Oxygen Demand (BOD) (44,000–65,000 mg/L), Chemical Oxygen Demand (COD) (75,000–99,000 mL/L), Total Solid (TS) (60,000–90,000 mg/L) along with strong odor and recalcitrant dark brown color [4]. The COD indicates that concentration of all organic compounds present in system which can be oxidized by using strong oxidizing agents and TOC indicates that the amount of organic present in the system [6]. Molasses based distillery wastewater contains, apart from high dissolved organic contents, high strength of nutrients in the form of nitrogen (1000–1200 mg/L), Potash (5000–12,000 mg/L) and Phosphate (500–1500 mg/L) that can be dissolved in water bodies.

Distillery Spent wash is highly pollution potential as well as hazardous if disposed directly into the environment. The high value of nitrogen, phosphates and COD of the distillery effluent may be result in eutrophication of natural water bodies. Highly colored compounds present in the distillery effluent reduces penetration of sunlight in the lakes, lagoons or rivers, which result in, reduce the dissolved oxygen concentration and photosynthesis activity [7]. In Table 8.1 shows typical characteristics of the molasses based distillery effluent.

8.1.2 IMPORTANCE OF TREATMENT

The limits for discharge of industrial wastewaters are setting up by world-wide environment regulatory authorities. The Central Pollution Control Board of India told to achieve zero discharge of spent wash of distillery industry [8].

Molasses based distillery effluent stream is a dark brown in color and highly organic effluent. It is most complex and strongest organic effluents due to high values of BOD, COD, TDS, TSS and low value of pH. The molasses based distillery effluent is the primary source of renewable energy because of it contains high concentration of organic loading. Most of the reported studies have concentrated only on removal of melanoidin/ color, BOD and COD from distillery effluent but not on the issues like enhancement of biodegradability [9]. However, there is an urgent need for development of a suitable treatment process for removal of melanoidin/ color, BOD and COD from distillery effluent along with biodegradability enhancement with lower investment [9].

TABLE 8.1 Summarized the Typical Characteristics of Molasses Based Distillery Wastewater (Spent Wash)

Parameter	Range
pH	3.8–4.4
Total Solids (mg/L)	60,000–90,000
Total Suspended Solid (mg/L)	2,000–14,000
Total Dissolved Solid (mg/L)	58,000–76,000
Total Volatile Solids (mg/L)	45,000–65,000
COD (mg/L)	70,000–98,000
BOD (for 5 days at 20^{0}C) (mg/L)	45,000–60,000
Total Nitrogen as N (mg/L)	1,000–1,200
Potash as K_2O (mg/L)	5,000–12,000
Phosphate as PO_4 (mg/L)	500–1,500
Alkalinity as $CaCO_3$ (mg/L)	8,000–16,000
Temperature (after heat exchanger) °C	70–80

8.1.3 REVIEW THE PRESENT STATE OF ART IN BRIEF

8.1.3.1 Conventional Methods

Up-flow Anaerobic sludge Blanket (UASB) reactor has been one of the beneficial and advantageous processes particularly in treatment distillery wastewater. The pilot scale UASB reactor shows that total COD removal efficiencies ranges between 84% and 98% were achieved [10].

Anaerobic digestion (AD) is also one of most suitable option for treatment of distillery effluent [11]. This is due to the fact that it contains high amount of biodegradable organic contaminants which can be converted into biogas. For generation of steam from boiler this biogas can be used thereby meeting the demand of energy of the treatment plant. The AD technique is effective for BOD as well as COD reduction. This process is incapable for removing of color associated with the distillery wastewater and which is the major challenge faced by AD [11, 12]. The wastewater from biogas digester is further treated to reduce its COD and BOD aerobically [8].

8.1.3.2 Microorganisms/Biological Treatment

Microorganisms present in wastewater are used in biological treatment for wastewater. The microorganisms decompose organic compounds in wastewater and convert it into energy and microbial biomass. In biological treatment process there is no separation of treated waste and microorganism. Bio-filtration is another method for wastewater treatment which is differing from biological treatment by the fact that there is separation of treated water and microorganism [5] for removal of heavy metals, dyes and hazardous organic pollutants from distillery wastewater by using Microbial bio-adsorbents [2]. The various microorganisms employed for the decolorization of distillery effluent as shown in Table 8.2.

8.1.3.3 Membrane Technology

The effective treatment method for distillery wastewater is the Membrane Bioreactor (MBRs). The MBRs offers complete retention of solid as well

TABLE 8.2 Microorganism Employed for the Decolorization of Distillery Effluent [13]

Name	Color Removal (%)
Bacteria	
Xanthomonas Fragariae	76
Bacillus Cereus	82
Acetobacteracetii	76.4
Pseudomonas Putida	60
Fungi	
Trametes versicolor	82
Geotrichum candidum	80
Mycelia sterilia	93
Aspergillus oryzae	75

as yielding of less sludge and best quality of effluent which is great advantages of MBRs. The MBRs application is limited by two reasons one is high membrane cost and other is the progressively fouling of membrane which leads to cleaning of membrane as well as replacement of membrane, thus contributing to the high operating costs [5].

8.1.3.4 Hybrid Technology/Integrated Technology

The hybrid UV photo degradation (AOP) and AD is most suitable treatment for effluent of distillery for the reduction of BOD as well as COD and Color of the effluent [12]. The overall color and COD removal for the integrated AD and AOP system as shown in Figures 8.1 and 8.2.

8.1.3.5 Biological Treatment Method

The biological treatment methods for molasses based distillery spent wash is either anaerobic or aerobic but combination of anaerobic and aerobic process used in the most of cases. If typically COD/BOD ratio will be 1.8 to 1.9 then it indicates that the suitability of effluent treatment of distillery wastewater by biological treatment [7].

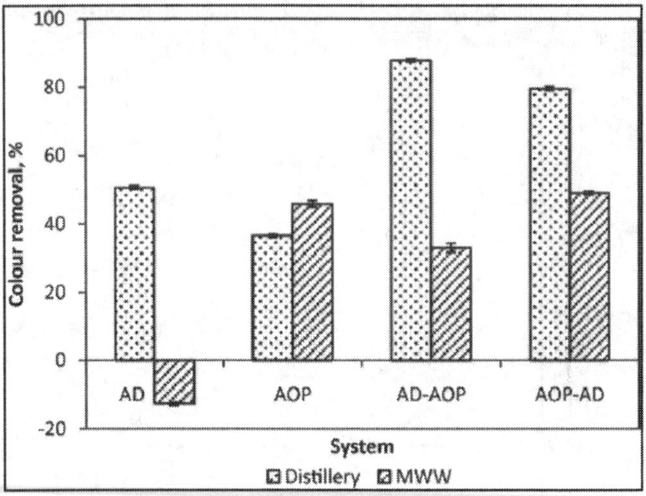

FIGURE 8.1 Overall color removals for the integrated AD and UV photo degradation (AOP) system.

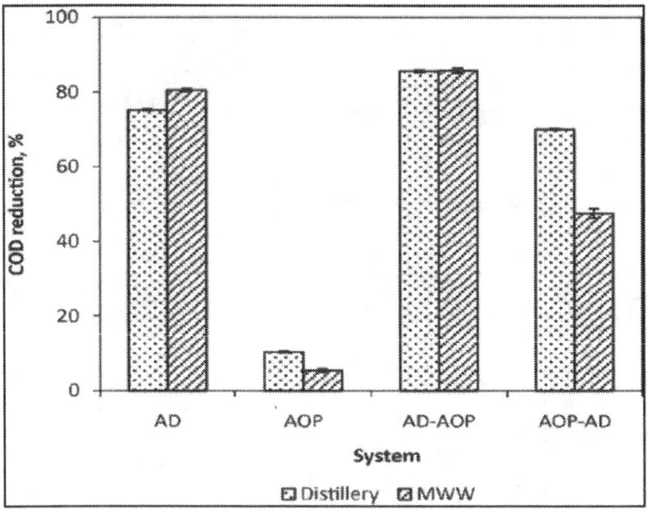

FIGURE 8.2 Overall COD reduction for the integrated AD and UV photo degradation (AOP) system.

8.1.3.6 Physico-Chemical Method

The Physico-chemical treatment method includes coagulation/flocculation, adsorption, and oxidation process like ozonation (O_3), hydrogen peroxide (H_2O_2) and Fenton's oxidation. These methods have also been used in combination for distillery spent wash treatment. The physico-chemical methods are employed for further reduction of the COD and color of the effluent after primary anaerobic treatment [7].

The decolorization of Melanoidins has been reported by various physico-chemical treatments system which is summarized in the Table 8.3. The majority of the color removed by this method is either by concentrating the color into sludge or by partially/complete breakdown of the color molecules.

The various physico-chemical treatment systems used for distillery spent wash treatment are tabulated in Table 8.3 with their efficiency.

i) **Adsorption:** Among all the physicochemical treatment methods, adsorption on activated carbon (AC) is mostly used for removal of specific organic pollutants and color of effluent. The well-known adsorbent is AC because of its high surface area, high adsorption capacity and high surface reactivity, and microporous structure.

ii) **Coagulation and Flocculation:** Mohana et al. [7] reported that effluent treatment by coagulation seems to be expensive process as expenses of chemical and sludge disposal. Thus, there is an urgent need for the development of cost effective alternative for post bio-methanated effluent.

iii) **Oxidation Processes:** For wastewater treatment ozone is a powerful oxidant. If ozone is dissolved in wastewater then it reacts with numbers of organic compounds in two ways:

- By direct oxidation as molecular ozone.
- By indirect reaction with formation of secondary oxidants like free radical particularly hydroxyl radicals (H_2O_2).

As hydroxyl radical and ozone are strongly oxidizing agents it can oxidize a number of organic compounds in wastewater [20].

Oxidation process based on Fenton's produces hydroxyl radical ($OH^•$), having high oxidation potential. The hydrogen peroxide (H_2O_2) and iron salt (Fe^{2+} or Fe^{3+}) is the mixture of Fenton's reagent which involves homogeneous reaction and which is environmentally acceptable, it produces hydroxyl radical thus ultimately decolorization of the effluent.

TABLE 8.3 Various Physico-Chemical Treatment Systems Used for Distillery Spent Wash with Their Efficiency

Treatment	% COD Removal	% Color Removal	References
Adsorption			
1) Chitosan, a biopolymer was used as anion exchanger	99	98	[14]
2) Chemically modified bagasse			[15]
– DEAE bagasse	40	51	
– CHPTAC bagasse	25	50	
3) Activated carbon prepared from agro industrial waste			[16]
– Phosphoric acid carbonized bagasse was used.	23	50	
4) Commercially available activated carbon	76 88	93 95	[16]
– AC (ME)			
– AC (LB)			
Coagulation–flocculation			[17]
(Flocculation of synthetic melanoidins was carried out by various inorganic ions)			
1) Polyferrichydroxysulphate (PFS)	NR	95	
2) Ferric chloride ($FeCl_3$)	NR	96	
3) Ferric sulphate ($Fe_2(SO4)_3$)	NR	95	
4) Aluminum sulphate ($Al2(SO_4)_3$)	NR	83	
5) Calcium oxide (CaO)	NR	77	
6) Calcium chloride ($CaCl_2$)	NR	46	
Oxidation processes			[18, 19]
Fenton's oxidation	88	99	
Ozonation	15–25	80	
Membrane technologies			[19]
Reverse osmosis	99.9	–	
Nanofiltration	97.1	100	

NR – Not Reported.

8.1.3.7 Recent Wastewater Treatment Technology

The various conventional methods are available for the treatment of industrial wastewater like biological treatment (aerobic and anaerobic), adsorption, and coagulation/flocculation. The main problem associated with biological treatment system is cost expensive and this treatment system is not able to completely degrade the organic effluent because of complex and refractory nature of the organic pollutant present in wastewater [21].

The other methods like adsorption, and coagulation/flocculation does not involve the chemical transformation but only transfer of waste components from one phase to another. Therefore, it is urgent need to develop other alternative cost effective treatment methods that can degrade the complex bio-refractory molecules. The alternative methods involving chemical transformation of organic pollutant by oxidation into CO_2 and H_2O as the end product. The hydroxyl radical oxidizes the organic pollutants which are present in wastewater [21].

8.1.3.8 The Alternative Method Termed as Advanced Oxidation Process (AOPs)

Gurol et al. [22] reported that advanced oxidation processes can be defined as the generation of highly reactive free radicals (especially hydroxyl radicals-$HO^{\bullet}$) in the water treatment processes at near ambient temperature and pressure. $HO^{\bullet}$ radicals are extraordinarily reactive species that attack most of the organic molecules present in the wastewater. The reaction kinetics is generally first order with respect to the concentration of hydroxyl radicals and to the concentration of the species cab be oxidized. Rate constants are generally in the range of 10^{8}–10^{11} $M^{-1}s^{-1}$, whereas the concentration of hydroxyl radicals lays between 10^{-10} and 10^{-12} M, thus a pseudo-first order constant between 1 and 10^{-4} s^{-1} is obtained. They also reported that advanced oxidation processes can be used for destroy the complex refractory organic matter present in the wastewater even after conventional treatment methods. In the treatment of pollutants these processes shows great potential either in high or low concentrations. The main goal of advanced oxidation processes is to mineralize the organic matter present in the wastewater into water, carbon dioxide and inorganic ions by degradation reactions with strongly oxidizing agent. The number Advanced Oxidation Processes (AOPs) have been used by many researcher for the production of hydroxyl radical for industrial wastewater treatment such as:

1. Cavitation.
2. Fenton reagent (Fe^{2+}/H_2O_2)
3. Ozone (O_3)
4. O_3/UV
5. Ozone/UV/H_2O_2

Cavitation is one of recent technique of AOPs which has been extensively used for pretreatment option for complex refractory organic matter present in industrial waste wastes water. The cavitation has ability to degradation of organic pollutant present in wastewater those other conventional methods.

Pinjari et al. [30], reported that the cavitation phenomena can be generated by monitoring the flow pattern of liquid hence is called as hydrodynamic cavitation and many researcher found that for same application cavitation phenomena is energy efficient than acoustic cavitation.

Satyawali and Balakrishanan [11], Migo et al. [31], Pia et al. [20], Pena et al. [19], Pala and Erden [18], Mohana et al. [7] reported that the numbers of approaches for distillery wastewater treatment like biological, Physico-chemical and phytoremediation. In all these applications, authors have concentrated only on the removal of color from distillery wastewater but not on the issue of enhancement of biodegradability.

8.2 CAVITATION

8.2.1 INTRODUCTION OF CAVITATION

Treatment of industrial wastewater is a key aspect of recent research interest because of increasing awareness of environment and more stringent environmental regulations worldwide. Some of research work done in the development and testing of new technique and their combinations with conventional biological and chemical methods. Cavitation is one of such recent technologies for industrial wastewater treatment but only a few studies have been reported in this regard. The cavitation technology can be suitable for degradation of molasses based distillery wastewater for lowering the toxicity levels of the effluent stream [6].

Cavitation is defined as formation of micro bubbles, its growth and subsequent collapse of it with small interval of time as milliseconds and releasing of large amount of energy at various locations in the reactor. Iskalieva et al. [23] reported that cavitation processes generates very high temperature and pressure like thousands of pressure in atmospheric and thousands of temperature in Kelvin. Cavitation processes can be considered as one of energy input methods. Apart from mode of generation of cavities, cavitation can be further classified as stable and transient cavitation. The formation of stable and transient cavitation is based on set of various operating parameters and constituents of liquid. It is very important to decide the set of operating parameter for obtaining maximum result with minimum energy input. The transient cavitation involves generation of larger bubbles size (maximum size of bubbles reached by cavities is hundred times than that of initial size) over the time scale of few acoustic cycles. The stable cavitation involves generation of smaller bubbles size over the time scale of many acoustic cycles.

8.2.2 HYDRODYNAMIC CAVITATION

The cavitation phenomena are called as hydrodynamic cavitation because cavitation generated by liquid flow pattern. The cavities formed in hydrodynamic cavitation due to liquid passing through the cavitating devices provided in the pipeline like venturi and orifice plate. When liquid pressure at vena-contract or throat falls below the vapor pressure of its liquid then the liquid flashes takes place with formation of number of micro bubbles with subsequently collapsing of them when pressure of liquid recovers in downstream side of cavitating devices.

When collapsing of vapor cavity then creation of hot spots takes place with generation of reactive free radicals with transient temperature of 10,000 K, and pressures of about 1000 atm. Under this extreme condition water molecules are dissociated into $OH^{\bullet}$ and $H^{\bullet}$ radicals. This $OH^{\bullet}$ radical then react with organic pollutant and oxidized/mineralized them [9].

Hydrodynamic cavitation (HC) technology can be considered as a pre-treatment method for molasses based distillery wastewater. By this method reduction of COD to TOC ratio and enhancement of Biodegradation Index (BI) (BOD_5 to COD ratio) as well as color reduction of distillery wastewater is also possible by this recent method [9]. The schematic representation of hydrodynamic cavitation reactor set-up along with cavitating device like venturi as shown in Figures 8.3 and 8.4.

Ozone is a strong oxidant agent and used in hydrodynamic cavitation which can oxidize both biodegradable and non-biodegradable organics present in the wastewater. When mass flow rate of ozone increases then both COD and color reduction also increases resulting from increasing the number of hydroxyl radical in the system. The ozonation system along with the iron oxide (Fe_2O_3) catalyst gave the highest efficiency in both COD and Color removal because of hydroxyl

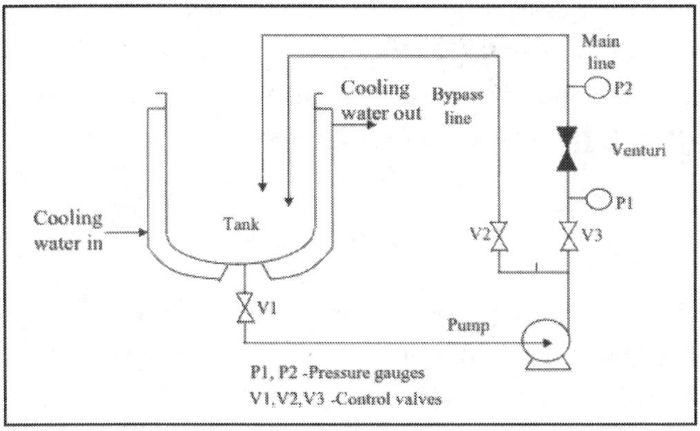

FIGURE 8.3 Schematic representation of hydrodynamic cavitation reactor set-up with venturi.

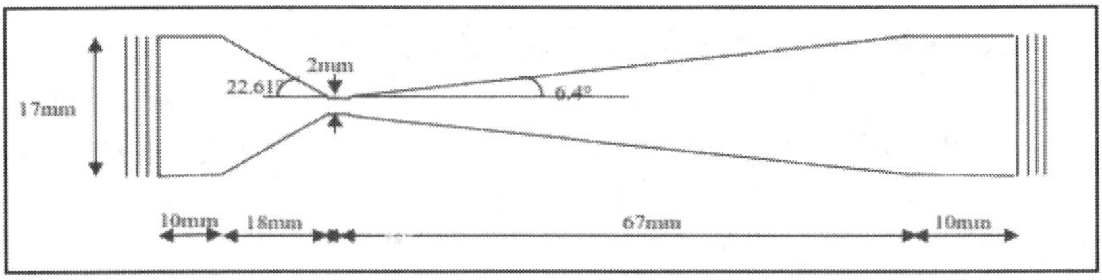

FIGURE 8.4 Schematic of venture.

free radical generated from catalyst is more reactive than that of ozone molecule itself [3].

Padoleya et al. [9] have reported that for degradation of pollutants in wastewater there are two mechanism using hydrodynamic cavitation, one is the thermal decomposition of volatile pollutant molecules and other is $OH^{\bullet}$ radicals reaction with pollutant at the interface of cavity – water. In case of non volatile pollutant, the mechanism for degradation of pollutant will be attack of $OH^{\bullet}$ radicals on pollutant molecules at the interface of cavity – water.

8.2.3 APPLICATION OF HYDRODYNAMIC CAVITATION TO DISTILLERY WASTEWATER

The wastewater used as molasses based distillery effluent called as spent wash obtained from a local site. For characterization of distillery wastewater effluents is first diluted by 2–3 times before passing through treatment to reduce the toxicity level [3]. The applications of hydrodynamic cavitation for distillery wastewater is to study the effect of inlet pressure for mineralization, effect of dilution for mineralization, effect of cavitation for color reduction, effect of cavitation for biodegradability, effect of time for Total Suspended Solids (TSS) and Volatile Suspended Solids (VSS), effect of temperature and pH.

8.2.4 ECONOMICAL ASPECTS FOR APPLICATION OF HYDRODYNAMIC CAVITATION TO DISTILLERY WASTEWATER

Nowadays, the Scientists and researchers have to emphasize the need for research on effective option for treatment of molasses based distillery wastewater with safe disposal. Padoleya et al. [9] have reported that cavitation technique is an energy efficient option used as pretreatment method in combination with other advanced oxidation processes or biological methods. As far as cost is concerned hydrodynamic cavitation technology required overall lower costs of treatment as compared to other method [6].

8.3 EFFECT OF DIFFERENT PARAMETERS OF CAVITATION ON TREATMENT OF DISTILLERY WASTEWATER

8.3.1 EFFECT OF INLET PRESSURE ON MINERALIZATION

Saharan et al. [24] reported that the effect of inlet pressure on reduction of COD and TOC by venturi was evaluated at an inlet pressure of 5 bar and 13 bar. The Molasses based distillery wastewater used for experimentation having average value of COD was 34000 mg/L. At different interval of time samples were collected for Analysis. The result indicates that percentage reduction of COD and TOC does not increase significantly with increase in inlet pressure as shown in Figure 8.5.

The reported result indicates that for the molasses based distillery wastewater, the maximum mineralization occurs at the initial stage of 50 min of circulation (85 passes) through hydrodynamic cavitation after which result indicates that the COD and TOC does not reduces significantly thus 50 min will be optimum treatment time.

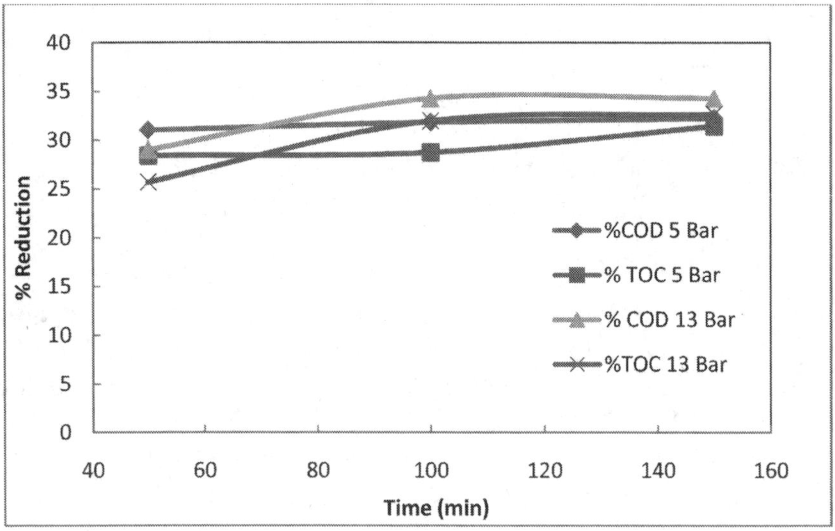

FIGURE 8.5 Effect of inlet pressure on mineralization.

8.3.2 EFFECT OF DILUTION ON MINERALIZATION

Saharan et al. [24], reported that the molasses based distillery wastewater used for experimentation to be diluted using tab water by 25% and 50 % concentration (V/V) having average value of COD was 34,000 mg/L at optimum inlet pressure of 5-bar using hydrodynamic cavitation. At different treatment time the various values of COD and TOC were measured. The reported result indicates that on mineralization of distillery wastewater dilution has no significant effect. Also at 50% dilution percentage reduction is marginally higher as shown in Figure 8.6.

But Chakinala et al. [25] reported that if increase in dilution then enhance the mineralization by using chemical added oxidants (Fenton's reagent). But obtained results are contrary to the literature reports.

8.3.3 EFFECT OF CAVITATION ON COLOR REDUCTION

Bhargava and Chandra [26] reported that the molasses based distillery wastewater, having dark brown in color, is not only due to complex

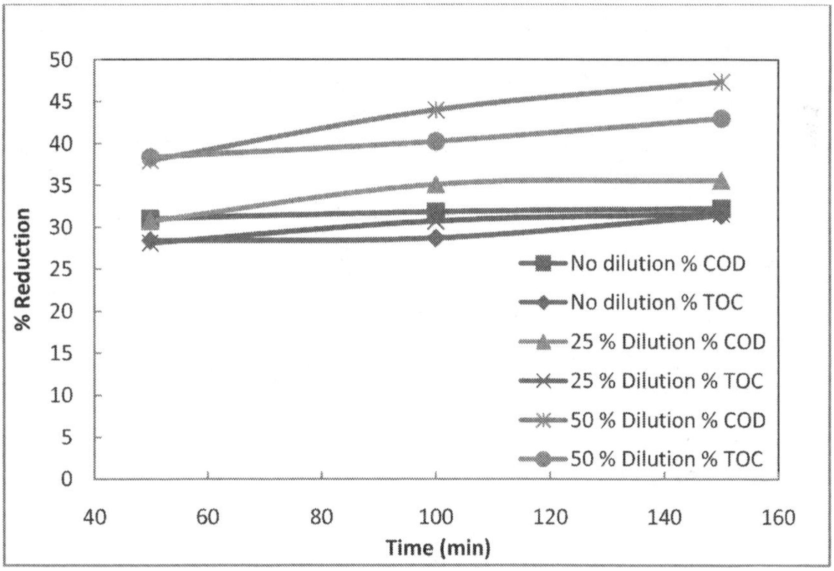

FIGURE 8.6 Effect of dilution for mineralization.

biopolymer called melanoidins, which is the product of Maillard reaction but also caramel colorants which are obtained from evaporation and concentration of sugarcane juice at very high temperature during distillation process of sugarcane molasses for absolute alcohol recovery. The recent pretreatment option by hydrodynamic cavitation has not only used for reduction of COD and TOC but also reduction of total color (indirectly toxicity) of molasses based distillery wastewater. At the optimum inlet pressure of 5 bars in hydrodynamic cavitation process indicates that the maximum color reductions of molasses based distillery wastewater were observed as 36%. The color reduction profile as shown in Figure 8.7.

For undiluted wastewater a maximum color reduction were observed as 34%, while for 25% dilution and 50% dilution of wastewater the maximum color reduction were observed as 41% and 48%, respectively.

Sreethawong and Chavadej [27] reported that the color removal from diluted distillery wastewater by hydrodynamic cavitation using ozonation process in presence of Fenton's reagent (Iron Oxide catalyst) and results indicates that in presence of Fenton's reagent the color and COD reduction was more. The reduction of color was 7% at

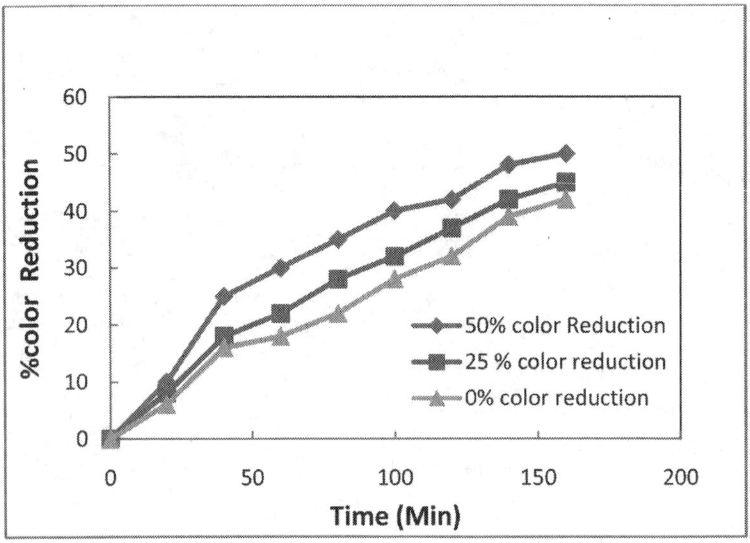

FIGURE 8.7 Effect of cavitation for color reduction.

25% dilution and 14% at 50% dilution for 150 min of treatment time and 34% at zero dilution.

8.3.4 EFFECT OF CAVITATION ON BIODEGRADABILITY

Metcalf and Eddy [32] reported that the ratio of BOD_5:COD can be expressed as Biodegradability Index (BI). The minimum of value of BI as 0.3 to 0.4 can be considered as ideal for good biodegradability of any wastewater. If BI ratio is ≥ 0.3 then process is suitable for coupling with aerobic treatment and if BI ratio is ≥ 0.4 then process is suitable for coupling with anaerobic treatment. The Figure 8.8 shows the variation of BI with respect to time. From results it can be suggested that for toxicity reduction lower inlet pressure (5 bars) is suitable whereas higher inlet pressure (13 bars) is suitable for enhanced the biodegradability.

Thus from results, hydrodynamic cavitation technique is not only suitable for reduction of toxicity of molasses based distillery wastewater but also pretreatment option for increase in biodegradability index of distillery wastewater.

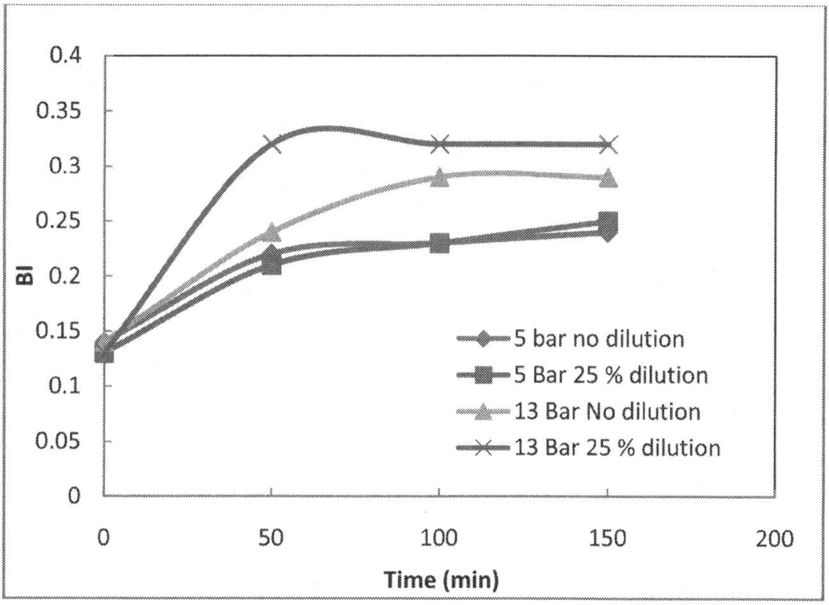

FIGURE 8.8 Effect of cavitation on Biodegradability Index (BI).

8.3.5 EFFECT OF TIME ON THE REDUCTION OF TOTAL SUSPENDED SOLIDS (TSS) AND VOLATILE SUSPENDED SOLIDS (VSS)

Suranjit Kumar Chanda [28] reported that the experiment were run for 90 min and results were presented, compared and discussed for three pre-treatment, for example, cavitation, ozone and combination of ozone and cavitation for molasses based distillery wastewater. The concentration of suspended solids measured as both TSS and VSS were reduced by the three techniques as shown in Figures 8.9 and 8.10. A greater reduction of both TSS and VSS was found in the case of the combined application of ozone and cavitation, compared to the other individual applications.

The effect of inlet pressure (515 KPa and 275 KPa) in the combined application of ozone and cavitation on suspended solids (TSS/VSS) reduction is shown in Figures 8.11 and 8.12.

From Figures 8.13 and 8.14 it can be seen that the slope of the curve decreased with respect to time and after 10 hours, TSS and VSS were reduced by 68% and 70%, respectively [28].

From the above discussion it can be concluded that the combined application of ozone and cavitation showed that marginally higher reduction of

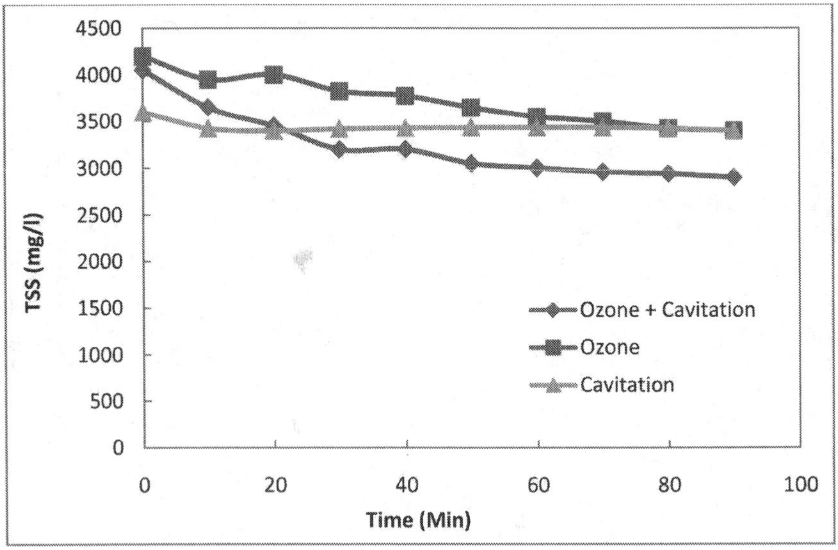

FIGURE 8.9 Reduction of TSS using different disintegration condition.

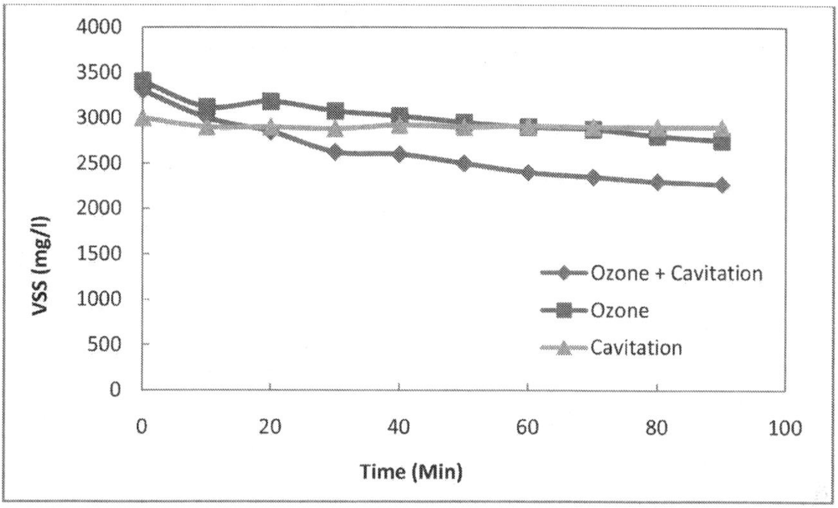

FIGURE 8.10 Reduction of VSS using different disintegration condition.

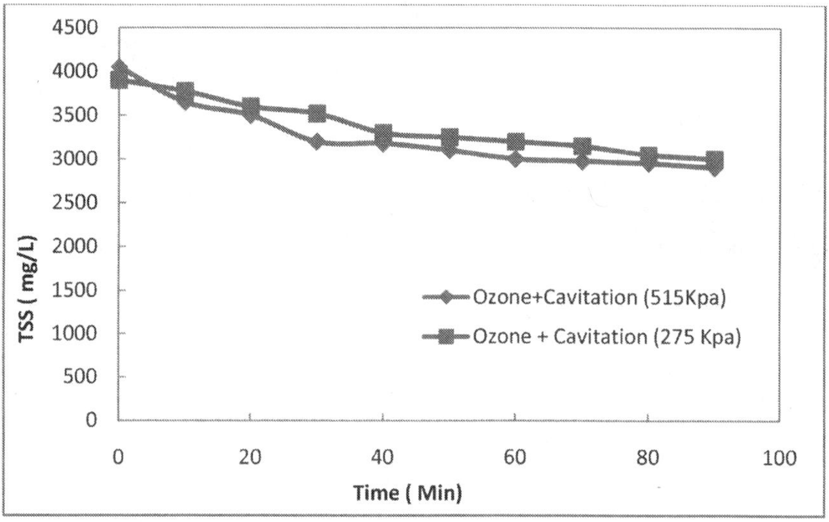

FIGURE 8.11 Reduction of TSS with different inlet pressures in combined ozone-cavitation treatment.

suspended solids (TSS/VSS) at higher inlet pressure than the individual technologies (i.e., ozonation, cavitation) and also that the rate of reduction decreased with respect to time.

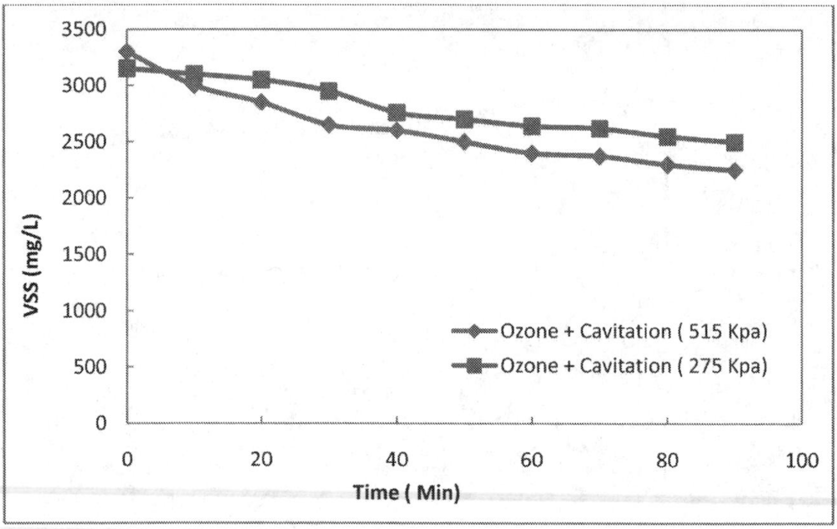

FIGURE 8.12 Reduction of VSS with different inlet pressures in combined ozone-cavitation treatment.

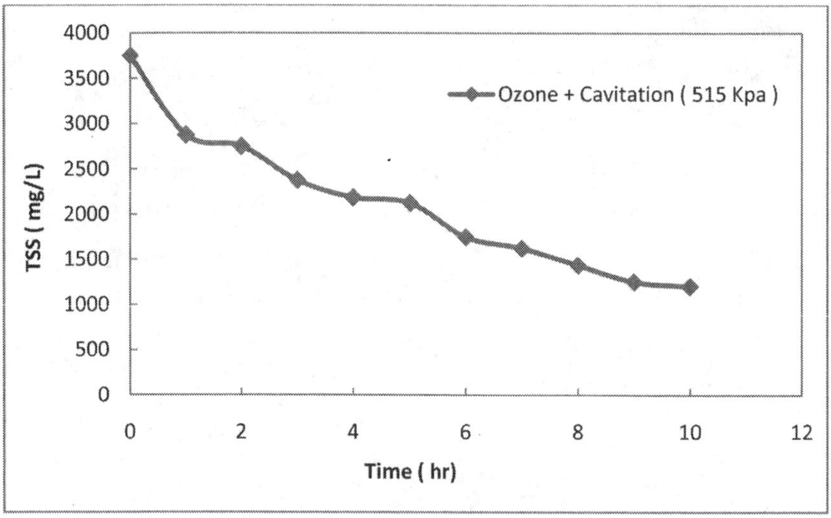

FIGURE 8.13 Reduction of TSS using combined application of ozone and cavitation.

8.3.6 EFFECT OF TEMPERATURE AND PH

The changes in temperature and pH are illustrated in Figures 8.15 and 8.16 for different disintegration conditions.

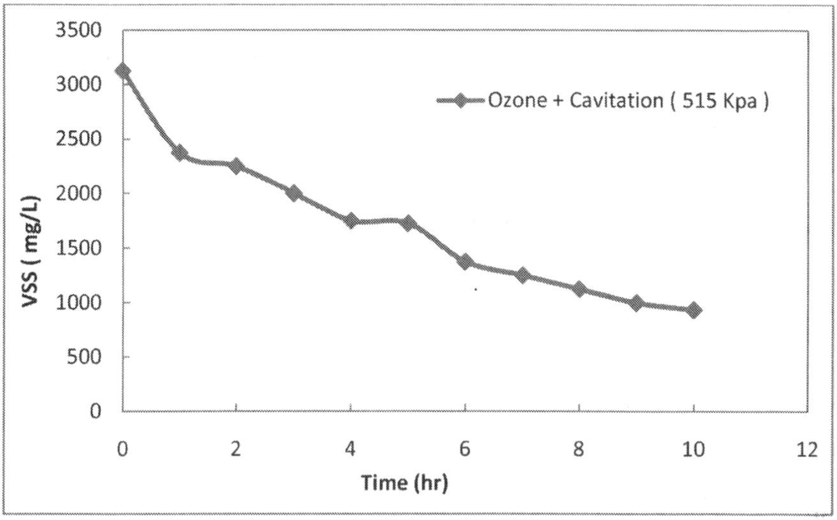

FIGURE 8.14 Reduction of VSS using combined application of ozone and cavitation.

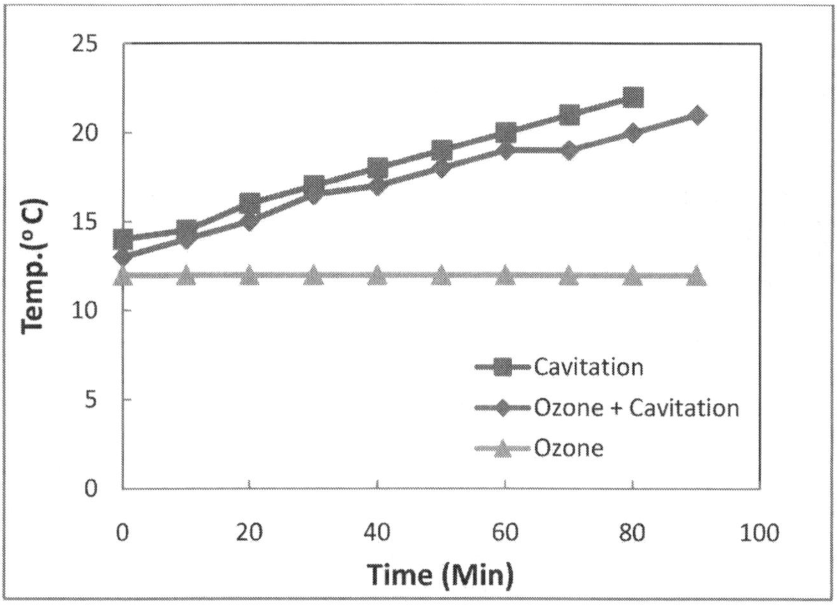

FIGURE 8.15 Changes in temperature for different disintegration conditions.

From Figure 8.15, it is observed that temperature increases for both the combined application of ozone and cavitation, and cavitation alone. This might be due to the collapse of cavities which causes the increase

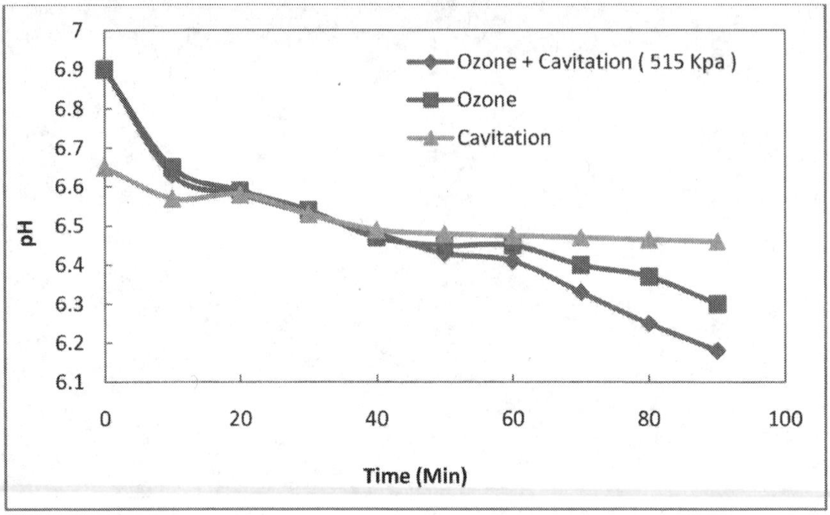

FIGURE 8.16 Changes in pH for different disintegration conditions.

of temperature [29]. On the other hand, no changes in temperature were observed with the application of ozone alone.

Figure 8.16 indicates that the pH changed from 6.9 to 6.18 due to the combined application of ozone and cavitation during the 90 min of experiments. Similar changes were observed due to the application of ozone [33]. Cavitation alone did not result in any significant impact on pH.

8.4 CONCLUSIONS

The hydrodynamic cavitation technique can be effectively used for enhancing the biodegradability index along with toxicity reduction (minimizing the COD/TOC ratio and color) of complex molasses based distillery wastewater. Lower inlet pressure (5 bars) is suitable for toxicity reduction of molasses based distillery wastewater while higher inlet pressure (13 bars) is suitable for enhancement of biodegradability index. The study shows that hydrodynamic cavitation as an advanced treatment or pretreatment technique for the molasses based distillery wastewater. The hydrodynamic cavitation can also be effective treatment option for complex bio-refractory pollutants and can serve as a cost effective option which can be used as an individual or in combination with other conventional treatment processes.

KEYWORDS

- **advanced oxidations processes (AOPs)**
- **biological treatment**
- **cavitation**
- **chemical treatment**
- **distillery effluent**
- **molasses**
- **physical treatment**

REFERENCES

1. CPCB, (2003). Annual Report: Central Pollution Control Board: New Delhi, India.
2. Shutosh Kumar Verma, Chandralata Raghu Kumar, & Chandrakant Govind Naik, (2011). "A novel hybrid technology for remediation of molasses-based raw effluents,". *Bioresource Technology, 102*, 2411–2418.
3. Thammanoon Sreetha Wong, & Sumaeth Chavadej, (2008) "Color removal of distillery wastewater by ozonation in the absence and presence of immobilized iron oxide catalyst," *Journal of Hazardous Materials, 155*, 486–493.
4. Saha, N. K, Balakrishnan, M. & Batra, V. S., (2005). "Improving industrial water use: Case study for an Indian distillery", *Resources, Conservation and Recycling, 43*, 163–174.
5. Pawar Avinash, et al., (2012). "Treatment of distillery wastewater using membrane technology", *International Journal of Advanced Engineering Research and Studies: IJAERS, 1*(3), 275–283.
6. Anand G. Chakinala, Parag R. Gogate, Arthur E. Burgess, & David H. Bremner, (2008). "Industrial wastewater treatment using hydrodynamic cavitation and heterogeneous advanced fenton processing", *Ultrason. Sonochem., 15*, 49–54.
7. Sarayu Mohana, Bhavik K. Acharya, & Datta Madamwar, (2009). "Distillery spent wash: Treatment technologies and potential applications," *Journal of Hazardous Materials, 163*, 12–25.
8. Parmesh Kumar Chaudhari, Indra Mani Mishra, & Shri Chand, (2007). "Decolorization and removal of chemical oxygen demand (COD) with energy recovery: Treatment of biodigester effluent of a molasses-based alcohol distillery using

inorganic coagulants", *Colloids and Surfaces A: Physicochem. Eng. Aspects, 296*, 238–247

9. Padoley, K. V., Virendra Kumar Saharan, Mudliar, S. N., Pandey, R. A., & Aniruddha B. Pandit, (2012). "Cavitationally induced biodegradability enhancement of a distillery wastewater", *Journal of Hazardous Materials, 219–220*, 69–74.
10. IlterTurkdogan-Aydınol, F., & Kaan Yetilmezsoy, (2010). "A fuzzy-logic-based model to predict biogas and methane production rates in a pilot- scale mesophilic UASB reactor reacting molasses wastewater", *Journal of Hazardous Materials, 182*, 460–471.
11. Satyawali, Y., & Balakrishnan, M., (2008) "Wastewater treatment in molasses-based alcohol distilleries for COD and color removal: a review", *Journal of Environmental Management, 86*, 481–497.
12. Seth Apollo, Maurice S. Onyango, & Aoyi Ochieng, (2013). "An integrated anaerobic digestion and UV photo catalytic treatment of distillery wastewater," *Journal of Hazardous Materials, 261*, 435–442
13. Pant, D., & Adholeya, A., (2007). "Biological approaches for treatment of distillery wastewater", *Bioresource Technology, 98*, 2321–2334.
14. Lalvo, I. G., Guerginov, I. I., Krysteva, M. A., & Farstov, K., (2000). "Treatment of wastewater from distilleries with Chitosan", *Water Res., 34*, 1503–1506.
15. Mane, J. D., Modi, S., Nagawade, S. Phadnis, S. P., & Bhandari, V. M., (2006). "Treatment of spent wash using chemically modified bagasse and color removal studies", *Biores. Technol. 97*, 1752–1755.
16. Satyawali, Y., & Balakrishnan, M., (2007). "Removal of color from biomethanated distillery spent wash by treatment with activated carbons", *Biores. Technol., 98*, 2629–2635.
17. Migo, V. P., Del Rosario, E. J., & Matsumura, M., (1997). "Flocculation of melanoidins induced by inorganic ions", *J. Fermen. Bioeng., 83*, 287–291.
18. Pala, A., & Erden, G., (2005). "Decolorization of a baker's yeast industry effluent by fenton's oxidation", *J. Hazard. Mater., 127*, 141–148.
19. Pena, M., Coca, M., Gonzalez, G., Rioja, R., & Garcia, M. T., (2003). "Chemical oxidation of wastewater from molasses fermentation with ozone", *Chemosphere, 51*, 893–900.
20. Bes-Pia, A., Mendoza-Roca, J. A., Alcaina-Miranda, M. I., Iborra-Clar, A., & Iborra-Clar, M. I. (2003). "Combination of physico-chemical treatment and Nanofiltration to reuse wastewater of printing, dyeing and finishing textile industry", *Desalination, 157*, 73–80.
21. Mohan M. Gore, & Prakash V. Chavan, (2013). "Hydrodynamic cavitation using degradation of reactive orange 4 dye", *Int. J. Chem. Sci., 11*(3), 1314–1320.
22. Gurol, M. O., & Vatistas, R., (1987). "Oxidation of Phenolic compounds by ozone and ozone + UV radiation: a comparative study", *Wat. Res., 21*, 895–900.
23. Asylzat Iskalieva, Bob Mbouyem Yimmou, Parag R. Gogate, Miklos Horvath, & Peter G. Horvath, (2012). "Cavitation assisted delignification of wheat straw: A review", *Ultrasonics Sonochemistry, 19*, 984–993
24. Saharan, V. K., & Pandit, A. B, (2013). "Hydrodynamic cavitation as a pretreatment tool for the biodegradability enhancement of distillery wastewater- Chapter 5: Hydrodynamic Cavitation based degradation of Bio-Refractory Pollutants, 140–161.

25. Chakinala, A. G., Gogate, P. R., Burgess, A. E., & Bremner, D. H., (2008). "Treatment of industrial wastewater effluents using hydrodynamic cavitation and the advanced Fenton process", *Ultrasonics Sonochemistry*, *15*, 49–54.
26. Bhargava, R. N., & Chandra, R. (2010). "Biodegradation of the major color containing compounds in distillery wastewater by an aerobic bacterial culture and characterization of their metabolites", *Biodegradation*, *21*, 703–711.
27. Sreethawong, T., & Chavadej, S., (2008). "Color removal of distillery wastewater by ozonation in the absence and presence of immobilized iron oxide catalyst", *Journal of Hazardous Materials*, *155*, 486–493.
28. Suranjit Kumar Chanda, (2012). "Disintegration of Sludge Using Ozone-Hydrodynamic Cavitation," The University of British Columbia, 1–87.
29. Benito, Y., Arrojo, S., Hauke, G., Vidal, P., de Conceicao Cunha, M., & Brebbia, C., (2005). "Hydrodynamic cavitation as a low-cost AOP for wastewater treatment: Preliminary results and a new design approach", *Water Resources Management*, *III*, 495–504.
30. Pinjari, D. V., & Pandit, A. B., (2010). "Cavitation milling of cellulose to nanosize", *Ultrasonics Sonochemistry*, *17*, 845–852.
31. Migo, V. P., Matsumura, M., Rosario, E. J. D., & Kataoka, H, (1993), "Decolorization of molasses wastewater using an inorganic flocculent", *J. Ferment. Bioengg.* *75*, 438.
32. Metcalf I. Eddy, (1979). "Wastewater Engineering: Treatment Disposal Reuse", 2nd ed., Tata McGraw Hill: New Delhi, India.
33. Bougrier, C., Battimelli, A., Delgenes, J. P. & Carrere, H., (2007). "Combined ozone pretreatment and anaerobic digestion for the reduction of biological sludge production in wastewater treatment", *Ozone: Science and Engineering*, *29*(3), 201–206.

CHAPTER 9

MICROWAVE- AND ULTRASOUND-ASSISTED SURFACTANT TREATED ADSORBENT FOR THE EFFICIENT REMOVAL OF EMULSIFIED OIL FROM WASTEWATER

P. AUGUSTA, P. KALAICHELVI, K. N. SHEEBA, and A. ARUNAGIRI

Department of Chemical Engineering, National Institute of Technology, Tiruchirappalli, Tamil Nadu, India,
Tel. +91 431 2503110; Fax: +91 431 2500133;
E-mail: kalai@nitt.edu

CONTENTS

ABSTRACT

Discharge of oily wastewater complying with both resource reuse suggestions and environmental regulations is a major challenge nowadays. The objective of this chapter is to elucidate the current status and prospect of development in oily wastewater treatment. The origin, characteristics and environmental impacts of the oil are herein outlined. The techniques for oil treatment including membrane separation, vacuum evaporation, chemical treatment, floatation, filtration, coalescence, deep bed filtration, electrical methods, adsorption, microwave and ultrasound technologies are elaborated. Among the various methods available, adsorption is the most promising choice of treatment owing to its robustness, efficacy and feasibility with the usage of an appropriate adsorbent. Hence, novel approaches to oil adsorption through surfactant modification, alkali treatment, microwave treatment, ultrasound application and/or combination of these approaches are proposed to maximize the oil removal efficiency and to minimize the adverse impacts of the disposal of oil into the environment. In addition, ecofriendly biosorbents such as, corn husk in terms of their adsorption capacities, process factors and pretreatment methods have been highlighted as an attractive alternative for the treatment of oil in wastewater.

9.1 INTRODUCTION

Oily wastewater has become prevalent as one of the primary environmental issues to date. The vast growth in the oil exploration and production together with the increase in industrial activities has paved way for oil pollution. The sources of oily wastewater encompasses a very broad range, the oil from the oil industry, oil refining, oil storage, transportation and petrochemical industries [1] and is provided in Table 9.1. Typical permissible limits for oil and grease discharges vary between 10–15 mg/L for

TABLE 9.1 Sources of Oily Effluents[1]

Industrial process	Oil concentration (mg/ L)
Petroleum refining	20–4000
Metal processing and finishing	100–20,000
Aluminum rolling	5000–50,000
Copper wire drawing	1000–10,000
Food processing (fish and seafood)	500–14,000
Edible oil refining	4000–6000
Paint manufacturing	1000–2000
Cleaning bilge water from ships	30–2000
Car washing	50–2000
Aircraft maintenance	500–1500
Leather processing (tannery effluents)	200–40,000
Wool sourcing	1500–12,500
Wood preservation	50–1500

mineral and synthetic oils whereas 100–150 mg/L for those from animal and vegetable sources.

Oils in wastewater exist in several forms such as fats, lubricants, cutting oils, heavy hydrocarbons, and light hydrocarbons. These oils can be classified into two types as free and emulsified oils. It is easier to treat the free oil fraction of the wastewater by physical techniques. However, it is difficult to process emulsions or the oil droplets dispersed in water phase because of their high stability in aqueous phase.

The amount of oily wastewater forming an emulsion is greatly increased in machine and tool works and other sources such as automotive, aerospace, packing, and mechanical engineering sectors. Emulsified oily wastewater contains oil (mineral, vegetable or synthetic), fatty acids, emulsifiers (anionic and nonionic surfactants), corrosion inhibitors (amines), bactericides and other chemicals, which endow a long-lasting and effective fluid.

Oil can be emulsified either mechanically or chemically. Mechanically emulsified oil involves electrostatically stabilized oil droplets ranging in size from 20–150 microns. These oils blend with water because of the shear resulting from the wastewater passing through a pump or due to splashing into a tank, which can disintegrate and disperse the oil droplets. Mechanical emulsion does not involve the influence of surfactants. Chemically emulsified oil entails the use of an emulsifier such as, a surfactant, detergent, or soap. The surfactants are hydrocarbon chains, the most simple ones being sodium laurel sulfate or stearic acid, which have a hydrophilic (water loving) and a lipophlic end (oil loving). The lipophilic end enters the oil droplet, while the hydrophilic end remains in the water. Since there is now a charge on the otherwise neutral oil droplet, when droplets approach each other they will repel each other and disperse. The size of the droplet is less than 20 microns; the color of the water is white. The white color is an indicator that the emulsion must be split to allow the removal of oil.

Emulsified oil is produced in various industrial applications where metal working operations use water soluble coolants, cutting, grinding oils and lubricants for machining. The crucial issue of oil emulsion is that it is prone to contamination after usage and it is deprived of its properties, hence, the replacement with a new emulsion results in oily waste. In the past, the waste emulsion was more often discharged into either sanitary sewers or public waterways without proper treatment, leading to environmental pollution and oil loss.

Treatment of oily wastewater emulsions consists of challenging tasks. First of all, stringent environmental regulations that impose very low oil concentrations (10 mg/L) in the discharge streams are tough to achieve. Secondly, the processing of such emulsified wastewater must be cost effective. As waste oils are significant resources, removal of the oil phase is necessary before discharge. Hence, the treatment of emulsified oil in wastewater is the need of date not only from an economic perspective but also in terms of culminating pollution of water resources.

9.2 IMPACT OF EMULSIFIED OIL IN WASTEWATER ON ENVIRONMENT

Oily wastewater involves high oil content, chemical oxygen demand (COD) and color. Free oil can be removed by gravitation and hence, it does not pose a serious issue. However, emulsified oil is a big issue due to its stability in the aqueous phase [2]. Oily wastewater mainly pollutes the environment in the ensuing facets [1] contaminating drinking water and groundwater resources,

[2] threatening human health; [3] polluting the atmosphere [4] upsetting crop productivity [5] deteriorating the natural landscape.

9.2.1 CONTAMINATING DRINKING WATER AND GROUND WATER RESOURCES

Besides creating an aesthetically unpleasant ambience, oil in wastewater blocks the penetration of sunlight, an indispensable element of the aquatic cycle. Subsequently, the photosynthesis process of aquatic plants gets affected and hence, the dissolved oxygen level in the water bodies is decreased. Moreover, oil may deposit on the gills of aquatic animals which will interrupt their breathing, leading to death. This in turn imposes needless stress on the already exhausted food chain to back up the ever-growing human population. The Ministry of Environment and Forests, Government of India has recommended the discharge wastewater quality with oil concentration not exceeding 5 mg/dm^3.

9.2.2 THREATENING HUMAN HEALTH

Oil compounds possess the potential for bioaccumulation as they are transportable along the food chain through oil-contaminated marine food. Studies on fuel extract from Erika oil spill demonstrated that consumption of oil-contaminated marine food could affect human health [3]. Hydrocarbon oily wastewater causes detrimental effects on human health, particularly to those who have bare-hand contact with the oil. Volatile organic compounds (VOC) and polycyclic aromatic hydrocarbons (PAH) metabolite levels were observed to be increased in urine samples of cleanup volunteers after cleanup. Moreover, there is a risk of skin cancer due to skin contact with used motor oils, chiefly due to PAHs.

9.2.3 POLLUTING THE ATMOSPHERE

Pollution of air due to VOCs is always attributed to the refinery oily sludge accumulated in lagoons or landfills. These air pollutants cause health risks to facility workers and nearby people.

9.2.4 UPSETTING CROP PRODUCTIVITY

The oil contaminated soils generate deficiency of nutrients, prevent seed germination, and are responsible for limited growth or death of plants on contact. Moreover, both the treated and untreated wastewater with oil used for irrigation bring in chemical pollutants, which may be deleterious to plants at higher concentrations.

9.2.5 DETERIORATING THE NATURAL LANDSCAPE

The oily sludge on discharge changes the physical and chemical properties of the terrestrial environment, resulting in morphological changes of the soil. Because of high viscosity, oily matter can enter the soil pores, adsorbed onto the mineral constituents of soil, or cover the soil surface. Hence, hygroscopic moisture would be reduced, hydraulic conductivity, and wet ability or water retention capacity of soils also would decrease. Specifically, higher molecular weight compounds and their degradation products exist near the surface of the soil and hydrophobic crusts are formed thereby limiting water availability and restricting water/air exchange. PHCs could also reduce the activity of soil enzymes (i.e., hydrogenase and invertase) and may be toxic to soil microorganisms.

9.3 TECHNIQUES AVAILABLE FOR TREATMENT OF EMULSIFIED OIL

9.3.1 MEMBRANE SEPARATION

Membrane technology is a prominent separation technique over the past few years. The pressure-driven membrane processes include microfiltration (MF), ultrafiltration (UF), nanofiltration (NF) and reverse osmosis (RO). These processes differ merely in the pore size of the membranes used, respectively. Microfiltration rejects particles in the range: 0.10–10 μm; ultrafiltration: 0.01–0.1 μm; nanofiltration: 0.001–0.01 μm, and reverse osmosis: less than 0.001 μm. Materials such as, polysulfone, polyamide, cellulose acetate, nylon, polytetrafluoroethylene, and polypropylene are used for membrane preparation. Membrane separation mainly consists of advantages such as, no addition of chemicals, lower energy requirements, ease of handling and better configuration for carrying out the process. However, fouling of the membrane is the crucial issue in membrane separation technology.

9.3.1.1 Ultrafiltration

Ultrafiltration is a proven technology for efficient recovery of solids and liquids. Its major areas of application include oily wastewater of several types, electropainting wastes, treatment and recovery of metal working fluids. Ultrafiltration exploits a semi permeable membrane to separate water, solvent and low molecular weight solutes from oil in oil-water emulsions. The oily wastewater is usually fed to the membrane unit and an operating pressure of 0.7–3.5 $kg/cm^2$is maintained. Oil droplets of high molecular weight are retained on the membrane and can be removed continuously.

Ultrafiltration membranes of pore sizes ranging between 0.001 and 0.1 mm are being employed to eliminate emulsified oil/water particles from wastewater [4, 5]. UF generates a liquid oil concentrate consisting of about 60% oil and solids, which can be subjected to hauling or incineration. The concentrate is usually only 3–5% of the original volume. Titanium dioxide (TiO_2) ultrafiltration membrane was able to achieve 99.5% of oil removal while treating the oil-water emulsions [6].

Ultrafiltration involves lower capital and operating costs. However, the membranes used for ultrafiltration do not withstand high temperatures and are not resistant to hazardous substances such as, oxidizing agents, some solvents and organic compounds. Moreover, ultrafiltration though being an apt technique to treat oily wastewaters, the membranes used is prone to extensive fouling due to adsorption and deposition of rejected oil and other substances on the surface of the membrane.

9.3.1.2 Reverse Osmosis

Reverse osmosis technology is prominently employed in the desalination of sea water, treatment of industrial wastewater and related fields due to its merits of technical, operational maturity, energy usage, investment and other aspects. It is efficient in removal of particles, dispersed and emulsified oil. Nevertheless, the method has certain setbacks such as, high pressure requirements, and high chances of membrane fouling even with trivial amounts of oil and grease.

9.3.2 VACUUM EVAPORATION

Vacuum evaporation is an apt alternative for the treatment of oily wastes, specifically when recycling of water is desired. The energy consumption is very low in the presence of vacuum or low pressures, such as 8 kPa, 10–40 kPa, or 40–90 kPa. However, the resulting aqueous effluent is of higher quality for vacuum evaporation when compared to other treatments, such as UF. Vacuum evaporation is favored for the treatment of oil-water emulsions, in spite of its operating costs, because of the high COD reduction that can be achieved (90–99.9%) [7].

9.3.3 CHEMICAL TREATMENT (DESTABILIZATION BY COAGULATION AND FLOCCULATION)

Coagulation–flocculation is a vital physicochemical/treatment step in industrial wastewater treatment to minimize the content of suspended and colloidal materials. Coagulation-flocculation comprises of addition of chemicals (coagulants) to the treatment tanks for speeding sedimentation (coagulants). Coagulants are inorganic or organic compounds such as aluminum sulphate, aluminum hydroxide or chloride or high molecular weight cationic polymers.

Chemical flocculation finds application when additional suspended solid particles cannot be removed by gravity separation method and when it is desirable to remove particulate sulfides in the water. It can also be used to remove emulsified oil. In the treatment of emulsified oil, the coagulants accelerate the disruption of emulsion by coalescing the oil droplets and the successive separation of the aqueous and oily phases through conventional settling or dissolved-air flotation. Aggregation zinc silicate (PISS) and anionic polyacrylamide (A-PAM) composite flocculant enhanced the oil removal efficiency up to 99% and the concentration of suspended solids was less than 5 mg/L [8]. However, this method has its own limitations such as, higher initial operating costs, generation of secondary pollutants and hence, the resultant processing problems. Thus, there arises the need for the development of novel cheap composite flocculants.

9.3.4 FLOTATION

Flotation is a gravity based separation process, driven by increasing density difference between continuous and dispersed phases. This is achieved by passing gas or air into the oily wastewater to enhance the development of air-solid or air-oil agglomerates. Flotation is categorized into three types: electro-flotation (EF), Induced (dispersed) air flotation (IAF) and dissolved air flotation (DAF). IAF and DAF have been used comprehensively in the removal of stable oil emulsions.

Applications of flotation, to date, at an industrial scale, have been reported for removal of light colloidal systems such as, emulsified oil from water, ions, pigments, ink and fibers from water. However, floatation is employed exclusively in the petrochemical industry for the separation of oil-water emulsions and treatment of oily metal-laden wastewater. DAF operations without additives removed merely 61% oil from the emulsified oil. Oil removal is more than 90% in oily wastewater treatment by flotation. Advantages claimed include clarity of the treated wastewater, minimization of organic losses and reduction of potential environmental problems. Disadvantages involve the low throughput, emission of H_2 bubbles, cost and maintenance of electrodes, and the high volume of sludge produced.

9.3.5 FILTRATION

Filtration is extensively used for removing oil and grease from oily wastewater, especially walnut shell filters. The solid content in oily wastewaters is usually low, but solid particles must be removed in order to avoid abrasion and deposition problems in different parts of the treatment plant. Two kinds of filters are in general employed in the plant. A magnetic filter is used first to remove metal filings, entrained with the cutting oil, and polarized particles. The effluent could then be passed either through a mesh or sand filter to remove the non-magnetic particles. A mesh filter is usually chosen, because of its ease of cleaning, with a 100 μm mesh size, resulting in a final stream with less than 0.1% of the initial solid content [9]. Filtration facilitates removal of suspended solids; separation of free, dispersed and emulsified oil. However, backwashing poses a serious drawback and it demands further treatment.

9.3.6 COALESCENCE IN PACKED BEDS

Coalescer, an enhanced filtration system, is prevalently used for liquid-liquid dispersion and emulsion. Coalescence of small droplets in fibrous bed coalescers is practiced. The principle behind is when dispersions flow through the bed, filter material captures the droplet because of wetting and retains it until a large number of droplets have been captured and then coalesces them to form large drops, later separated by gravity settlers. In a study, the optimum conditions for oil removal using bed coalescers were determined to be the initial oil concentration (30% (v/v)), bed height (100 mm), pH 6, and the emulsion flow velocity (127×10^{-3} dm^3/min) at which the efficiency of oil removal was assessed as 83.4% [10]. The benefits of coalescers are easy installation, maintenance and automatization while its demerit includes the need for bed periodical replacement of bed based the concentration of particle in effluent and slow process of oil removal.

9.3.7 DEEP BED FILTRATION

Deep-bed filtration is a usual technique for removal of fine particles in drinking water or wastewater. A deep bed filter consists of a deep bed of sand (or similar material) through which the oil/water mixture flows. Coalescence through depth filter is one of the vital emulsion separation mechanisms using porous and fibrous filter media. In depth filtration, coalescence of two or more oil droplets is expected to generate a single droplet. Cellulose-based-deep-filter exploits pecan/walnut shell as the filter medium, and it can separate particles of 2 μm and above at 98% efficiency [11]. Furthermore, the decrease in COD was higher than 95% while using DBF and, therefore, the processed effluent might be treated in a conventional wastewater treatment plant. The easy availability of sawdust and calcium sulphate makes this process a suitable technique for treatment and removal of waste cutting oils. The advantage of this technique is that operation at higher liquid loading rates is possible. However, it is essential to prefilter every particle to prevent significant damage to porous medium, which prove to be quite expensive, and in many cases impractical.

9.3.8 ELECTRICAL METHODS

Electrical methods can be applied instead of conventional mechanical-chemical techniques that lead sometimes to irregular average removal degrees of all important pollutants. Experiments are typically performed in an electrocoagulation/electroflotation cell without diaphragm with monopolar electrodes in order to establish the optimal operational conditions (e.g., electroflotation/electrocoagulation time, electric current intensity or electricity quantity) for removal efficiencies of more than 60% COD, 95% oil, 80% turbidity and 85% zinc ions [12].

9.3.8.1 Electrocoalescence

Electrostatic coalescence is ineffective type of clarification of emulsion. The electrostatic demulsification process leads to destabilization or coalescence of surface charge of drop causing breakage of droplet and coalescence. The applied voltage polarizes the water droplet; therefore, film of emulsifier around the water droplet is destabilized. Consequently, larger drops are formed by aggregation of smaller droplets so that they can separate out. Nevertheless, electrocoalescence is affected by break-up and development of secondary droplets whenever the strength of electric field is too high or droplets have become too large in size. The electrocoalescence process is still not completely comprehended, partially due to the complex electric and hydrodynamic interactions.

9.3.8.2 Electrofloatation

Electroflotation process produces bubbles by water electrolysis to release hydrogen and oxygen at the cathode and anode, respectively. The oil in the form of oil-in-water emulsion is present in the effluent. When electric current is passed through the electrodes, the emulsion quickly destabilizes and the oil is separated out as follows: an electrostatic attractive force attracts the negatively charged oil droplets towards anode and the charges on the droplets are neutralized, developing large droplets. Eventually, an oil layer is formed on the top that can be removed.

On contrary to conventional flotation, EF owns several advantages. Firstly, it is well known for the fast removal of pollutants. Secondly, it enables simultaneous flotation and coagulation, with less amount of sludge production. Thirdly, the EF equipment is very compact and hence requires less space for installation.In addition, automation is easy as dose can be controlled by simply altering the current, and the process involves low capital and operating costs. pH control is a limitation of EF and it affects various flotation parameters, especially collector-mineral interaction and bubble size. Moreover, the nascent gases released from the electrodes affect the unabsorbed reagent as well as the surface compound.

9.3.8.3 Electrocoagulation

Electrocoagulation is a modification of conventional coagulation method in which coagulant is produced electrolytically on passing electric current. Aluminum and iron electrodes are generally effective for electrocoagulation. Electrocoagulation has been effectively employed to treat various wastewater contaminants, including phenol, oil, boron, petroleum hydrocarbons, fluoride, black liquor, indigo carmine.

Electrocoagulation yields clear, colorless and odorless water with zero discharge. It is capable of handling diverse waste effluents with numerous contaminants. The electrolytic processes are governed electrically and not mechanically, thus, less maintenance is required. Instead, sometimes, oxide films forming on the cathode reduce the efficiency of electrocoagulation. Moreover, this technique necessitates a high conductivity of the wastewater and replacement of anodes and also, the use of electricity may be expensive.

9.3.9 ADSORPTION

Adsorption is a promising technique to overcome the inherent limitations of the techniques discussed above. Oil adsorption denotes the attraction of oil to the sorbent surface. The attractive forces during adsorption can be classified as the chemical (covalent or hydrogen bonds), electrostatic (ion – ion, ion– dipole), and physical (Coulombic, Kiesom energy, Debye energy, London dispersion energy) forces.

The most common adsorbents are activated carbon (AC), clay, silica, alumina and zeolite. AC is the most successful adsorbent across the globe, has a high surface area, highly porous structure, a high crystalline form and mechanical strength. Still, adsorbent-grade activated carbon is expensive and both regeneration and disposal of the used carbons seem to be difficult. The main disadvantages of adsorption include its labor intensiveness and poor removal of fine emulsion (<100 μm). Accordingly, several researchers have investigated the possibility of employing inexpensive materials.

Natural oil sorbents are derived either from plant and animal residues (organic sorbent) or minerals (inorganic sorbent). These organic and inorganic sorbents are capable of absorbing oil between 3–15 and 4–20 folds of their weight, respectively. Nevertheless, inorganic sorbents have disadvantages such as, low oil sorption capacities and low buoyancy [13]. However, organic fibrous sorbents are known for their higher sorption capacity. Moreover, oil recovery is high and, it is possible to collect the organic sorbent and dispose easier than other types of sorbents. In general, raw natural sorbents have exceptional adsorption capacity, similar density as that of synthetic sorbents, chemical-free, and very much biodegradable.

Popular natural oil sorbents include kapok fiber, sugarcane bagasse, rice husk, barley straw, cotton, wool, wood residues, and various plant and animal materials. These sorbents are either utilized as receiver made into sheets, booms, pads, filter and fiber assemblies. Plant materials possess large surface area and bigger pores, which can adsorb organic pollutants by physical and chemical mechanisms, akin to charcoal [14, 15]. These materials were found to be promising candidates for adsorption of oil by-products by several researchers.

Literature reveals that oil adsorption is attributed by the functional group properties of the sorbent [16, 17]. Significant functional groups causing oil adsorption are O-H, C = O, and C-O [18–20]. The binding sites for oil on agricultural waste or raw materials are provided by different functional groups such as, carboxyl, hydroxyl, sulfate, phosphate, ether and amino groups [21]. Surface morphology of the sorbent is another important feature for oil adsorption. For instance, various better performing natural sorbents such as, kapok fiber and populous seed consist of hollow structures, which make a higher surface area accessible for oil to bind [22]. However, raw agricultural wastes exert low sorption capacity [23] and, hence, modification of the raw material becomes essential to improve its performance. Various agricultural wastes used for emulsified oil treatment and their corresponding potential of oil removal are illustrated in Table 9.2.

9.3.10 ROLE OF MICROWAVE IN WASTEWATER TREATMENT

The microwave (MW) frequency exists between 300 MHz to 300 GHz, a frequency either near 900 MHz or close to 2450 MHz is in general applied for industrial purposes. Unlike conventional heating, the microwave energy is not transferred by conduction or convection, but is easily transformed into heat inside the particles by dipole rotation and ionic conduction. The short treatment time as well as less energy consumption is the main advantage of MW.

Focusing on wastewater treatment, MW energy is used for the treatment of ballast water with harmful organisms, remediation of phenol, treatment of petroleum refinery wastewater, safe discharge of highly-contaminated pharmaceutical wastewater. MW is successful in degrading several organic compounds such as, pesticides, ammonia nitrogen and organic dyes. One mole of photon from MW has energy of 0.4–40 J at a frequency range of 1–100 GHz. Yet, the energy of MW is not enough to disturb the chemical bonds of various organic compounds. The efficiency of MW for pollutant degradation can be augmented by coupling MW with photo-Fenton and Fenton like processes, photolysis, photolysis in the presence of H_2O_2 and photocatalysis.

TABLE 9.2 Adsorp tion Potential of Various Biosorbents

Sources of agricultural waste	Type of wastewater	Concentration of emulsified oil	Percentage of oil removal	Adsorption capacity	References
Barley straw	Oily wastewater-(synthetic emulsified oil-canola oil from aqueous solution)	A stock solution of emulsified oil wastewater was prepared by mixing 11.5 g of canola oil with 1000 mL of water and 12.5 g of an emulsifier (Ajax, Colgate-Palmolive, Australia).	90.7 ± 0.9	576.0 ± 0.3 mgg^{-1} at 25°C.	21
Corn husk	Coolant wastewater (emulsified oil)	4670 mg/L.	92.73	N/A	24
Sugarcane Bagasse	Oil by-product (to simulate the wastewater contaminated by oil by-products and creation a stable oil emulsion mixture, commercial oil by-product was dispersed in water)	600 ppm	N/A	20 g oil by-product per 1 g sorbent.	25
Egg shell	Oil-in-water mixture was prepared by mixing crude oil with distilled water	1600 mg/L	100	N/A	26
Cornhusk	Engine oil	10% (v/v)	99.66	N/A	27
Mango seed kernel powder	Produced water	30 ppm	93.33	N/A	28

MW has also been used in combination with adsorbents to increase the efficiency of removal of several pollutants and also to minimize the reaction time or else only to modify the surface of the adsorbent thereby increasing its adsorption efficiency. For instance, modification of adsorbent due to microwave radiation results in an improved efficiency of degradation of dye and heavy metals. The surface chemistry of coal based activated carbon fibers has been effectively modified by microwave treatment. Moreover, microwave activation of activated carbon generated from tobacco stems, waste tea, commercial coal-tar pitch, acrylic textile fibers, textile fibers, coffee grounds, acrylic textile fibers, oil palm stone chars have proven to perform better than standard activated carbon.

9.3.11 ROLE OF ULTRASOUND IN WASTEWATER TREATMENT

Ultrasound implies a longitudinal wave above the frequency of 20 kHz. It is well known that ultrasound power finds potential applications in cleaning and plastic welding, in versatile fields such as electrochemistry, food technology, nanotechnology, chemical synthesis, dissolution and extraction, dispersion of solids, phase separation, water and sewage treatment. The ultrasonic wave produces compressions and rarefactions while passing through the treatment medium. The rarefaction cycle generates a negative pressure and produces microbubbles. The microbubbles grow larger, collapse and produce shock waves, yielding a very high temperature and pressure within a few microseconds.

Ultrasound exhibits various effects in solid–liquid systems including the improvement of mass transfer rate, increase in surface area by creating several microcracks on the solid surface and cleaning the surfaces of solid particle. Of late, sonochemical treatment is employed in wastewater treatment for oxidation, degradation, adsorption and desorption of pollutants. The oxidation of aqueous pollutant, 2,4-dihydroxybenzoic acid and phenol by ultrasound has taken place in the bubble–bulk interface region. The degradation potential of ultrasound treatment has been investigated for dyes such as, azodye acid orange 7 with Fe^0/granular activated carbon system, 2-chlorophenol, acid orange 8 and acid red 1 using titanium dioxide particles, methylene blue by applying TiO_2 pellets, pentachlorophenate in the presence of air, oxygen and argon. US is currently being utilized to remove surfactants from wastewater as it decomposes the surfactant, promoting adsorption. Treatment of sewage sludge with US at 200 kHz have resulted in sludge disintegration as well as the removal of the linear alkylbenzenesulfonates simultaneously [29]. Moriwaki et al. [30] have reported the removal of perfluorooctanesulfonate and perfluorooctanoic acid in aqueous solutions aided by US.

US has also brought out the adsorption of phenol and chlorophenol onto granular activated carbon, phenol on resin and heavy metal on core-surfactin shell. The two modifications of ultrasound treatment includes exposure of both biosorbent and wastewater to ultrasound or biosorbent alone in the presence of chemical environment followed by its loading in batch/column system as ultrasound assisted bio-sorbent for pollutant removal.

9.4 ECOFRIENDLY SURFACTANT REFORMED ADSORBENT

Biological adsorbents are renewable, produced in large volumes and economic on comparison with the commercial adsorbents. Moreover, they can be used without or with little processing (washing, drying, grinding) and hence, production costs are reduced along with excluding the energy costs involved in thermal treatment. Various biosorbents were employed for the removal of stable emulsified oil from wastewater including modified barley straw, chitosan and walnut shell, natural wool-fibers, recycled wool based nonwoven material and sepiolite, chitosan, and hydrophobized and expanded vermiculite, bentonite and zeolites.

Corn husk is one of the biosorbents containing cellulose, which is composed of repeating units of β-D-glucose. The polar hydroxyl groups on the cellulose take part in chemical reactions and hence, adsorb heavy metals from effluents. Corn husk has thus proved to be a promising adsorbent for the removal of Cd(II), Pb(II) and Zn(II) ions both with and without modification using EDTA [31]. Moreover, the surface properties of the functional groups present on cellulose could be changed by assimilation of other functional groups, which greatly influences the adsorption potential.

Raw agricultural wastes in general exert low sorption capacity and, hence, modification of the raw material is vital to stimulate its performance. Modification of the adsorbent is also crucial to curb the release of minerals [32]. The improvement criteria generally involves washing, drying and comminution of plant materials, followed by impregnation, carbonization and activation. On the other hand, the revelation of modification towards MW and ultrasound (US) generates a high possibility to enhance batch adsorption efficiency [27].

Some investigations have been carried out to demonstrate the efficiency of surfactant modification of inorganic and organic adsorbents for adsorption of emulsified oil [33, 34]. Surfactant modified adsorbents also depicted better removal of various contaminants such as metals, dyes and organics. Very few studies have reported the surfactant modification of agricultural waste/byproducts mostly mineral surfaces have been modified by surfactants. Modification of adsorbents by cationic surfactants provides added advantages over anionic, zwitter ionic and nonionic classes of surfactants.

Hexadecylpyridinium chloride monohydrate (CPC, C21H38NCl) is the commonly used surfactant for oil adsorption. The cationic heads of the surfactant offer an electrostatic force for adsorption on negatively charged surface [35], thus generating a hydrophobic layer on the solid surface, which moreover permits an organic contaminant partition into the layer. This is referred to as adsolubilization; a blend of adsorption and solubilization [36]. The use of surfactant treated clays (organoclay) for separation of emulsified oil in wastewater has been investigated. Surfactant with long alkyl chain such as CPC is vital since short alkyl chain is poor remover of organic contaminants. Moreover, the planar and polar head groups in CPC produce larger hemimicelle aggregates, enabling them to get more adsorbed onto solid surface. Oei et al. [37] showed that that CPC-modified barley straw could be effective for dye adsorption as well as for emulsified oil removal.

Modification of agricultural waste by surfactants is performed as follows: Raw agricultural waste adsorbent is prepared by washing with distilled water and drying in shady daylight followed by deep drying in hot air oven at 65°C. The chemical modifications can be then performed by addition of sodium hydroxide, followed by cationic surfactant, cetylpyridinium chloride solution (CPC, $C_{21}H_{38}NCl$). Pretreatment with 0.05 M NaOH is carried out for15 min in respective equipment (RS at 160 rpm, MW at 119 W and ultrasonicator at 20 kHz) for good progress of surfactant on the surface of adsorbent. Then, excess of solution is poured out and the adsorbent is washed with distilled water and dried in hot air oven overnight. Eventually, the washed and dried pre-treated adsorbent can be soaked in2.5 m mol/L of cationic solution in the respective equipment for 30 min to enable surface modification [27].

9.5 IMPREGNATION OF SURFACTANT ON ADSORBENT SURFACE THROUGH ADVANCED TECHNIQUES

9.5.1 *MICROWAVE ASSISTED TECHNIQUE (MAT)*

Agricultural wastes from renewable resources are economic and biodegradable. However, use of raw adsorbents causes several issues, such as, lower adsorption potential for anionic pollutants, secondary pollution, high COD and biological chemical demand (BOD) as well as total organic carbon (TOC) due to release of organic compounds. Hence, treatment or modification of these biosorbents before usage in adsorption becomes vital. Cationic surfactants are usually used at large to modify a biosorbent for removal of anionic ions or oil pollutants. Cationic surfactants such as, cetylpyridinium chloride can prevent the release of minerals and causes the biosorbent to be water repellent facilitating the maximum adsorption. Critical micelle concentration of CPC is 0.8 M mol. Barley straw modified by hexadecylpyridinium chloride was used to adsorb anionic and reactive dyes from aqueous solution. Cationic surfactant modified coir pith was used to remove anionic dyes and thiocyanate anion. Hexadecylpyridinium bromide was employed to modify wheat straw to remove anionic dyes from solution. Moreover, sawdust modified by cetyltrimethyl ammonium bromide for adsorption of congored and peanut husk (PH) modified using CPB to adsorb anionic dyes were reported.

Treatment of biosorbents with aqueous sodium hydroxide solutions disrupts the covalent bonds between lignocellulosic components and hydrolyzes hemicellulose and depolymerizes lignin. Hence, morphological, molecular and supramolecular properties of cellulose may be greatly affected, leading to alterations in crystallinity, pore structure, accessibility, stiffness, unit cell structure and coordination of fibrils in cellulosic fibers

[38]. Furthermore, alkali treatment enhances the mechanical and chemical properties of cellulose including structural durability, reactivity and ion-exchange potential. Alkali treatment also eliminates natural fats and waxes from the surfaces of cellulose fibers and exposes the reactive functional groups such as, –OH.

While considering the modification of adsorbent by microwaves, the resistance of the adsorbent decreases and adsorption increases until surface sites are occupied by pollutants [39]. In microwave assisted preparation process, the dried biosorbent is in general treated with alkali and a surfactant followed by microwave treatment. For instance, corn husk has been modified by microwave treatment after impregnation with NaOH and CPC [27].

9.5.2 ULTRASOUND ASSISTED TECHNIQUE (UAT)

Due to the vast increase in the use of ultrasound for environmental remediation, several researchers have focused on the influence of ultrasound on sorption and desorption processes. The influence of ultrasound on removal of organic pollutants from aqueous solutions can be illustrated physically or chemically. Physically, the ultrasonic waves can clean the surface of solid absorbent, decrease the particle size and enhance mass transfer. Chemically, cavitation occurs during ultrasonication whereas the cavity collapses asymmetrically near the solid boundary and causes high-speed jets of liquid, which hit the surface strongly. Hence, the point of contact gets damaged heavily and results in novel exposed, and highly reactive surfaces.

Ultrasonic waves minimize the thickness of liquid films close to the solid phase, improve the mass transfer, and enable the diffusion of adsorbate through the interface. The adsorption of some dyes such as disperse blue 2BLN, acid black 210 and acid brown 348, direct blue 78 from their aqueous solutions, have been reported using exfoliated graphite under ultrasound irradiation. The leaves of olive tree (*Oleaeuropaea*) were used as adsorbent for the removal of cadmium in the presence and absence of ultrasound and by with simultaneous ultrasonication and stirring. The adsorption was greater in the presence of ultrasound.

The maximum adsorption capacity of the adsorbent for heavy metals in the presence of ultrasound was greater than that in its absence. The shear forces produced during cavitation are chiefly responsible for the improvement of adsorption of heavy metals, Pb(II) and Cu(II) removal by saffron corm in the presence of ultrasound. Due to the higher specific area, the adsorption capacity of the ultrasound activated carbon from hazelnut shells for removal of Cu(II) ions was two folds greater than that of the raw plant material.

The adsorption of fatty acids by activated carbon and acetic acids by weak basic ion exchangers were high when ultrasonic radiation was applied. This might have occurred by pushing of molecules into micropores mediated by US waves, achieving more active sites. However, the impact of ultrasound on adsorption and desorption has been controversial. It was observed that the desorption rate of activated carbon increased by sonication. It has also been demonstrated the decrease in the adsorption of phenol in the presence of ultrasound. However, Schueller and Yang [40] viewed an improvement of the adsorption kinetics since ultrasound enhanced the mass transfer through cavitation and acoustic streaming. In some cases, the adsorption equilibrium is shifted towards lower sorption concentrations by US waves, which may be likely due to thermal changes taking place near the adsorbent particles.

The surface modification of biosorbents was conducted under the influence of ultrasound in the presence of cationic surfactants. Alkaline impregnation is for prechemical activation and ultrasound irradiation is particularly used in this step to improve the diffusion of NaOH into the pores of the cellulosic material [41]. Corn husk was subjected to ultrasonication and chemical treatment with NaOH and CPC for emulsified oil removal. The ultrasound assisted biosorbents performed better than the raw adsorbents for oil removal [27].

9.6 CHARACTERIZATION OF THE MODIFIED ADSORBENT

FTIR spectroscopy provides a preliminary quantitative analysis of major functional groups present in adsorbents. For example, in a study reported by Ibrahim et al. [42] for removal of oil using surfactant modified raw barley straw, two bands at 2922 and 2853 cm^{-1} can be due to asymmetric and symmetric stretching vibration of methylene C–H adsorption bands arising from the alkyl chain of the surfactant, CPC. In another study reported by Augusta et al. [27] reformation of the cornhusk biosorbent by CPC has been confirmed by the FTIR spectra with a strong fall at 3,332 cm^{-1} due to the vibration of OH stretch, and bands near 2,920 and 2,854 cm^{-1} denoting asymmetric and symmetric stretches of methyl C–H group (derivatives of CPC) [43]. A deep fall indicating an OH vibration was observed for raw biosorbents such as corn husk [27] while the depth decreased after modification due to the increase in hydrophobicity [44]. FTIR of oil loaded adsorbent confirms the adsorption of oil such as, two strong peaks at 2922 and 2853 cm^{-1} suggesting the adsorption of oil to the hydrophobic, alkyl chain layer on the surface of surfactant modified raw adsorbent [42]. The transmittance of both US and MW modified adsorbents also get increased thus confirming modification by the surfactant [27]. Continuous hot bubbling during US stimulates the movement of surfactant and biosorbent, helping adsorption and producing a strong transmittance spectra [27]. FTIR spectra of raw, MW assisted and US assisted cornhusk are shown in Figure 9.1.

The deformation of the surface of the modified adsorbents due to the release of internal volatile matter and increase in pore volume are clearly evident in SEM Analysis. Modified barley straw has shown irregular shapes in SEM with particle sizes less than 500 nm. The surface displays a porous structure with round holes on the surface [42]. SEM images of CPC reformed cornhusk has demonstrated a more or less smooth surface with uniform circular spots for the raw adsorbent. However, MW and US reformed cornhusk did not contain regular circular spots, as folding led to the reformation of irregular surface area [27]. The strong transmittance of acoustic waves disturb, erode and increase the surface roughness of the biosorbent. Circular spots are revealed on SEM images of ultrasound assisted adsorbents which have arisen from bubbles at temperature of 4000–6000 K and pressure of 100–200 MPa. SEM images of raw, MW assisted and US assisted cornhusk are illustrated in Figure 9.2.

BET surface area and pore volume of barley straw decreased following surfactant modification [42]. This was because of the attachment of

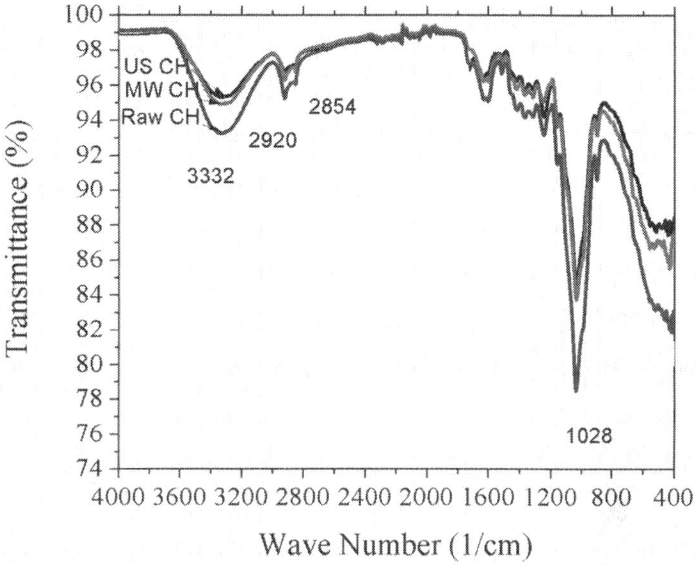

FIGURE 9.1 FTIR Spectra of Raw, MW assisted, and US assisted corn husk.

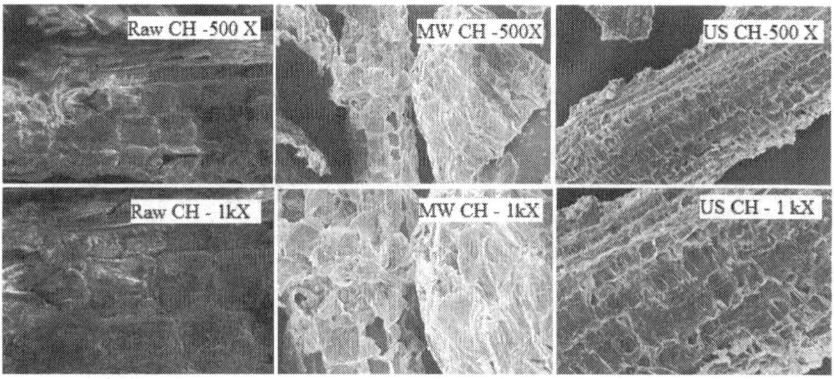

FIGURE 9.2 SEM images of Raw, MW assisted, and US assisted corn husk.

surfactant groups to the internal framework of raw barley straw leading to the narrowing of pore channels [21]. Adsorption of CPC + ions caused a reduction in acidic groups on surfactant modified barley straw while the uptake of CPC + ions resulted in an increase in basic groups. Namasivayam and Sureshkumar [45], reported the same on coir pith modification.

9.7 EFFECT OF VARIOUS PROCESS PARAMETERS ON OIL REMOVAL

High cost of activated carbon production is the major challenge for commercial manufacturers. Hence, inexpensive low cost precursors such as agricultural by-products and wastes are need of date for adsorption processes. The biosorption of oil might be influenced by several parameters such as, pH, temperature, contact time, sorbent dosage and shaking speed. These factors determine the overall biosorption by affecting the uptake of oil, selectivity and amount of oil removed. Hence, optimization of these factors to achieve maximum oil adsorption becomes necessary.

9.7.1 pH

Among process factors, pH plays a vital role in controlling the removal of oil. Changes in pH modify surface properties and binding sites of sorbent [42], and emulsion breaking [46]. Various researchers have investigated the effect of pH on oil removal. It is observed that changes in pH affects the various sorbents differently. Palm oil and crude oil removal efficiency by chitosan was found to be increased at strongly acidic and basic pH [46, 47]. Recycled wool-based nonwoven material behaved likewise in removal of motor oil [48]. Acidic pH promotes reaction between oil molecules and NH_{2-} functional group in sorbent [46]. At acidic pH, a large number of protons are available and saturate the sorbent sites, increases the cationic properties of sorbent, [49] thus reducing its hydrophobic properties. Saponification causing hydrolysis of oil in sorbate is responsible for high removal efficiency at strongly basic conditions. Corn cob modified with lauric acid and ethanediol revealed maximum oil adsorption at pH5 while minimum at pH9 [50].

Several studies showed an increase in oil removal at basic pH. For instance, natural wool fiber removed maximum amount of motor oil at pH 10, and minimum at pH 5 [48]. Ionization of the sorbate is the reason for low oil absorption at acidic pH [42]. Some other studies showed no significant effect of pH on oil removal efficiency. For instance, difference in removal of mineral oil was less than 15% at pH between 3 and 10.5. This is because of chemical stability of oil and absence of ionizable or hydrophilic groups, which could be affected by pH [51].

Canola oil and mineral oil removal using surfactant-modified barley straw showed minimum removal efficiency at pH 2 and removal efficiency increased in parallel to increasing pH. However, oil removal did not change significantly at pH between 6 and 10 [51, 52]. Removal efficiency of emulsified oil using modified and raw corn husk increased with increase in pH. At pH 6, 81.76, 62.76 and 50 % of removal was attained for cationic surfactant modified, NaOH treated and raw corn husk. Meanwhile at pH 8, 89.11, 71.70 and 59.10% were achieved [24]. The increased oil sorption at a relatively high pH may be due to the decreased number of protons, thus maintaining the hydrophobicity of the adsorbent surface. Nevertheless, removal efficiency of emulsified engine oil by microwave and ultrasound assisted cornhusk decreased with the increase in pH. The acidic media stimulated the interaction of oil molecules with cationic surfactant ions on the adsorbent surface. Still, the interaction differs for different modification procedures adopted. MW assisted cornhusk was found to perform better than the US assisted. The maximum adsorption capacity at pH 2 for US and MW assisted corn husk were 1,618 and 1,658 mg/g, respectively [27]. The effect of pH on percentage of removal of emulsified oil is presented in Figure 9.3.

9.7.2 TEMPERATURE

Adsorption of oil is in general lowered at temperatures greater than 80°C. Brownian motion of oil molecules increases with increasing temperature. Hence, more energy is needed for the oil molecule to adhere onto the sorbent surface [51]. Furthermore, at such temperatures of more than 80°C, the randomness of oil adsorption–desorption is observed to increase, resulting in less probability of oil attachment on sorbent surface. For instance, natural wool fibers and recycled non-woven fibers showed higher motor oil removal at 20°C. The efficiency of oil removal by both the fibers was as low as only 2.5% at temperatures between 80 to 95°C [48]. This may be because of degradation of fiber structure at higher temperatures [53], thereby losing their functionality.

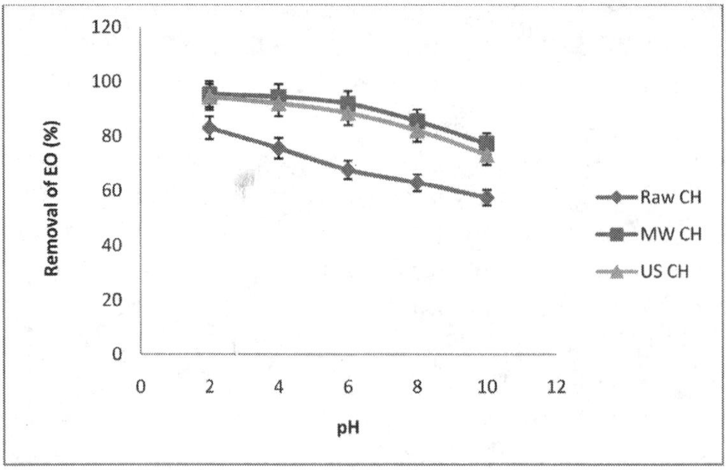

FIGURE 9.3 Effect of pH on oil removal by Raw, MW assisted, and US assisted corn husk.

Efficiency of motor oil sorption by the inorganic sorbent sepiolite also decreased with increase of temperature. This could be due to occurrence of higher physical adsorption [53]. However, adsorption of oil onto zeolite was 18.4% at 80°C, which may be due to the molecular sieve like nature of zeolite [54]. Bentonite also revealed higher motor oil removal efficiency with increase in temperature and it can be ascribed to the higher chemisorptions [55, 56]. A trivial effect of temperature on adsorption at lower temperatures of 23°C and 33°C was evident for removal of emulsified oil by surfactant modified barley straw [21]. However, oil sorption was slightly increased at a relatively higher temperature of 43°C. This could be ascribed to the increase in movement of oil molecules and thus leading to more intense interactions between sorbent and sorbate [57] thereby enhancing the rate of diffusion of adsorbate across the surface of the adsorbent [48]. In addition, the adsorption capacity of raw, microwave and ultrasound assisted corn husk was similar between 25°C and 30°C. Beyond 30°C, the assisted adsorbents behaved differently and attained saturation at 40°C. MW assisted adsorbent showed higher oil adsorption than US, which may be due to the porosity developed during chemical modification of corn husk. Though there was oil movement due to the porosity generated, moment of molecules was low in US assisted corn husk, hence showing lower adsorption than MW [27]. The effect of temperature on percentage of removal of emulsified oil is demonstrated in Figure 9.4.

9.7.3 CONTACT TIME

The effect of contact time between sorbate and sorbent is more often investigated in batch or batch flow adsorption processes. Increase in contact time leads to increased interaction between oil–sorbent, hence increasing the efficiency of oil removal [47]. Nevertheless, contact time is crucial only at the beginning of oil adsorption, and is less vital near equilibrium at which only a small increase in oil sorption is witnessed [42, 47]. This may be due to limited number of sorbent surfaces available for entrapment of oil near equilibrium [46]. Previous studies on oil removal have reported substantial removal of oil before 30–60 min, based on the type and dosage of sorbent. In a batch-flow column adsorption study (0–1400 min), contact time had influenced oil removal only during the initial 60 min, with 99% removal [51]. Similar results were obtained for oil adsorption by vegetable fibers performed between 0–1500 min [58]. Nonetheless, significant increase in removal of oil from palm oil mill effluent using chatoyant was found only during the first 30 min [47].

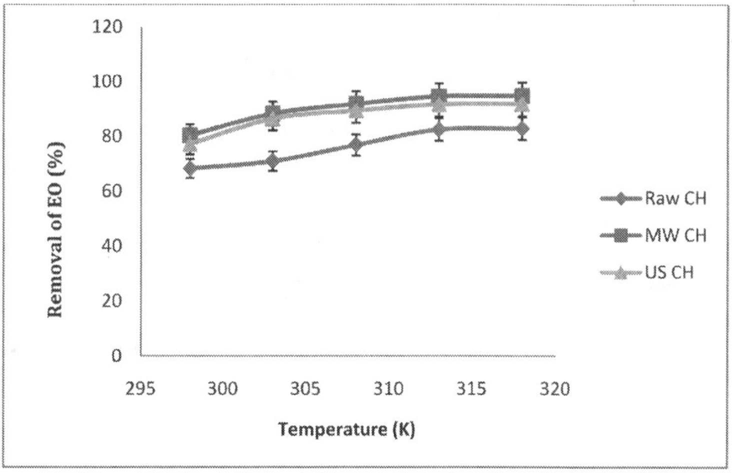

FIGURE 9.4 Effect of temperature on oil removal by Raw, MW assisted, and US assisted corn husk.

It is explicit from experiments that the adsorption capacity of surfactant modified adsorbents also increases with increase in contact time till reaching steady phase. The modified adsorbents exhibit an active rapid phase, a gradual phase and further increase in time results in a steady phase. Active rapid phase occurs because of the existence of unoccupied bare sites on the adsorbent surface. However, the number of bare sites decreases with increase in time [59] leading to a moderate gradual phase and finally a saturated steady phase is reached. However, the equilibrium time for removal of oil by surfactant modified barley straw was dependent on initial concentration of oil. Equilibrium was reached earlier at 20 min for lower concentration of oil while at 40 min for a relatively high concentration and hence, it was determined that the equilibrium time for oil sorption could be fixed as 60 min [42]. On the other hand, raw, MW assisted, US assisted cornhusk showed active rapid phase near 40 min and gradual phase near 100 min, after which a steady phase was reached. Hence, the contact time was set as 120 min for emulsified oil removal [27]. Figure 9.5 displays the effect of contact time on percentage of removal of emulsified oil.

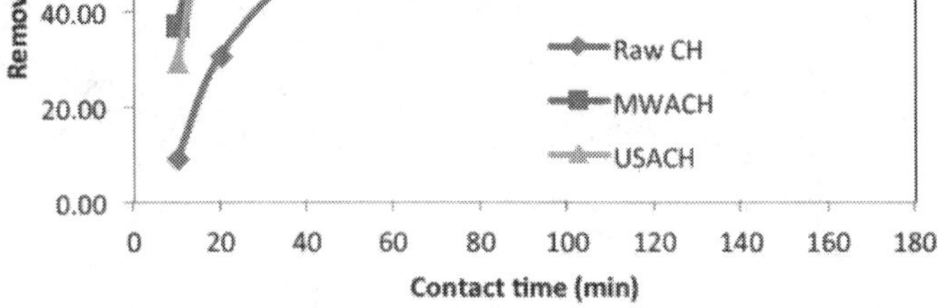

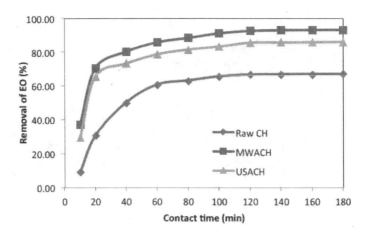

FIGURE 9.5 Effect of contact time on oil removal by Raw, MW assisted, and US assisted corn husk.

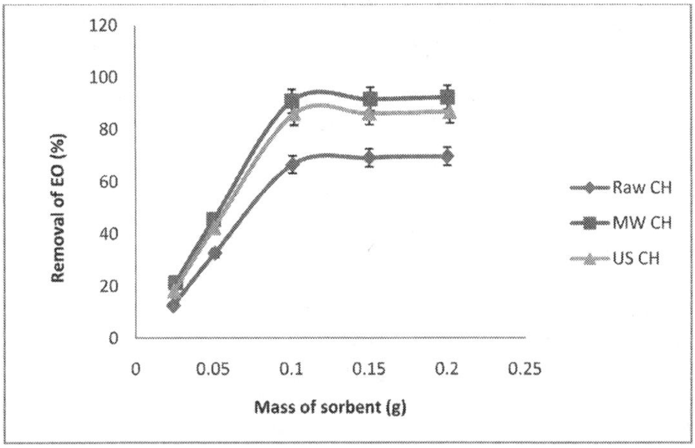

FIGURE 9.6 Effect of sorbent dosage on oil removal by Raw, MW assisted, and US assisted corn husk.

9.7.4 SORBENT DOSAGE

Oil removal efficiency decreases with increase in dosage of the sorbent as the number of binding sites available for adsorption increases at higher dosage of adsorbent [47]. However, the adsorption potential decreases with an increase in dosage, this phenomenon is associated with the increase of unsaturated oil binding sites [21]. Moreover, efficiency of oil removal by is decreased by saturation effect when the maximum adsorption has been achieved [48]. For instance, as the dosage increased, sorption of emulsified oil by raw and treated corn husk also increased but after a certain point, removal was not affected by dosage. Hence, 10 mg/L of adsorbent was chosen for the entire study. 86.4, 69.09 and 46.41% of oil removal at 5 mg/mL and 98.01, 76.70 and 58.04% removal at 30 mg/mL were achieved, respectively for cationic surfactant modified, NaOH treated and raw corn husk [24]. Similar adsorption patterns were obtained during the study of effect of dosage on removal of oil by surfactant modified barley straw and for microwave and ultrasound assisted corn husk [27].

Activated carbon, MW assisted cornhusk, US assisted corn husk and raw corn husk at the dosage of 10 g/L exhibited adsorption capacities of 1703 mg/g, 1573 mg/g, 1467 mg/g and 1148 mg/g, respectively

[27]. The influence of sorbent dosage on percentage of removal of emulsified oil is manifested in Figure 9.6. Pore size and pore volume of the adsorbent are critical for adsorption along with its surface area. This could be confirmed from US cornhusk with less surface area but a higher pore size to display better adsorption potential. However, MW assisted cornhusk had a relatively higher pore size, pore volume and surface area compared to raw and US assisted, which could be ascribed to its higher adsorption potential. In addition, higher surface area, pore volume and pore size together contribute to a good emulsified oil-holding capacity of the adsorbent. It is also clear that lower initial concentration of emulsified oil enables a higher percentage of removal. Similar findings were also reported for removal of oil from palm oil effluent by rubber powder [47].

9.7.5 SHAKING SPEED

Speed of agitation also influences the adsorption of emulsified oil. An increase in speed of shaking results in an initial increase of oil sorption followed by a decreased sorption of oil at higher speeds. This phenomenon is associated with the increase in mass transfer with the increase in speed. Nonetheless, a vigorous shaking may more probably leads to desorption of more oil from the adsorbent, resulting in the breakage of bonds between oil and sorbent. For instance, an increase in agitation speed from 90 to 190 rpm caused an enhanced oil uptake by surfactant modified barley straw. However, the oil uptake was reduced slightly at higher speeds of 230 and 315 rpm. Hence, shaking speed was fixed as 130–190 rpm to facilitate maximum adsorption [42]. Similarly, maximum oil removal of 92.73, 73.92 and 56.33 % were achieved for cationic surfactant modified, NaOH treated and raw corn husk at 150 rpm [24]. However, the adsorption capacity decreased till 200 rpm and then started to increase after 220 rpm whereas the sorption potential of MW assisted corn husk did not decrease till 200 rpm. The decrease in adsorption till 200 rpm might be due to the possibility of dispersion of oil droplet back into the solution and resulting in weak bonds between adsorbent and adsorbate. The adsorption capacity of US assisted cornhusk only was greater at 220 rpm. At high speeds, oil molecules may get attached and diffuse, resulting in low adsorption as reported for removal of oil from palm oil effluent by rubber powder [47]. Maximum adsorption capacity is usually achieved at low speeds, hence, the lowest speed of 140 rpm was preferred for maximum adsorption. The maximum adsorption capacities attained for raw, MW and US assisted were 1294 mg/g, 1584 mg/g and 1435 mg/g, respectively at 140 rpm. Due to the hot spot formed in the interior of MW assisted adsorbent, at low speeds, even distribution of oil molecules into the adsorbent had occurred, causing a higher oil uptake by MW than US and raw cornhusk [27]. The effect of shaking speed on percentage of removal of emulsified oil is demonstrated in Figure 9.7.

In conclusion, MW assisted cornhusk showed higher adsorption potential than US assisted and raw cornhusk and performed nearly similar to activated carbon. The difference in adsorption capacities of MW and US assisted adsorbents was negligible after optimization of the process variables. Modification of the adsorbent had taken place on the surface and within the particles in US and MW as revealed by SEM images thus supporting the results obtained for adsorption capacity [27]. Hence, chemical modification of adsorbent together with the application of MW or US can be proposed for other adsorbents to enable maximum adsorption.

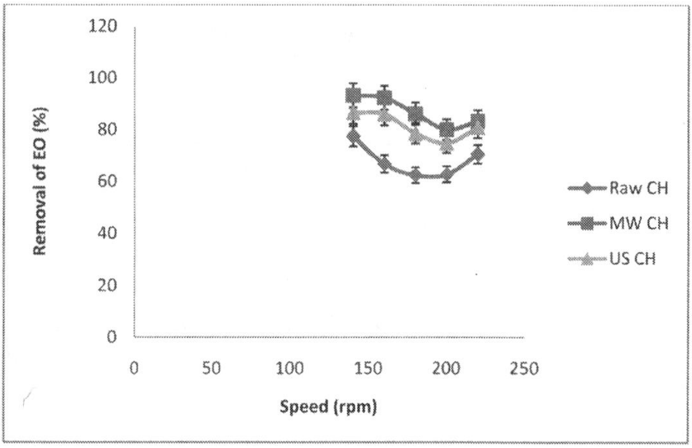

FIGURE 9.7 Effect of shaking speed on oil removal by Raw, MW assisted, and US assisted corn husk.

9.8 ISOTHERM STUDY OF ADSORPTION OF OIL

In general, effective separation of solute from solution relies on the separation of two phases at equilibrium. The equilibrium is attained when the quantity of solute adsorbed is equal to the amount being desorbed. The amount of solute taken up by the adsorbent is a function of concentration of adsorbate and temperature. Adsorption isotherms determine the quantity of adsorbate as a function of concentration at constant temperature. Several previous studies have employed Langmuir, Freundlich, Dubinin-Radushkevish (D-R) or Temkin isotherm models to elucidate the oil adsorption phenomenon of biosorbents.

The Langmuir adsorption isotherm describes monolayer adsorption of oil on homogenous surface of the adsorbent. The Langmuir equation is given by

$$q_e = \frac{q_0 C_e}{(1+K_\mathrm{L} C_e)} \tag{1}$$

where q_0 denotes the maximum adsorption capacity (mg/g), K_L is a constant associated with s of adsorption (L/mg), and C_e denotes the equilibrium concentration of oil in liquid phase (mg/L). On linearization, this equation becomes

$$\frac{C_e}{q_e} = \frac{C_e}{q_0} + \frac{1}{q_0 K_\mathrm{L}} \tag{2}$$

$$R_\mathrm{L} = \frac{1}{1+K_\mathrm{L} C_0} \tag{3}$$

On plotting C_e/q_e vs. C_e, q_0 and K_L can be determined from the intercept and slope, respectively. Adsorption is favored if R_L value lies between 0 and 1.

Freundlich isotherm deals with multilayer adsorption involving interactions between adjacent molecules on heterogeneous surface and is expressed as follows

$$\log q_e = \log K_f + \frac{1}{n} \log C_e \tag{4}$$

where, C_e indicates the concentration of oil at equilibrium in mg/L, and q_e refers to the amount of adsorbent adsorbed at equilibrium time in mg/g. K_f and n are the Freundlich constants with n indicating the extent of favorability of the adsorption process and K_f representing the adsorption capacity of the sorbent. Plotting log q_e vs. logCe provides the values of n and K_f as slope and intercepts, respectively.

Dubinin–Radushkevish (D–R) model can calculate the adsorption energy and is given by

$$\ln q_e = \ln q_0 - K_{\mathrm{DR}}\varepsilon^2 \tag{5}$$

where, K_{DR} is sorption mean free energy mol^2/J^2 and ε indicates Polanyi potential derived and plot of lnq_e vs. ε^2 yields a slope K_{DR} and intercept lnq_0.

The Flory–Huggins model determines the degree of coverage of the adsorbent surface and is expressed as,

$$\log\frac{\theta}{c_e} = \log k_{\mathrm{FH}} + \alpha_{\mathrm{FH}} \log\ (1-\theta) \tag{6}$$

where, θ is degree of surface coverage, α_{FH} implies the number of adsorbents in adsorption sites, k_{FH} is the equilibrium constant of adsorption, and plot of log (θ/c_e) vs. log ($1-\theta$) provides a slope and α_{FH} intercept k_{FH}.

The Temkin isotherm model assumes that the decrease in heat of adsorption is linear instead of logarithmic as illustrated in the Freundlich equation.

$$q_e = A + B\ln(C_e) \tag{7}$$

where, A and B are Temkin constants.

However, in case of multisolutes adsorption such as in oil adsorption, the adsorption isotherms can be applied only when all the solutes under investigation follow these isotherms. Yet, not all organic molecules follow these isotherms as single solutes [60] Hence, it becomes necessary to know the adsorption coefficient of each single solute before applying these isotherms [60]. Adsorption isotherms on oil adsorption by various biosorbents available in literature are provided in Tables 9.3 and 9.4.

TABLE 9.3 Reported Langmuir and Freundlich Isotherms on Oil Adsorption by Various Biosorbents

Sorbent	Oil type	Langmuir		Freundlich		Ref.
		Maximum sorption at equilibrium, q_0(mg/g)	Langmuir constant K_L (l/g)	Adsorption intensity (1/*n*)	Freundlich constant, K_f	
Chitosan	Crude oil	2000	5.96	0.27	1.4 g/g $(l/g)^{1/n}$	46
Chitosan	Palm oil mill effluent	Powder: 3420 Flake: 1970	Powder: 2.18 Flake: 1.2	Powder: 0.388 Flake: 0.457	Powder: 2.31 Flake: 2.08	47
Barley straw (surfactant-modified)	Canola oil	576	3.78	0.235	84.4 l/mg	21
Barley straw (surfactant-modified)	Canola oil	613.3	5.2	0.21	112.1 l/mg	42
Barley straw (surfactant-modified)	Mineral oil	584.2	14	0.168	165.0 l/mg	42
Sugarcane bagasse	Gasoline	8.3604 mL/g	60.7344 mL/mL	0.3431	14.9450	61
Sugarcane bagasse	Heptane	2.7771 mL/g	266.4997 mL/mL	0.1250	3.4663	61
Corn husk (raw)	Emulsified engine oil	1.658	0.067	1.58	–2.253	27
Corn husk (MW assisted)	Emulsified engine oil	1.988	0.057	3.65	–1.692	27
Corn husk (US assisted)	Emulsified engine oil	1.912	0.042	2.273	–1.650	27

In case of gasoline adsorption by sugarcane bagasse, all the four isotherms namely Langmuir, Freundlich, D-R and Temkin were fitted well till the concentration of pollutant was low. The Freundlich isotherm exhibited the best performance while describing gasoline adsorption than other isotherms applied.

TABLE 9.4 Reported Dubinin–Radushkevish (D-R), Flory–Huggins and Temkin Isotherms on Oil Adsorption by Various Biosorbents

Sorbent	Oil type	D-R		Flory–Huggins		Temkin		Re.
		Maximum sorption at equilibrium, q_0(mg/g)	K_{DR} (mol²/J²)	α_{FH}	K_{FH}	A (mL/g)	B (mL/g)	
Sugarcane bagassse	Gasoline	10.99164	0.2658	N/A	N/A	10.60758	1.45493	61
Sugarcane bagassse	Heptane	3.0595	0.0051	N/A	N/A	3.28293	0.20404	61
Corn husk (raw)	Emulsified engine oil	0.359	–0.049	–1.332	0.478	N/A	N/A	27
Corn husk (MW assisted)	Emulsified engine oil	–0.238	–0.067	–1.445	0.591	N/A	N/A	27
Corn husk (US assisted)	Emulsified engine oil	–0.244	–0.049	–1.373	0.543	N/A	N/A	27

This can be ascribed to the Freundlich isotherm, which considers the heterogeneity of the solid and does not restrict the amount adsorbed to the monolayer. However, the performance was not found to be satisfactory for higher concentrations. For higher concentrations of pollutants, D-R and Temkin isotherms displayed better performances than Langmuir but still worse than Freundlich. The same behavior was observed for n-heptane but the fit for all the four isotherms tested was very poor [61]. Most of the oil adsorption processes obeyed Freundlich isotherm rather than Langmuir isotherm [47, 62, 63].

On contrary, surfactant modified barley straw and MW and US assisted cornhusk showed higher regression coefficients, R^2 for Langmuir model indicating a better fit of the experimental data to the Langmuir model [21, 27]. In case of MW and US assisted cornhusk used for engine oil removal, $1/n$ values were higher for MW assisted cornhusk compared to US assisted and raw, thus indicating a higher adsorption potential for the MW assisted cornhusk [27]. The value of adsorption energy K_{DR} obtained in D-R model for all the three raw, MW and US cornhusk was below 40 implying the role of physisorption. In addition, α_{FH} value in Flory–Huggins model, which represents the molecules occupied on the surface was found to be high for MW assisted cornhusk indicating its higher adsorption potential [27].

9.9 KINETIC STUDY OF ADSORPTION OF OIL

Adsorption kinetics provides information about adsorption capacity and rate of biosorption occurrence. It becomes vital to determine the appropriate kinetic model that best fits the experimental data. In fact, adsorption is governed by mechanisms such as mass transfer, diffusion control, chemical reaction and particle diffusion. In bulk diffusion, adsorption is assumed to be fast and not rate-determining. Instead, in batch reactor, bulk diffusion and adsorption steps are rate-limiting. The adsorbate molecules are in general transferred from the bulk of the solution into the solid phase by intraparticle diffusion, which is almost the rate limiting parameter in several adsorption processes. To elucidate the adsorption process, the experimental data need to be compared with first-order, second-order and intra-particle diffusion model. The linearized forms of Lagergren first-order and second-order model are expressed in Eqs. (8) and (9).

$$\log(q_e - q_t) = \log q_e - K_1 t \tag{8}$$

$$\frac{t}{q_t} = \frac{1}{K_2 q_e^2} + \frac{t}{q_e} \tag{9}$$

where, q_e and q_t specify the quantity of emulsified oil adsorbed at equilibrium and time t in (mg/g) and K_1 and K_2 are the rate constants for pseudo-first-order and pseudo-second-order, respectively, in l/min and g/mg min. The plot of log (q_e-q_t) vs. t and t/qt vs. t yields K_1 and K_2 value. Intra-particle diffusion model suggested by Weber and Morris is provided in Eq. (10):

$$q_t = K_{id} t^{1/2} + I \quad (10)$$

where, k_{id} designates the intra-particle diffusion rate constant (mg/g min^2). The plot of q_t vs $t^{1/2}$ provides an intercept I, which specifies the thickness of the boundary layer, which is known to be prominent in reformed adsorbents and the slope gives k_{id} value.

Only limited studies have reported the kinetics of oil adsorption as several organic components are present in oil. The kinetic parameters of pseudo-first-order, pseudo-second-order and intra-particle diffusion models for removal of emulsified oil by raw, MW and US assisted cornhusk are provided in Table 9.5. R^2 values for first order model was very low, hence assuring that adsorption was not of first order type. The second order equation is based on the adsorption capacity and agrees with chemisorption being rate limiting q_e values obtained using second order equation for raw as well as MW and US assisted cornhusk conformed well with the experimental q_e data and R^2 values were near to unity [27]. In addition, K_2 values were higher than K_1 for all the three adsorbents tested. This has rendered the pseudo-second order model best in describing the adsorption of emulsified oil [27]. Similar results were observed for the removal of oil from palm oil mill effluent by chitosan powder and flakes [47]. I values obtained in intra-

TABLE 9.5 Kinetics of Adsorption of Emulsified Oil by Raw, MW Assisted and US Assisted Corn Husk [27]

Kinetics model	Parameters	Cornhusk		
		Raw	MW	US
Pseudo first order	$q_{e, cal}$	1.398	1.148	0.612
	K_1	0.017	0.018	0.008
	R^2	0.880	0.720	0.865
Pseudo second order	$q_{e, cal}$	1.769	1.589	1.607
	K_2	0.0002	0.178	0.171
	R^2	0.977	0.997	0.998
Intraparticle diffusion	I	0.243	1.271	1.272
	K_{id}	0.115	0.027	0.027
	R^2	0.978	0.978	0.947

particle diffusion model were also higher for MW and US assisted cornhusk than raw cornhusk. The nearly equal values of *I* for the MW and US assisted cornhusk have justified the oil uptake capacity of the latter being similar to the former irrespective of surface area, pore volume and pore size [27].

Hence, it can be concluded that the surfactant modification coupled with microwave or ultrasound treatment is a good method to enhance the capacity of biosorbents to adsorb oil from oil-water emulsion.

9.10 CONCLUSION

Oil treatment and discharge is a vital issue in petroleum industry. This chapter introduced the origin and characteristics of oil in wastewater. The toxic and adverse environmental effects of oil and the need for its effective treatment were demonstrated. Several methods available for the treatment and removal of oil were explored. Adsorption was reviewed as the best choice for oil removal and agricultural products such as corn husk were recommended as viable alternatives for oil adsorption for their simplicity, remarkable oil removal features, eco-friendly nature and easy availability. In particular, MW and US assisted adsorbents could present higher adsorption potential than raw adsorbent. Optimization of process variables further enhanced the oil uptake capacity of the reformed adsorbents. In fact, the porosity created by microwave and ultrasound is the reason behind the increase in surface modification of the adsorbent during alkali and surfactant treatment leading to an increase in oil holding capacity. Hence, it can be concluded that upgradation of the biosorbents through surfactant modification together with microwave and ultrasound application could further improve their oil adsorption potential.

KEYWORDS

- **adsorption**
- **corn husk**
- **microwave**
- **oil**
- **ultrasound**

REFERENCES

1. Patterson, J., & Patterson, J., (1985). Industrial Wastewater Treatment Technology; Butterworth: Boston.
2. Nag, A., (1995). Utilization of charred sawdust as an adsorbent of dyes, toxic salts and oil from water, *Process Safety Environ. Prot: Trans. Inst. Chem. Eng. Part B, 73*, 299–306.
3. Aguilera, F., Méndez, J., Pásaro, E., & Laffon, B., (2010). Review on the effects of exposure to spilled oils on human health. *J. Appl. Toxicol., 30,* 291–301.
4. Chakrabarty, B., Ghoshal, A., & Purkait, M., (2010). Cross-flow ultrafiltration of stable oil-in-water emulsion using polysulfone membranes. *Chemical Engineering Journal, 165*, 447–456.
5. Ohya, H., Kim, J., Chinen, A., Aihara, M., Semenova, S., Negishi, Y., et al., (1998). Effects of pore size on separation mechanisms of microfiltration of oily water, using porous glass tubular membrane. *Journal of Membrane Science, 145*, 1–14.
6. Falahati, H., & Tremblay, A., (2011). Flux dependent oil permeation in the ultrafiltration of highly concentrated and unstable oil-in-water emulsions. *Journal of Membrane Science, 371*, 239–247.
7. Benito, J., Cambiella, A., Lobo, A., Gutiérrez, G., Coca, J., & Pazos, C., (2009). Formulation characterization and treatment of metalworking oil-in-water emulsions. *Clean Techn Environ Policy,12,* 31–41.
8. Zeng, Y., Yang, C., Zhang, J., & Pu, W., (2007). Feasibility investigation of oily wastewater treatment by combination of zinc and pam in coagulation flocculation. *Journal of Hazardous Materials, 147*, 991–996.
9. Benito, J., Ríos, G., Ortea, E., Fernández, E., Cambiella, A., Pazos, C., et al., (2002). Design and construction of a modular pilot plant for the treatment of oil-containing wastewaters. *Desalination, 147*, 5–10.
10. Kundu, P., & Mishra, I., (2013). Removal of emulsified oil from oily wastewater (oil-in-water emulsion) using packed bed of polymeric resin beads. *Separation and Purification Technology, 118*, 519–529.
11. Schulz, J. C., (1996). Nutshell filter technology.*Fluid Particle Separation Journal, 9*, 14–20.
12. Zaharia, C., Surpateanu, M., Cretescu, I., Macoveanu, M., & Braunstein, H., (2005). Electrocoagulation/electroflotation: Methods applied for wastewater treatment.*Environmental Engineering Management Journal, 4,* 463–472.
13. Cojocaru, C., Macoveanu, M., & Cretescu, I., (2011). Peat-based sorbents for the removal of oil spills from water surface: application of artificial neural network modeling. *Colloids and Surfaces A: Physicochemical and Engineering Aspects, 384*, 675–684.
14. Gurgel, L., Freitas, R., & Gil, L., (2008). Adsorption Of Cu(II), Cd(II), And Pb(II) from aqueous single metal solutions by sugarcane bagasse and mercerized sugarcane bagasse chemically modified with succinic anhydride. *Carbohydrate Polymers, 74*, 922–929.
15. Lillo-Ródenas, M., Marco-Lozar, J., Cazorla-Amorós, D., & Linares-Solano, A. (2007). Activated carbons prepared by pyrolysis of mixtures of carbon precursor/alkaline hydroxide. *Journal of Analytical and Applied Pyrolysis, 80*, 166–174.

16. Abdullah, M., Rahmah, A., & Man, Z., (2010). Physicochemical and sorption characteristics of Malaysian *Ceiba Pentandra* (L.) *Gaertnasa* natural oil sorbent. *Journal of Hazardous Materials, 177*, 683–691.
17. Srinivasan, A., & Viraraghavan, T., (2010). Oil removal from water using biomaterials. *Bioresource Technology, 101,* 6594–6600.
18. Ribeiro, T., Rubio, J., & Smith, R., (2003). A dried hydrophobic aquaphyte as an oil filter for oil/water emulsions. *Spill Science & Technology Bulletin, 8*, 483–489.
19. Khan, E., Virojnagud, W., & Ratpukdi, T., (2004). Use of biomass sorbents for oil removal from gas station runoff. *Chemosphere, 57*, 681–689.
20. Said, A., Ludwick, A., & Aglan, H. (2009). Usefulness of raw bagasse for oil absorption: a comparison of raw and acylated bagasse and their components. *Bioresource Technology, 100*, 2219–2222.
21. Ibrahim, S., Wang, S., & Ang, H., (2010). Removal of emulsified oil from oily wastewater using agricultural waste barley straw. *Biochemical Engineering Journal, 49*, 78–83.
22. Likon, M., Remškar, M., Ducman, V., & Švegl, F. (2013). Populus seed fibers as a natural source for production of oil super absorbents. *Journal of Environmental Management, 114*, 158–167.
23. Weber, W., McGinley, P., & Katz, L., (1991). Sorption phenomena in subsurface systems: concepts, models and effects on contaminant fate and transport. *Water Research, 25*, 499–528.
24. Augusta, P., & Kalaichelvi, P., (2013). Study on removal of emulsified oil from coolant wastewater using corn husk. *Proc. of Int. Conf. on Emerging Trends in Engineering and Technology*, 317–320.
25. Hamid, S., Javad, T., & Reza, R., (2014). Oil by-product removal from aqueous solution using sugarcane bagasse as absorbent. *International Journal of Emerging Science and Engineering (IJESE), 2*, 2319–6378.
26. Muhammad, I., El-Nafaty, U., Makarf, Y., & Abdulsalam, S., (2015). Oil removal from produced water using surfactant modified eggshell. *International Conference on Environmental, Energy and Biotechnology, 85*, 14.
27. Augusta, P., & Kalaichelvi, P., (2015). Investigation on microwave and ultrasound-assisted cornhusk for the removal of emulsified engine oil from water. *Desalination and Water Treatment, 57 (28)*, 13120–13131.
28. El-Nafaty, U., Sirajuddeen, A., & Muhammad, I., (2014). Isotherm studies on oil removal from produced water using mango seed kernel powder as sorbent material. *Chemical and Process Engineering Research, 23*, 55–65.
29. Gallipoli, A., & Braguglia, C., (2012). High-frequency ultrasound treatment of sludge: combined effect of surfactants removal and floc disintegration. *Ultrasonics Sonochemistry, 19*, 864–871.
30. Moriwaki, H., Takagi, Y., Tanaka, M., Tsuruho, K., Okitsu, K., & Maeda, Y., (2005). Sonochemical decomposition of perfluorooctanesulfonate and perfluorooctanoic acid. *Environmental Science & Technology, 39*, 3388–3392.
31. Igwe, J. C., & Abia, A. A., (2006). A bioseparation process for removing heavy metals from wastewater using biosorbents, *Afr. J. Biotechnol., 5*, 1167–1179.
32. Altundogan, H., (2005). Cr(VI) removal from aqueous solution by Iron (III) hydroxide-loaded sugar beet pulp. *Process Biochemistry, 40*, 1443–1452.

33. Moazed, H., & Viraraghavan, T., (2005). Removal of oil from water by bentoniteorganoclay. *Pract. Period Hazard Toxic Radioact. Waste Manage., 9,* 130–134.
34. Fanta, G., Abbott, T., Burr, R., & Doane, W., (1987). Ion exchange reactions of quaternary ammonium halides with wheat straw. Preparation of oil-absorbents. *Carbohydrate Polymers, 7,* 97–109.
35. Tiberg, F., Brinck, J., & Grant, L., (1999). Adsorption and surface-induced self-assembly of surfactants at the solid–aqueous interface. *Current Opinion in Colloid & Interface Science, 4,* 411–419.
36. Adak, A., Pal, A., & Bandyopadhyay, M., (2006). Removal of phenol from water environment by surfactant-modified alumina through adsolubilization. *Colloids and Surfaces A: Physicochemical and Engineering Aspects, 277,* 63–68.
37. Oei, B., Ibrahim, S., Wang, S., & Ang, H., (2009). Surfactant modified barley straw for removal of acid and reactive dyes from aqueous solution. *Bioresource Technology, 100,* 4292–4295.
38. Široký, J., Blackburn, R., Bechtold, T., Taylor, J., & White, P., (2011). Alkali treatment of cellulose II fibers and effect on dye sorption. *Carbohydrate Polymers, 84,* 299–307.
39. Wan Ngah, W., & Hanafiah, M., (2008). Removal of heavy metal ions from wastewater by chemically modified plant wastes as adsorbents: A review. *Bioresource Technology, 99,* 3935–3948.
40. Schueller, B., & Yang, R., (2001). Ultrasound enhanced adsorption and desorption of phenol on activated carbon and polymeric resin. *Industrial & Engineering Chemistry Research, 40,* 4912–4918.
41. Şayan, E., (2006). Ultrasound-assisted preparation of activated carbon from alkaline impregnated hazelnut shell: An optimization study on removal of Cu^{2+} from aqueous solution. *Chemical Engineering Journal, 115,* 213–218.
42. Ibrahim, S., Ang, H., & Wang, S., (2009). Removal of emulsified food and mineral oils from wastewater using surfactant modified barley straw. *Bioresource Technology, 100,* 5744–5749.
43. Okiel, K., El-Sayed, M., & El-Kady, M., (2011). Treatment of oil/water emulsions by adsorption onto activated carbon, bentonite and deposited carbon. *Egyptian Journal of Petroleum, 20,* 9–15.
44. Chen, D., Chen, J., Luan, X., Ji, H., & Xia, Z., (2011). Characterization of anionic/cationic surfactants modified montmorillonite and its application for the removal of methyl orange. *Chemical Engineering Journal, 171,* 1150–1158.
45. Namasivayam, C., & Sureshkumar, M., (2008). Removal of Chromium(VI) from water and wastewater using surfactant modified coconut coir pith as a biosorbent. *Bioresource Technology, 99,* 2218–2225.
46. Sokker, H., El-Sawy, N., Hassan, M., & El-Anadouli, B., (2011). Adsorption of crude oil from aqueous solution by hydrogel of chitosan based polyacrylamide prepared by radiation induced graft polymerization. *Journal of Hazardous Materials., 190,* 359–365.
47. Ahmad, A., Sumathi, S., & Hameed, B., (2005). Adsorption of residue oil from palm oil mill effluent using powder and flake chitosan: equilibrium and kinetic studies. *Water Research, 39,* 2483–2494.
48. Rajaković-Ognjanović, V., Aleksić, G., & Rajaković, L., (2008). Governing factors for motor oil removal from water with different sorption materials. *Journal of Hazardous Materials, 154,* 558–563.

49. Yang, C., (2007). Electrochemical coagulation for oily water demulsification. *Separation and Purification Technology, 54*, 388–395.
50. Ji, Z., Lin, H., Chen, Y., Dong, Y., & Imran, M., (2015). Corn cob modified by lauric acid and ethanediol for emulsified oil adsorption. *Journal of Central South University, 22*, 2096–2105.
51. Simonovic, B., Arandjelovic, D., Jovanovic, M., Kovacevic, B., Pezo, L., & Jovanovic, A., (2009). Removal of mineral oil and wastewater pollutants using hard coal. *CI&CEQ, 15*, 57–62.
52. Pal, R., (2011). Rheology of simple and multiple emulsions. *Current Opinion in Colloid & Interface Science, 16*, 41–60.
53. Radetic, M., Jocic, D., Jovancic, P., Rajakovic, L., Thomas, H., & Petrovic, L., (2003). Recycled-wool-based nonwoven material as a sorbent for lead cations. *Journal of Applied Polymer Science, 90*, 379–386.
54. Ellis, J., Korth, J., & Peng, L., (1995). Treatment of retort waters from Stuart oil shale using high-silica zeolites. *Fuel, 74*, 860–864.
55. Barrer, R., (1989). Clay minerals as selective and shape-selective sorbents. *Pure and Applied Chemistry, 61,* 1903-1912.
56. Babel, S., (2003). Low-cost adsorbents for heavy metals uptake from contaminated water: A review. *Journal of Hazardous Materials, 97*, 219–243.
57. Argun, M., Dursun, S., Ozdemir, C., & Karatas, M. (2007). Heavy metal adsorption by modified oak sawdust: thermodynamics and kinetics. *Journal of Hazardous Materials, 141*, 77–85.
58. Annunciado, T., Sydenstricker, T., & Amico, S., (2005). Experimental investigation of various vegetable fibers as sorbent materials for oil spills. *Marine Pollution Bulletin, 50*, 1340–1346.
59. Bhattacharyya, K., & Sen Gupta, S., (2006). Pb(II) uptake by kaolinite and montmorillonite in aqueous medium: influence of acid activation of the clays. *Colloids and Surfaces A: Physicochemical and Engineering Aspects, 277*, 191–200.
60. Sheintuch, M., & Rebhun, M., (1988). Adsorption isotherms for multisolute systems with known and unknown composition. *Water Research, 22,* 421–430.
61. BrandÃ£o, P., Souza, T., Ferreira, C., Hori, C., & Romanielo, L., (2010). Removal of petroleum hydrocarbons from aqueous solution using sugarcane bagasse as adsorbent. *Journal of Hazardous Materials, 175*, 1106–1112.
62. Mysore, D., Viraraghavan, T., & Jin, Y., (2005). Treatment of oily waters using vermiculite. *Water Research, 39*, 2643–2653.
63. Banerjee, S., Joshi, M., & Jayaram, R., (2006). Treatment of oil spills using organo-fly ash. *Desalination, 195*, 32–39.

CHAPTER 10

EXTERNAL MASS TRANSFER STUDIES ON ADSORPTION OF METHYLENE BLUE ON PSIDIUM GUAVA LEAVES POWDER

R. W. GAIKWAD, S. L. BHAGAT, and A. R. WARADE

Department of Chemical Engineering, Pravara Rural Engineering College, Loni, 413736, Ahmed Nagar, (MS), India, E-mail: rwgaikwad@yahoo.com

CONTENTS

ABSTRACT

In this work elimination of methylene blue from wastewater by using Psidium guava leaves powder was investigated. Typical adsorption isotherm, Langmuir and Freundlich were used to signify the experimental data. The external mass transfer coefficient have been resolved for the procedure affected by various variables together with pH, preliminary concentration, amount of adsorbent, contact time and speed of agitation. Up to 100% dye removal is seen at lower dye concentration at 12.5 mg/L, adsorbent dose at 500 mg, pH 5 and agitation speed at 100 rpm.

Furusuwa–Smith model is used to assess the effect of external mass transfer coefficient, β, on the operating variables. The value of β, were then related to dimensionless mass exchange numbers ($Sh/Sc^{0.33}$). The dimensionless mass transfer numbers were found to differ with preliminary concentration, $C^{0.11}$and Psidium guava leaves powder amount, $M^{2.6}$. The equilibrium data fitted very well in Freundlich and Langmuir isotherm.

10.1 INTRODUCTION

More than 10,000 dyes have been widely used in fabric, paper, rubber, plastics, leather and cosmetic, drug-based, and food businesses [1]. The photosynthetic activity reduces due to colored wastes into the receiving water bodies [2].

As dyes are designed to resist breakdown with time, exposure to sunlight, water, soap, and oxidizing agent cannot be easily removed by ordinary wastewater treatment processes due to their complex structure and manmade origins [3].

Methylene blue is cationic dyes. Due to methylene blue, peculiar sickness like multi plied heart rate, vomiting, upset, Heinz body formation, cyanosis, jaundice, quadriplegia, and tissue necrosis in humans have been took place [4].

A variety of ordinary methods such as physical, chemical, and organic processes have been attempted for the elimination of dyes from water-based media [5–7].

A physico-chemical method of treatment like adsorption is the simple and cheap to eliminate the dyes from wastewater [8]. The adsorption is carried out to find other low-cost adsorbents [9].

The waste dumping problem can be solved by activated carbon generated from these unwanted material. Farming wastes comprises fruit peel off [10], jute dietary fiber [11], rice shells [12], soy meal hull [13], rice husk [14], triggered day abyss [15], as well as bamboo bedding and sheets airborne debris [16]. Almost all of the carbons which are activated tend to be produced by the two-stage procedure, for example, carbonization then activation. In the present study, an adsorbent derived from Psidium guava leaves (PGL) has been used to eliminate methylene blue. The effects of operating parameters such as initial dye concentration, adsorbent dosage, pH, agitation speed, contact time, particle size, and temperature were studied. The consequences of working factor such as pH, preliminary concentration, dosage, contact time and stirring speed, and effect of mass transfer on adsorbent have been studied. The point of the present study was to survey the parameters which impact the external mass transfer coefficient amid the adsorption of methylene blue on (PGL) powder.

10.2 MATERIALS AND METHODS

Every chemicals utilized as a part of this study were of scientific grade. Methylene blue was provided by Merck India private limited. The common character of methylene blue ($C_{16}H_{18}N_3SCl$) are molar mass = 319.86, C.I. No. = 52015. Stock arrangement of Methylene blue was arranged by dissolving 200 mg of Methylene blue/Liter in two fold distillated water. The structure of Methylene blue is shown in Figure 10.1.

The pH of the solution was adjusted to a desired value using HCL and NaoH. The PGL powder utilized as an adsorbent. The normally dried-up guava leaves were obtained locally from the greenery enclosure of Loni town of Maharashtra state (India). The PGL were initially washed to evacuate soil and after that dried, powdered and sieved through a 200–250 sieve. The powdered tree leaves were again warmed on a hot plate for just about 15 min at 60–70°C for careful drying. The powder is then stored in firmly packed glass bottle. The characteristics of PGL powder are given in Table 10.1.

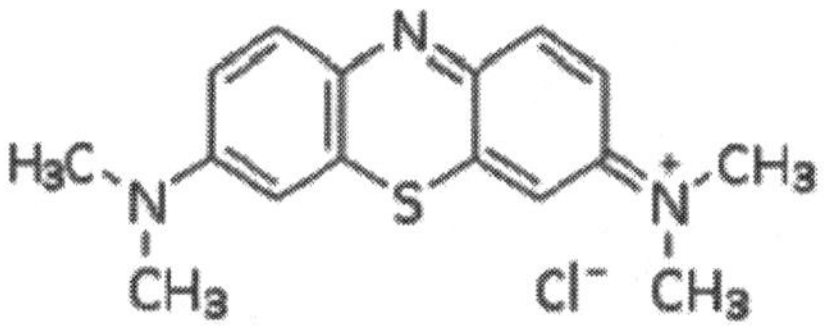

FIGURE 10.1 Structure of methylene blue.

TABLE 10.1 A Physical Characteristic of Psidium Guava Leaves Powder

Sr. No.	Characteristics	Value
1	Bulk density, gm × mL^{-1}	0.333
2	Surface area, m^2 × gm^{-1}	1.70
3	Average particle size, microns	200
4	Matter soluble in water, %	Nil
5	Matter soluble in 1M HCL, %	Nil
6	Matter soluble in 0.001M NaOH, %	Nil
7	pH	6.2

The temperature of the suspension was adjusted at the end of reaction time, after pH adjustment. After pH conformity the temperature of the arrangement was additionally balanced toward the end of response time. The suspension was separated through the Whatman filter paper No. 42 and the filtrate was taken to examinations for the methylene blue present in the suspension by UV spectrophotometer (Systronics-118) at 665 nm. So as to have a reasonable representation of the impact of various parameters on the adsorption of methylene blue, the tests were conducted at various preliminary concentrations, quantity of PGL powder. For the period of the adsorption of methylene blue from the suspension, two factors namely; concentration of dye in the solution and time of contact between methylene blue and PGL powder, play an important role. A quick transfer of methylene blue cuts through the equilibrium period. The time taken to accomplish the equilibrium is vital to foresee the effectiveness and the possibility of the PGL powder. The investigations were led at two diverse preliminary concentrations of methylene blue, for example, 12.5 mg/L and 25 mg/L of suspension. The quantity of PGL powder is kept constant at 500 mg and at 25°C with a pH 5.

The quantity of equilibrium adsorption of methylene blue is calculated by means of equation

$$q_e = (C_0 - C_e)V/W \qquad (1)$$

where, q_e is the methylene blue adsorption by PGL powder mg/g, C_0 is the preliminary methylene blue concentration; C_e is the methylene blue concentration (mg/L) after the batch adsorption process; W is the weight of adsorbent (gm); and V is the volume of methylene blue solution.

The percentage removal of methylene blue is defined as the ratio of difference in methylene blue concentration before and after adsorption ($C_0 - C_e$) to the preliminary concentration of the methylene blue of the aqueous solution of the dye (C_0) and was calculated by means of equation

$$\% \text{ methylene blue removal} = (C_0 - C_e) \times 100/C_0 \quad (2)$$

10.3 RESULTS AND DISCUSSION

10.3.1 IMPACT OF pH

The pH of the methylene blue suspension perform an important function in the total adsorption process, mainly on adsorption capability [17]. Figure 10.2 demonstrates the impact of pH on the adsorption of methylene blue onto the PGL powder at 12.5 mg/L which is a preliminary concentration and the amount of adsorbent (500 gm/L) at 50 rpm agitation speed and at 25°C. If pH of the dye solution was changed from 3 to 8, the adsorption capacity of PGL powder was raised. During the adsorption, at a lower pH, the adsorbent surface is positively charged, supporting adsorption of anionic pollutant. Since methylene blue is a cationic, positive charge which possessed the achievable adsorption position competing methylene blue molecules, bringing about a lower adsorption of methylene blue; while at a higher pH, negative charged surface encourages adsorption of cationic contaminants.

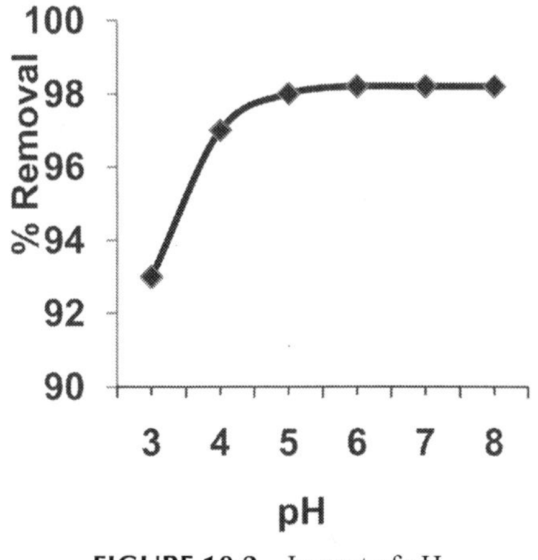

FIGURE 10.2 Impact of pH.

10.3.2 IMPACT OF PRELIMINARY METHYLENE BLUE CONCENTRATION

The impact of preliminary methylene blue concentration on the adsorption of methylene blue onto the PGL powder was considered. Initially the rate of adsorption increases due to difference in bulk flow and when equilibrium is established the nature of the graph become a straight line as there is decrease in mass transfer as shown in Figure 10.3. The equilibrium time is 4 min. The time taken to attain equilibrium is independent of the concentration. It was found that 93% removal for 25 mg/L of dye solution and 94.6% removal for 12.5 mg/L of dye solution occur.

10.3.3 IMPACT OF PSIDIUM GUAVA LEAVES POWDER QUANTITY

PGL powder quantity is a vital parameter due to its strong effect on the capability of an adsorbent at given preliminary concentration of Methylene blue. Effect of PGL powder quantity on elimination of Methylene blue was checked by changeable adsorbent quantity from 100 g/1000 mL to 800 g/1000 mL. The adsorption of Methylene blue increased with the PGL powder quantity and arrive at on equilibrium value after 500 g of adsorbent quantity (Figure 10.4). The percentage of Methylene blue elimination increased with increasing amount of PGL powder, however the ratio of Methylene blue adsorbed to PGL powder (mg/g) decreased with increasing amount of PGL powder.

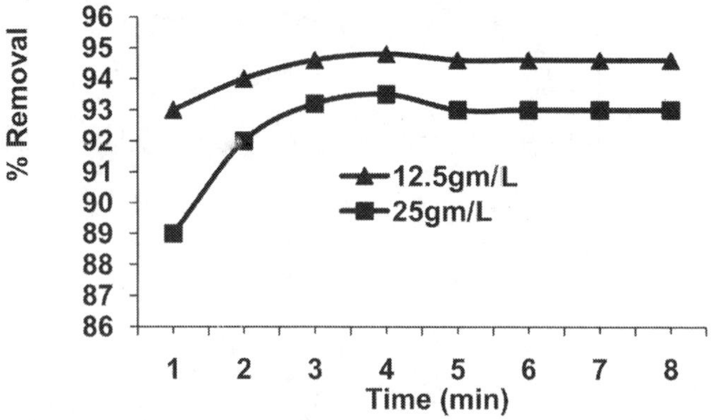

FIGURE 10.3 Impact of preliminary methylene blue concentration.

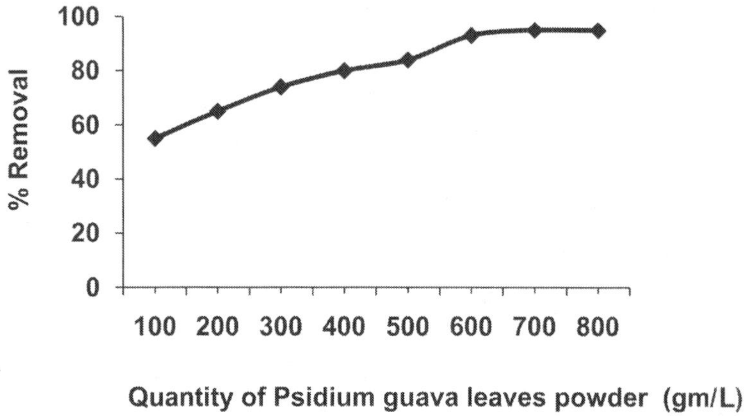

FIGURE 10.4 Impact of adsorbent dose.

10.3.4 IMPACT OF CONTACT TIME

Contact time is likewise a vital element influencing elimination; the majority of adsorption happens in the starting 30 min and rise gradually later. These investigations have been completed at variety of time of contact, for example, 10 to 100 min. Figure 10.5 shows that adsorption of methylene blue onto the PGL powder reached to the equilibrium in 70 min.

It was accounted for that in the middle of adsorption of methylene blue, at first the methylene blue molecule achieve the limit layer; then they need to diffuse into the adsorbent surface; lastly, they need to diffuse into the permeable structure of the PGL powder. As a result, this occurrence is time consuming [18].

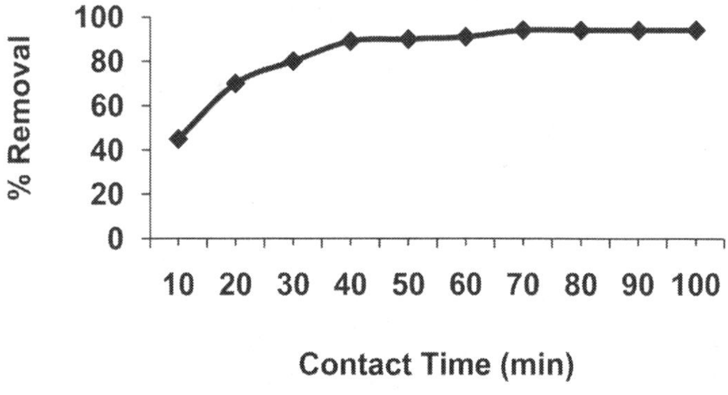

FIGURE 10.5 Impact of contact time.

10.3.5 IMPACT OF STIRRING SPEED

The stirring is an imperative process in adsorption which affects the diffusion of the solute in the mass of the suspension and creation of the exterior edge coat. The impact of sorption of methylene blue on PGL powder was considered at various stirring speeds.

The outcome of this study is revealed in Figure 10.6. It was found that the adsorbed methylene blue by PGL powder improved with the rise in stirring speed. By additional stirring speed, there was a more rise in sorption as every single site was accessible for additional sorption. With stirring, the outer mass transfer or exchange coefficient fosters, result in faster adsorption of the methylene blue.

10.3.5 ISOTHERM MODELING

The adsorption isotherm stand for the link between the quantity adsorbed by strong adsorbent and the measure of adsorbate continued in the suspension at a time [19]. Isotherm such as Langmuir and Freundlich, were utilized to depict the adsorption process. Langmuir isotherm [20] alludes to homogeneous monolayer adsorption where as the straight type of the

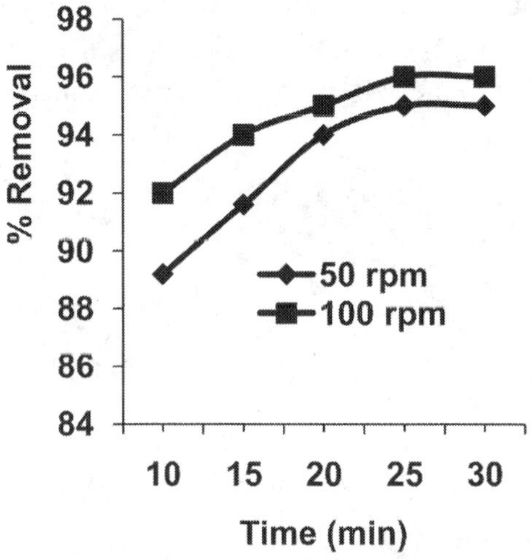

FIGURE 10.6 Impact of agitation speed.

Freundlich isotherm [21] is inferred by accepting a mixed surface of adsorption capability and adsorption strength with a inconsistent allocation of heat of adsorption.

The Langmuir isotherm equation for study is specified as

$$\frac{C_e}{q_e} = \frac{1}{Q^0 b} \frac{C_e}{Q^0 b} \tag{3}$$

Here, q_e is the quantity of methylene blue in the PGL powder at equilibrium (mg/g), C_e is the concentration at equilibrium (mg/L) and Q^0 and b are the Langmuir constants indicating adsorptive capacity and energy of adsorption, respectively.

The Eq. (3) can be linearized to get the maximum capacity, q_{max} by plotting a graph of C_e/q_e vs. C_e. The Eq. (3), can be linearized to get the most extreme limit, by plotting a chart of C_e/q_e vs. C_e.

Freundlich isotherm is given by

$$q_e = K_f C_e^{1/2} \tag{4}$$

On rearranging this equation we get

$$\log q_e = \log K_f + 1/n \log C_e \tag{5}$$

Here K_f and $1/n$ are Freundlich isotherm constants linked with the adsorption capability and adsorption strength correspondingly. A figure of log $1/q_e$ vs. log $1/C_e$ produce a straight line with a slop of $1/n$ and intercept log K_f. Both the isotherms Langmuir and Freundlich, of methylene blue are depicted in Figures 10.7 and 10.8. The investigational results specified that the isotherms of methylene blue adsorption on PGL powder abide by Langmuir and Freundlich models.

10.3.6 SURFACE MASS TRANSFER

The particular adsorption curve within Figure 10.3 supported to look for the external mass transfer applying correlation proposed by Furusuwa and Smith [22], and that is specified by mathematical statement (6) as takes after:

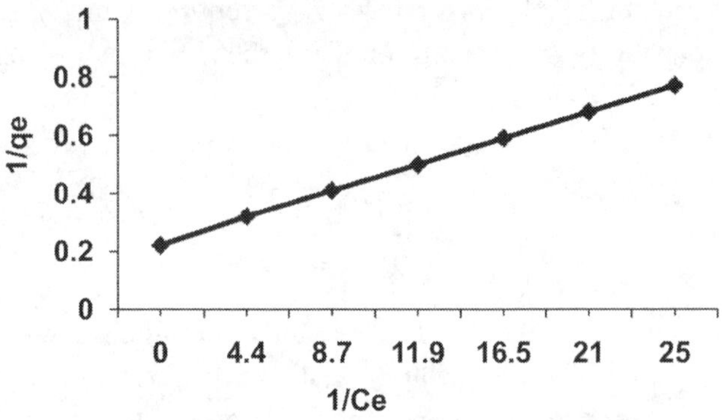

FIGURE 10.7 The Langmuir isotherm.

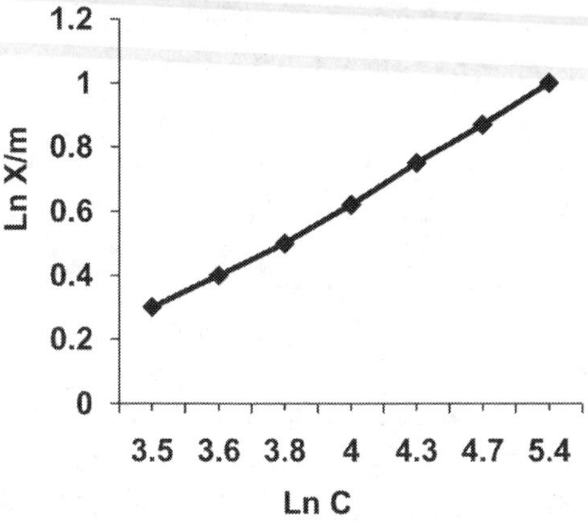

Figure 10.8 The Freundlich isotherm.

$$d(C/C_o)/dt = \beta S \tag{6}$$

In which β mathematical statement (6) implies this external mass transfer coefficient along with S signifies this external outside area of associated with PGL powder of particle totally liberated suspension and the facts ended up earlier discussed simply by McKay et al. [23].

Therefore, the actual bulk transfer coefficient, β, might be measured in the slope of curvature utilizing mathematical statement (6). The actual determined mass exchange coefficients were then correlated as mass exchange dimensionless numbers (Sh/Sc 0.33) along with were plotted against Log C as appeared in Figure 10.9. In which S_h along with S_c represent Sherwood along with Schmidt range, respectively. Through Figure 10.8, it turned out witnessed that this mass transport speed reduces with rise in preliminary concentration. A comparable declaration had been documented through McKay et al. [23] for agitated adsorption programs. This lowering in mass transfer rate using preliminary concentration can be clarified out based on molecular relationship as well as interionic groups of methylene blue, that may reduce the activity coefficient in the methylene blue as well as viable diffusivity [24].

The curvature in Figure 10.8 fits the mathematical statement as takes after:

$$Sh/Sc^{0.33} = 3.1\ \mathrm{Log}(C)^{0} \tag{7}$$

Figure 10.4 demonstrates the plot between % elimination versus time at various quantity of PGL powder. From Figure 10.4, it was viewed that the methylene blue elimination rate gets improved with addition of quantity of PGL powder. This is because with improving adsorbent quantity, effective site necessary for more uptakes associated with solute transfer to take place raise. Likewise, Figure 10.4 affirms that the sorption rate

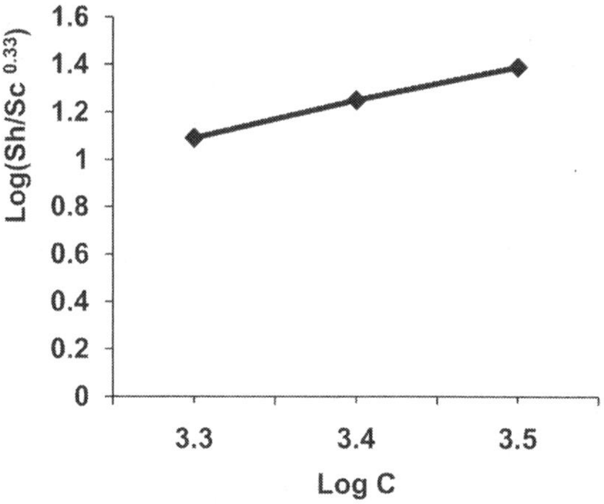

FIGURE 10.9 Impact of mass transfer on methylene blue concentration.

raise with expansion in PGL powder. This is because of the expansion in methylene blue exchange rate per unit time onto unit surface zone of PGL powder with rising PGL powder quantity. As clarified before the mass exchange coefficient, β, were ascertained from the slant of curvature in Figure 10.3 utilizing mathematical statement (1). The determined values of β, were linked as dimensionless mass exchange numbers ($S_h/S_c^{0.33}$) as appeared in Figure 10.10. From Figure 10.10, it was viewed that the mass exchange rate diminishes with rise in PGL powder quantity. This is on account of with rising PGL powder quantity. the concentration rise necessary for mass transfer to take place, reduce with rise in PGL powder quantity. Thus, the solute exchange rate per unit time onto unit surface territory divide, this causes a decline in mass exchange rate. The curvature in Figure 10.10 suit fits the mathematical statement (8) as takes after:

$$Sh/Sc^{0.33} = 1.2\ \text{Log}(M)^{2.6} \tag{8}$$

10.4 CONCLUSION

The treatment of methylene blue wastewater utilizing PGL powder was explored under various test conditions. From this investigation it can be inferred that PGL powder can be utilized as an effective permeable

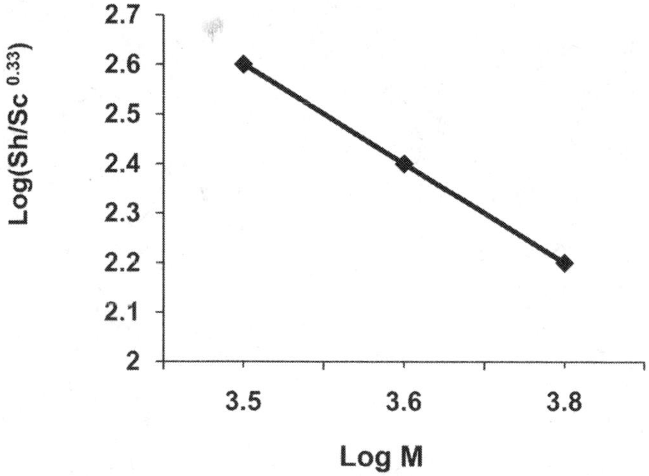

FIGURE 10.10 Impact of mass transfer on dosage.

adsorbent for treatment of wastewater. The quantity of methylene blue eliminated by PGL powder rise with escalating these parameters: quantity, pH, contact time, stirring speed. Isotherm such as Freundlich and Langmuir were utilized to articulate the adsorption process of methylene blue elimination. This process follows the Freundlich. The outcome shows that this adsorption procedure is physical. The external mass transfer rate was found to decline with quantity of PGL powder and rise with preliminary methylene blue concentration.

KEYWORDS

- **adsorption**
- **external mass transfer**
- **isotherms**
- **mass transfer coefficient**
- **methylene blue**
- **Psidium guava leaves powder**

REFERENCES

1. Mondal, S. (2008). Methods of dye removal from dye house effluent: An overview, *Environmental Engineering Science*, *25*(3), 383–396.
2. Rastogi, K., Sahu, J. N., Meikap, B. C. & Biswas, M. N., (2008). Removal of methylene blue from wastewater using fly ash as an adsorbent by hydrocyclone, *Journal of Hazardous Materials,158*(2–3), 531–540.
3. Wang, L., Zhang, J., & Wang, A., (2008). Removal of methylene blue from aqueous solution using chitosan-g-poly(acrylic acid)/montmorillonite super adsorbent nanocomposite, *Colloids and Surfaces A*, *322*(1–3), 47–53.
4. Ahmad, R. & Kumar, R. (2010). Adsorption studies of hazardous malachite green onto treated ginger waste, *Journal of Environmental Management*, *91*(4), 1032–1038.
5. Barka, N., Qourzal, S., Assabbane, A., Nounah, A., & Ait-Ichou, Y., (2010). Photocatalytic degradation of an azo reactive dye, Reactive Yellow 84, in water using an industrial titanium dioxide coated media, *Arabian Journal of Chemistry,3*(4), 279–283.
6. Elahmadi, M. F., Bensalah, N., & Gadri, A., (2009). Treatment of aqueous wastes contaminated with Congo Red dye by electrochemical oxidation and ozonation processes, *Journal of Hazardous Materials,168*(2–3), 1163–1169.

7. Khadhraoui, M., Trabelsi, H., Ksibi, M., Bouguerra, S., & Elleuch, B. (2009). Discoloration and detoxicification of a Congo red dye solution by means of ozone treatment for a possible water reuse, *Journal of Hazardous Materials*, *161*(2–3), 974–981.
8. Tehrani-Bagha, A. R., Nikkar, H., Mahmoodi, N. M., Markazi, M. & Menger, F. M., (2011). The sorption of cationic dyes onto kaolin: kinetic, isotherm and thermodynamic studies, *Desalination*, *266*(1–3), 274–280.
9. Crini, G. (2006). Non-conventional low-cost adsorbents for dye removal: a review, *Bioresource Technology*, *97*(9), 1061–1085.
10. Sivaraj, R., Namasivayam, C. & Kadirvelu, K. (2001). Orange peel as an adsorbent in the removal of acid violet 17 (acid dye) from aqueous solutions, *Waste Management*, *21*(1), 105–110.
11. Senthilkumaar, S., Varadarajan, P. R., Porkodi, K., & Subbhuraam, C. V., (2005). Adsorption of methylene blue onto jute fiber carbon: kinetics and equilibrium studies, *Journal of Colloid and Interface Science*, *284*(1), 78–82.
12. Bulut, Y. & Aydin, H., (2006). A kinetics and thermodynamics study of methylene blue adsorption on wheat shells, *Desalination*, *194*(1–3), 259–267.
13. Arami, M., Limaee, N. Y., Mahmoodi, N. M. & Tabrizi, N. S., (2006). Equilibrium and kinetics for the adsorption of direct acid dyes from aqueous solution by soy meal hull, *Journal of Hazardous Materials,135*(1–3), 171–179.
14. Malik, P. K. (2003). Use of activated carbons prepared from sawdust and rice-husk for adsorption of acid dyes: a case study of acid yellow 36, *Dyes and Pigments*, *56*(3), 239–249.
15. Banat, F., Al-Asheh, S., & Al-Makhadmeh, L., (2003). Evaluation of the use of raw and activated date pits as potential adsorbents for dye containing waters, *Process Biochemistry*, *39*(2), 193–202.
16. Kannan, N., & Sundaram, M. M. (2001). Kinetics and mechanism of removal of methylene blue by adsorption on various carbons: A comparative study, *Dyes and Pigments*, *51*(1), 25–40.
17. Wang, S., Boyjoo, Y., & Choueib, A., (2005). A comparative study of dye removal using fly ash treated by different methods, *Chemosphere*, *60*(10), 1401–1407.
18. Senthilkumaar, S., Varadarajan, P. R., Porkodi, K., & Subbhuraam, C. V. (2005). Adsorption of methylene blue onto jute fiber carbon: kinetics and equilibrium studies, *Journal of Colloid and Interface Science*, *284*(1), 78–82.
19. Hu, Z., Chen, H., & Yuan, S., (2010). Removal of Congo red from aqueous solution by cat tail root, *Journal of Hazardous Materials*, *173*, 292–297.
20. Langmuir, I., (1916). The constitution and fundamental properties of solids and liquids, *Journal of American Chemical Society*, *38*(11), 2221–2295.
21. Freundlich, H. M. F., (1906). Over the adsorption in solution, *Journal of. Phys. Chem.*, *57*, 385–470.
22. Furusuwa, T., & Smith, J. M., (1974). Intraparticle mass transport in slurries by dynamic adsorption studies. *AiChE, 20*, 88.
23. Mckay, G., Allen, S. J., Mcconvey, I. F., & Otterburn, M. S. (1981). Transport process in the sorption of colored ions by peat particles. *J. Colloid. Interface. Sci. 80*, 323.
24. Nassar, M. M., & Magdy, Y. H. (1999). Mass transfer during adsorption of basic dyes on clay in fixed bed. *Indian. Chem. Engg. 40*, 27.

CHAPTER 11

THE ROLE OF TIO_2 NANOPARTICLES ON MIXED MATRIX CELLULOSE ACETATE ASYMMETRIC MEMBRANES

SHIRISH H. SONAWANE,[1] ANTONINE TERRIEN,[2] ANA SOFIA FIGUEIREDO,[3,4] M. CLARA GONÇALVES,[5,6] and MARIA NORBERTA DE PINHO[3,5]

[1]*Department of Chemical Engineering, National Institute of Technology, Warangal, 546004, India, E-mail: shirishsonawane09@gmail.com, Tel: +91-870-2462626*

[2]*ETU, University of Nantes, France*

[3]*CEFEMA, Instituto Superior Técnico, Universidade de Lisboa, Lisbon, Portugal, E-mail: marianpinho@ist.utl.pt*

[4]*Departamental Area of Chemical Engineering, Instituto Superior de Engenharia de Lisboa, Instituto Politécnico de Lisboa, Lisbon, Portugal*

[5]*Department of Chemical Engineering, Instituto Superior Técnico, Universidade de Lisboa, Lisbon, Portugal*

[6]*CQE Centro de Química Estrutural, Instituto Superior Técnico, Universidade de Lisboa, Lisbon, Portugal, E-mail: clara.goncalves@ist.utl.pt*

CONTENTS

ABSTRACT

In this work new anti-fouling mixed matrix membranes with increased permeability and better surface properties have been produced. Mixed matrix membranes were obtained by incorporating ultrasound assisted sol-gel TiO_2 nanoparticles into cellulose acetate (CA) membranes.

TiO_2 nanoparticles, within the range of 20 to 30 nm, were added to cellulose acetate/acetone/formamide casting solutions with weight ratios acetone/formamide varying from 1.44 to 2.77, to prepare ultrafiltration/nanofiltration membranes over a wide range of hydraulic permeabilities. Bond formation between TiO_2 nanoparticles and cellulose acetate matrix was confirmed by FTIR through the disappearance of the peak centered at 2600 cm^{-1}. Zeta Potential evidenced changes in the membrane surface chemistry. The addition of ultrasound assisted sol-gel TiO_2 nanoparticles to the casting solution changed the membranes porosity structure, as observed by SEM. Further, the pure water flux increased with the acetone concentration and TiO_2 nanoparticles content. Although the presence of TiO_2 nanoparticles rendered the cellulose acetate membranes less electronegative, TiO_2 nanoparticles play an important role in the adsorption and separation of the cations present in the water.

11.1 INTRODUCTION

Simplicity and efficiency in process separation place membranes in an outstanding point when it comes to chemical industries, dairy and food processing, desalination, and drinking and industrial water treatments [1, 2]. Membrane filtration processes are commonly used industrially in the separation, concentration and purification of mixtures as an alternative to costly and more complex unit operations such as distillation, centrifugation, and extraction.

Most of the membrane separation techniques utilize polymer membranes, being cellulose acetate (CA) membrane one of the most promising candidates due to its high flexibility, low cost and excellent water permeation properties [3, 4]. Mixed matrix membrane is a recent research area in separation technologies, where inorganic nanoparticles and/or additives are mixed/incorporated into the organic polymer matrix in order to improve their performance in terms of selectivity, separation, and flux [4–6]. The presence of polyethylene or polypropylene glycol promotes the hydrophilic character of the neat matrix. M. Ali et al. [6] studied the addition of ZnO inorganic nanoparticles along with polyethylene glycol 600 (PEG 600) into CA membranes. The authors observed that the presence of 10 wt.% PEG into the ZnO/CA membranes increases the permeation flux up to 1.2 times the permeation flux of the plain ZnO/CA membranes on the pervaporation of a feed composed by isopropyl alcohol (71 wt.%) in water.

Titanium oxide is one of the most important ceramic materials used as bactericide, UV protector and self-cleaning agent. In paper industry, for example, TiO_2 nanoparticles allow the tune of the refractive index; in water treatment applications, solar glasses, paints and cements TiO_2 nanoparticles play a photocatalyst effect; and in solar cell panels the role of host sensitizers [7–10].

Vatanpur et al. [11] reported the introduction of different types of TiO_2 nanoparticles into polyethersulfone (PES) membranes and correlates the nanoparticles morphology, size and structure with the membranes flux performance. The maximum flux was observed for the addition of 4 wt.% of TiO_2 nanoparticles (P25 Degussa: 20% rutile and 80% anatase). The flux recovery percentage was observed to increase with the decrease in TiO_2 nanoparticles size.

Lee et al. [12] studied polyamide thin-film nanofiltration membranes containing in-situ polymerized TiO_2 nanoparticles. This method allows higher membrane loading (~5 wt.% of TiO_2 nanoparticles), assembling more robust PA-TiO_2 nanocomposite structures, along with high and more stable rejection values (higher than 95%) with respect to $MgSO_4$, and permeation flux of 9.1 L/m^2h. Elemental analysis by XPS demonstrated that substantial amounts of TiO_2 nanoparticles remained on the surface of the membranes after nanofiltration operation for 2 days.

Sairam et al. [13] studied the use of mixed matrix membrane containing TiO_2 nanoparticles for the pervaporation system, where isopropyl alcohol is separated from water at normal temperature. Addition of small amount of filler nanoparticles into the cross linked PVA membrane matrix has been instrumental in increasing the membrane selectivity to infinity values. The addition of nanoparticles as fillers has decreased the extent of swelling and the flux properties, increasing the membranes' selectivity.

Yang et al. [14] studied the influence of TiO_2 nanoparticles structure on TiO_2/polysulfone ultrafiltration (UF) membranes. TiO_2 nanoparticles presence changed the rheological properties of casting solution from Newtonian to non-Newtonian viscous behavior. At 2 wt.% TiO_2 content, the composite membranes held excellent water permeability, hydrophilicity, mechanical strength and good anti-fouling ability with almost unchanged retentions. High TiO_2 contend (>2 wt.%) caused nanoparticle aggregation and deterioration of PSF/TiO_2 membranes performances.

Bendixen et al. [15] studied the membrane nanoparticle interactions in an asymmetric flow by incorporation TiO_2 nanoparticles in different membrane materials (regenerated cellulose, polyethersulfone and polyvinyl difluoride). Nanoparticles zeta potential and the membrane structures play an important role in the separation efficiency.

In this work ultrasound assisted sol-gel TiO_2 nanoparticles were incorporated in CA asymmetric membranes for the first time. Ultrasonic synthesis reduced the TiO_2 nanoparticle size and consequently enhanced its photo catalytic activity; no matter the solvent is pure water or ethanol/water mixed solutions [16]. TiO_2/CA mixed matrix membranes are prepared and their chemical properties and water permeation properties are studied as function of the TiO_2 nanoparticles contend. The influence of TiO_2 nanoparticles size and shape on the CA asymmetric membrane composites permeability has been studied.

11.2 MATERIALS AND METHODS

11.2.1 MATERIALS

Titanium tetraisopropoxide (Merck), acetic acid glacial (Panreac), 2-propanol (Aldrich), acetone (Labchem), cellulose acetate (39.8% acetyl content, MW ~30,000), formamide (Aldrich) and polyvinylpyrolidone (98% pure, BDH) were analytical grade reagents used without further purification.

11.2.2 TIO_2 NANOPARTICLE SYNTHESIS

The ultrasound assisted sol-gel synthesis of TiO_2 was described in Refs. [17–19]. Briefly, 10 mL of titanium precursor titanium tetra isopropoxide (TTIP) is added drop wise to 60 mL of 2-propanol. The resulting clear solution is then added drop wise to 10 mL of glacial acetic acid solution (solution A). Solution A is then subjected to acoustic cavitations using an ultrasonic bath (ACE 22 kHz 120 W O/P) at 50% amplitudes delivering required O/P. The irradiation was carried out for 20 min in a 10 s pulse mode, for example, 10 s ON and 5 s OFF mode. The energy input was calculated using the calorimetric method. Separate solution of polyvinyl pyrollidone (PVP) was prepared in propanol with 0.1 g of PVP in 10 mL of propanol. The synthesized TiO_2 particles were added into the PVP solution and mixed it well. After 30 min, the resulting solution is filtered, dried. Figure 11.1 shows TiO_2 nanoparticles synthesis flowchart. PVP encapsulation renders hydrophobic character to the TiO_2 nanoparticles.

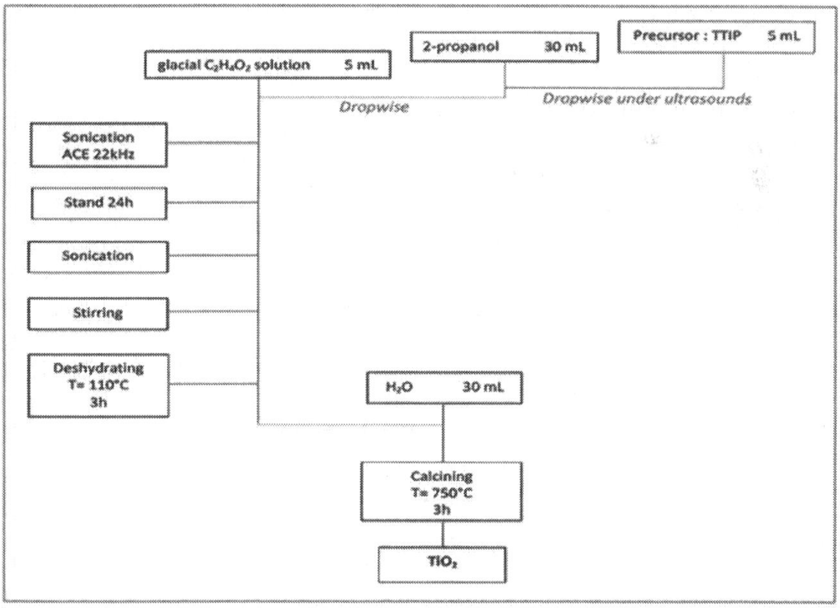

FIGURE 11.1 Synthesis of TiO_2 nanoparticles using sol-gel ultrasound cavitation technique.

11.2.3 PREPARATION OF TIO_2 CELLULOSE ACETATE MIXED MEMBRANE

The TiO_2 mixed CA asymmetric membranes were prepared by phase inversion using the wet process described by Kunst and Sourirajan [20]. As reported in Table 11.1, the five different CA casting solution compositions of the unmodified CA membranes (series C) just differ in the solvent ratios acetone/formamide, where the casting solution with higher content in formamide will promote a more porous membrane. Table 11.1 also presents the five different casting solutions compositions of the TiO_2 CA mixed membranes (series T). The TiO_2 nanoparticles (0.1 g, TiO_2 nanoparticles) were initially dispersed in a mixture of polyvinyl pyrolidone (0.2 mL of 0.5 M, PVP) and acetone (3 mL acetone) in an ultrasound bath for 10 min. The addition of TiO_2 nanoparticles into acetate/acetone/formamide casting solutions was carried out in ultrasound batch. The nanocomposite casting solution was poured on the glass plate and spread using a casting knife with 10 μm thickness to form a film of membrane. The coagulation was carried out in ice-cold water bath and the membrane was removed from water. The details of the casting conditions are reported in the Table 11.2.

TABLE 11.1 Composition of Casting Solutions of the CA Nanocomposite Membranes

Membrane	C1	T1	C2	T2	C3	T3	C4	T4	C5	T5
Casting solution (wt.%)										
Cellulose acetate	17.0	16.8	17.0	16.8	17.0	16.8	17.0	16.8	17.0	16.8
Acetone	49.0	48.5	51.0	50.5	53.0	52.5	56.0	55.5	61.0	60.4
Formamide	34.0	33.7	32.0	31.7	30.0	29.7	27.0	26.7	22.0	21.8
TiO_2 nanoparticles	—	1.0	—	1.0	—	1.0	—	1.0	—	1.0

TABLE 11.2 Film Casting Conditions for Neat CA and CA Nanocomposite Membranes

Casting conditions	
Temperature of casting solutions ((C)	20–25
Solvent evaporation time (min)	0.5
Gelation Medium	Ice cold deionized water
Gelation Time (h)	2

11.2.4 CHARACTERIZATION OF TIO_2/CA MEMBRANES

Size, size distribution and morphology of TiO_2 nanoparticles have been observed by conventional transmission electron microscope (TEM, Hitachi H-8100, equipped with a LaB6 filament, and EDS Noran System Six detector from Thermo Noran). The samples (TiO_2 nanoparticles suspended in ethanol) were suspended over a copper grid, followed by fast evaporation in air.

Surface morphology of the asymmetric TiO_2/CA mixed matrix membranes was carried out using FEG-SEM, JSM 7001F model, equipped with and EDSD from Oxford, INCA 250 model. The small pieces of membrane sample were snapped under liquid nitrogen to give a generally consistent and clean cut. The membranes samples sputter-coated with thin film of gold before characterization.

FTIR spectra of the asymmetric CA and mixed matrix TiO_2/CA membranes were recorded by a Thermo-Scientific Nicolet 5800 FTIR spectrometer, in transmission mode, in the region 4,000–500 cm^{-1}.

Zeta potential versus pH of the different TiO_2/CA nanocomposite membrane was carried out using an electrokinetic Analyzer (SurPASS, Anton Paar GmbH, Austria). Zeta potential value was determined using 0.001 M KCl as a background electrolyte solution and the solution pH was varied from 3 to 11 by adding small amounts of HCl and NaOH 0.1M. The operating pressure ranged from 0 to 300 milibar and the temperature was about 25°C.

11.2.5 CALCULATIONS OF PERMEATE WATER FLUX CALCULATIONS FOR NANOCOMPOSITE MEMBRANE

The permeation experiments were carried out with deionized water, to determine the membrane pure water flux, J_{pw}. The pure water permeation flux was determined by weighing the permeate volume collected (Mp) in a given time (t) and per membrane area (A_m) as $J_{pw} = Mp/(A_m.t)$. The hydraulic permeability (Lp) is obtained by the slope of the linear variation between the pure water permeate flux (J_{pw}) as a function of the transmembrane pressure (ΔP) defined as $Lp = J_{pw}/\Delta P$. The range of the transmembrane pressure used was 1, 2 and 3 bar with a feed flow rate of 180 L/h. The permeation experiments were performed in an installation, schematized in Figure 11.2, consisting of a feed tank, a pump, a flow meter, five permeation cells, a valve and two manometers (placed before and after the permeation cells). The system pressure and the flow is assured by the pump, and the pressure is adjusted by the valve. The five permeation cells are flat plate cells with two detachable parts separated by a porous plate (membrane support) with a membrane surface area of 13.2×10^{-4} m^2 described previously by Afonso and De Pinho [21]. Before the experiments membranes were compacted for 2 h with deionized water at a transmembrane pressure of 3 bars, and the stabilization time for each experiment run was 30 min.

11.3 RESULTS AND DISCUSSION

The variation of the solvent systems, over different acetone/formamide ratios yields membranes covering the wide range of ultrafiltration,

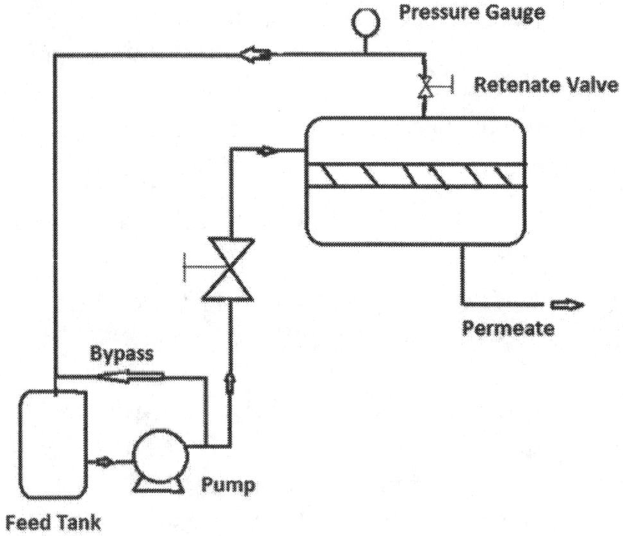

FIGURE 11.2 Experimental setup for testing of water permeation rate through nanocomposite TiO_2/CA membrane. (a) FTIR spectra of CA membranes and nanocomposite TiO_2/CA membranes. (b) FTIR spectra of nanocomposite TiO_2/CA membranes with different compositions.

nanofiltration, and reverse osmosis as reported in literature [22, 23]. The mixed matrix TiO_2/CA asymmetric membranes prepared from casting solutions reported in Table 11.1 exhibited different membrane porosities. Higher content of formamide in the casting solutions will originate membranes less dense and with large pore size while casting solutions with lower amount of formamide produce denser membranes with smaller pore size [24].

The mixed matrix TiO_2/CA asymmetric membranes prepared in the present work display hydraulic permeabilities ranging from 11.3 $kg.m^{-2}.h^{-1}.bar^{-1}$ to 54.2 $kg.m^{-2}.h^{-1}.bar^{-1}$.

TiO_2 nanoparticles (encapsulated in PVP) exhibited a hydrophobic character. The incorporation of TiO_2 nanoparticles in CA membranes play important role in ionic change performance.

Due to acoustic cavitations there is high energy collision of bubbles which has effect onto the particle size of TiO_2, % crystalline and % rutile form of TiO_2. It has been reported that the sonochemical synthesis is more effective process for the production nanoparticles as it shows the improved properties in terms of stability of structure. Due to the exothermic nature of the hydrolysis reaction, 2-propanol was used as heat removal medium. The dissipation of the heat has large impact onto the crystal structure of the TiO_2 nanoparticles, hence it has to proceed under strict control [17–19].

11.3.1 FTIR STUDIES OF TIO_2 NANOPARTICLES AND NANOCOMPOSITE CA MEMBRANES

FTIR has been used to identify functional groups in the TiO_2 nanocomposite CA membranes. For TiO_2/CA asymmetric membranes, specific interactions between the cellulose acetate polymers and TiO_2 nanoparticles may lead to change or shift of the fundamental IR peaks.

Figure 11.3a shows the FTIR of neat CA membranes (without TiO_2 nanoparticles) (series C). The FTIR spectra of neat CA membranes, with different compositions, exhibited major prominent peaks in the range of 1720 to 1800 cm^{-1}, assigned to the carbonyl stretching (C=O). The range of 3200 to 3600 cm^{-1} corresponds to the stretching of hydroxyl groups (-OH); the band in the range of 2500 to 3000 cm^{-1} corresponds to the methylene

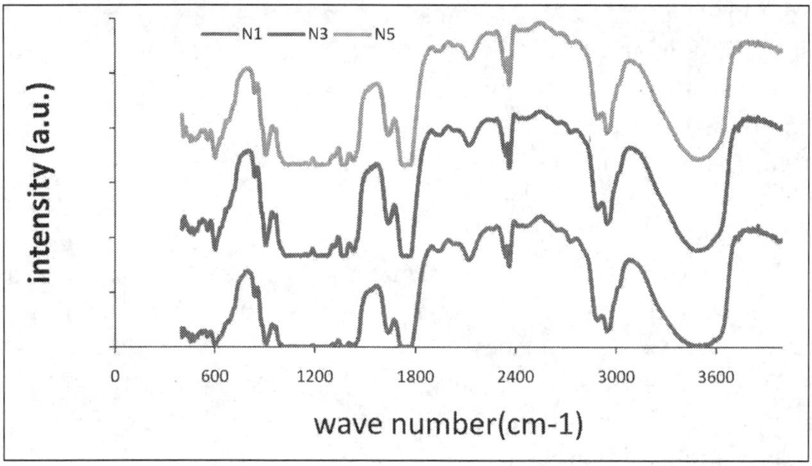

FIGURE 11.3(a) FTIR spectra of CA membrane with different compositions of CA/formamide/acetone.

group (-CH_2), specifically C-H stretching. The absorption bands in the range of 1680 to 1545 cm^{-1} corresponds to C-O stretching; the bands centered at 1455 and 1370 cm^{-1} are assigned to CH_2 scissoring (H-C-H) and H_2O bending vibration (H-O-H), respectively [25]. The important intense bands due to the -OH, -CH and C = O stretching are found at 3500–3100 cm^{-1}, 2944 cm^{-1}, 2889 cm^{-1}, and 1744 cm^{-1}, respectively, while the C-O single bond stretching modes are located at 1228 cm^{-1} and 1044 cm^{-1} [26]. The FTIR band at 3500 cm^{-1} is assigned to -OH stretching, strongly intense in the spectrum of wet membranes. The bending of H-O-H for the absorbed water is found at 1636 cm^{-1}.

FTIR spectra of the five different TiO_2/CA membrane compositions, with 1 wt.% of TiO_2 nanoparticles (series T), are reported in the Figure 11.3b, where molecular interaction of TiO_2 nanoparticles and CA matrix are evidenced. The FTIR spectra of TiO_2/CA membranes show a broad peak at 3619 cm^{-1} assigned to the stretching vibration of O-H group. The strong peak around 232 cm^{-1} represents the stretching vibrations of carboxylic acid while strong peak near to 1522 cm^{-1} evidence the presence of C = C bond. The peak at the 710 cm^{-1} is assigned to C-H vibrations. The FTIR spectrum of TiO_2 nanocomposite CA membranes shows the stretching of O-H, C≡C, and C = C represents at the 2100 cm^{-1} to 2130 cm^{-1} and 1520 to 1522 cm^{-1}, respectively. Some of the strong peaks rose at 3415

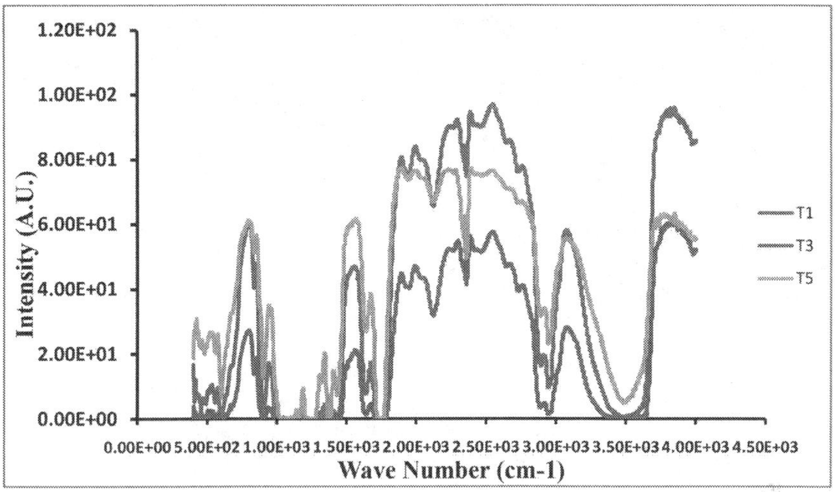

FIGURE 11.3(b) FTIR spectra of CA membrane containing TiO_2 nanoparticles at different compositions. FTIR spectra of nanocomposite TiO_2/CA membranes with different compositions.

cm^{-1} and 2880 cm^{-1} are due to the presence of the TiO_2 nanoparticles in the membrane. Major changes in the TiO_2/CA membranes attributed to the presence of TiO_2 nanoparticles occurred at 2100–2600 cm^{-1}. The disappearance of the peak centered at 2600 cm^1 is attributed to bond formation between the OH groups present on the surface of CA matrix and TiO_2 nanoparticles [27–29].

11.3.2 MORPHOLOGY STUDIES OF NANOPARTICLES AND NANOCOMPOSITE CA MEMBRANE

TEM images of ultrasound assisted sol-gel TiO_2 nanoparticles are shown in Figure 11.4a. Spindle shape TiO_2 nanoparticles are in the 20–30 nm range size.

Figure 11.4b reports SEM images of TiO_2/CA membranes. The white spot on the surface indicates homogeneous distributions of TiO_2 nanoparticles on the membrane nanocomposite. Further the 10,000x magnification SEM images indicate that the pore size of the membrane is largely affected by the presence of the TiO_2 nanoparticles. The size of the pores is in the range of 0.1–0.2 micrometers. The presence of only 1% TiO_2

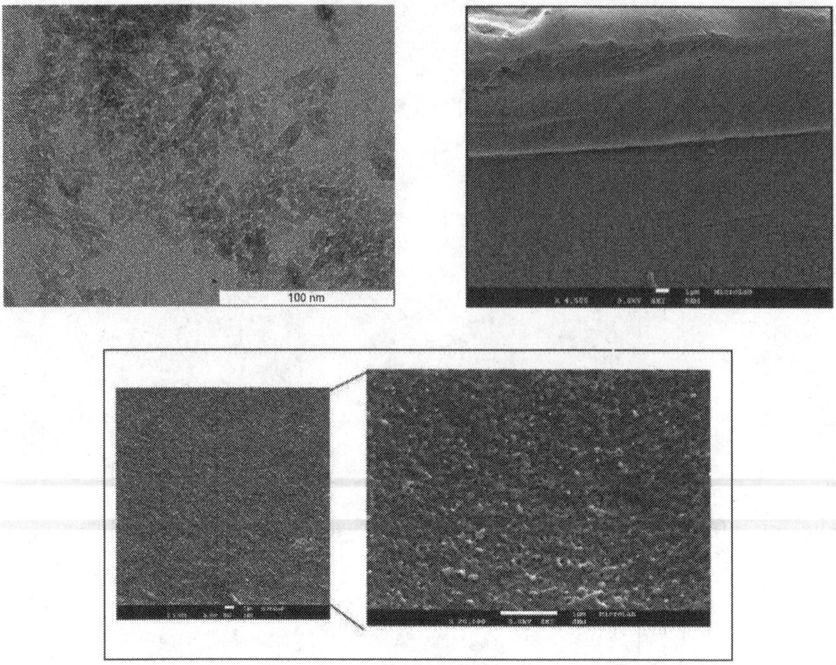

FIGURE 11.4(a) TEM Image of the TiO_2 nanoparticles.

nanoparticles has a large impact on the pore size and pore distributions of the composite CA membrane. The homogeneous particles distribution in the nanocomposite membrane indicates that PVP encapsulating agent play an efficient role in the adherence of the TiO_2 nanoparticles to the cellulose acetate membrane. Figure 11.4b proves the homogeneous distribution of TiO_2 nanoparticles onto the surface.

Figure 11.4c shows SEM images of T5 TiO_2 nanocomposite CA membrane. SEM images indicate that the membrane is not dense in nature and pore size is higher than the T1 composition, confirming the TiO_2 nanoparticles impact on pore size and porc distribution. Thc porc size is higher than 1 to 2 micrometers (in case of the formulation T5).

11.3.3 ZETA POTENTIAL ANALYSIS OF NANOCOMPOSITE CA MEMBRANE

The zeta potential for three different TiO_2/CA nanocomposite membranes is reported in the Figures 11.5a–11.5c. CA membrane, as well as most polymeric

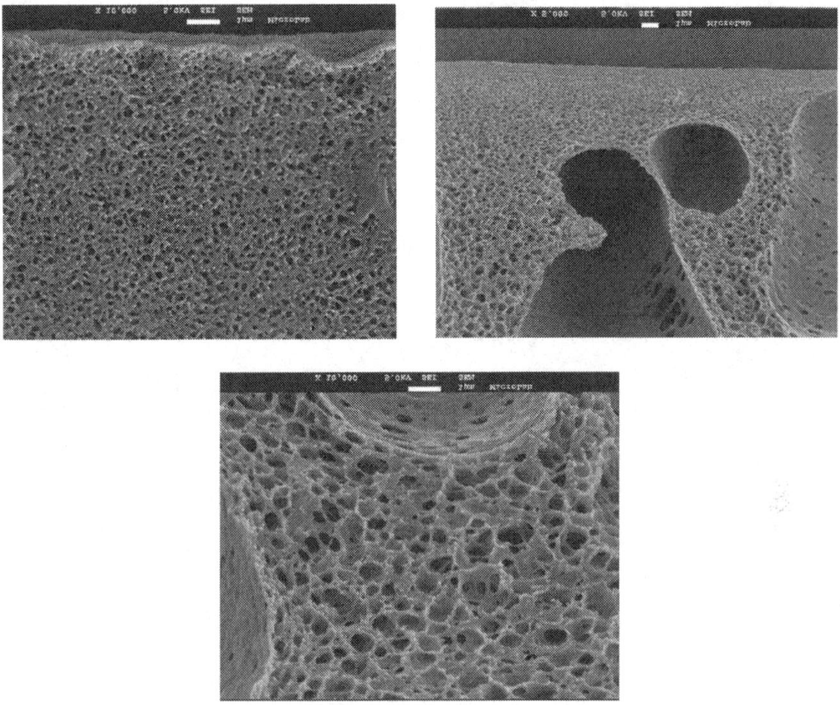

FIGURE 11.4(b) I, II, III SEM images of the CA membranes containing the TiO_2 nanoparticles (formulation T1 reported in Table 11.2)

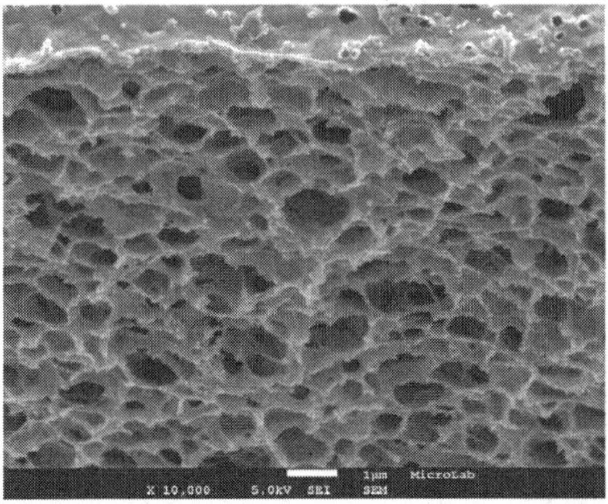

FIGURE 11.4(c) SEM image of the CA membranes containing the TiO_2 nanoparticles (formulation T5 reported in Table 11.2).

membranes, when brought into contact with an aqueous solution acquires an electric surface charge. As can be observed, for all membranes, the zeta potential is negative in all the pH range studied, and becomes more negative with pH. For TiO_2 nanocomposite membrane T1, the value of zeta potential at pH

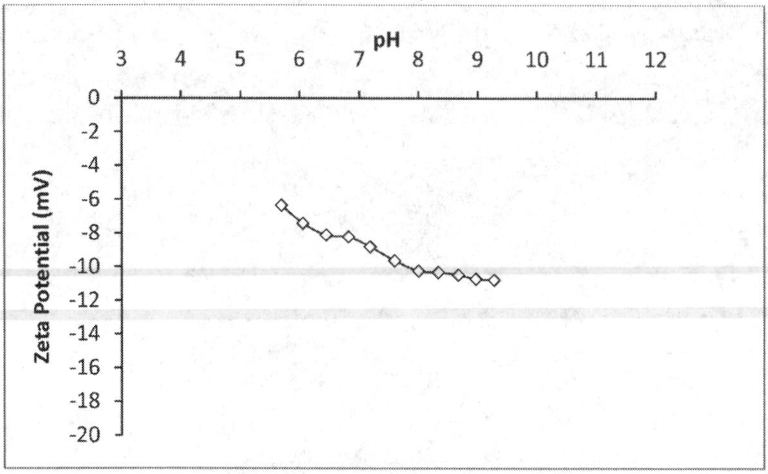

FIGURE 11.5(a) Zeta potential analysis at different pH of the CA membranes containing TiO_2 nanoparticles for the formulation T1.

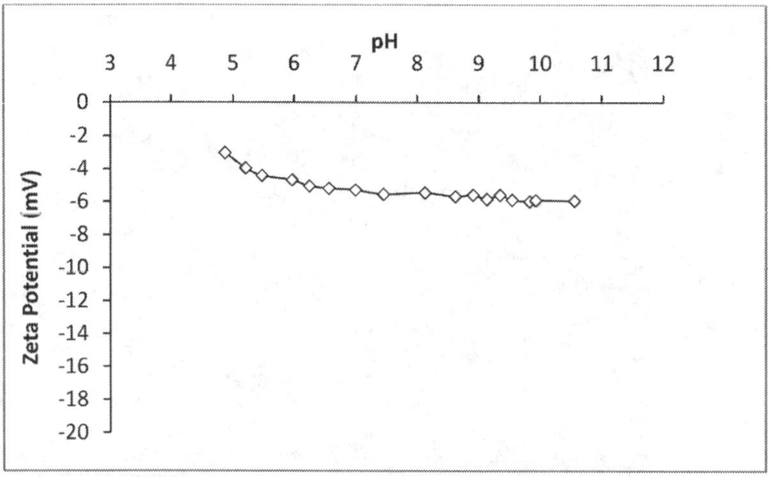

FIGURE 11.5(b) Zeta potential analysis at different pH of the CA membranes containing TiO_2 nanoparticles for the formulation T3.

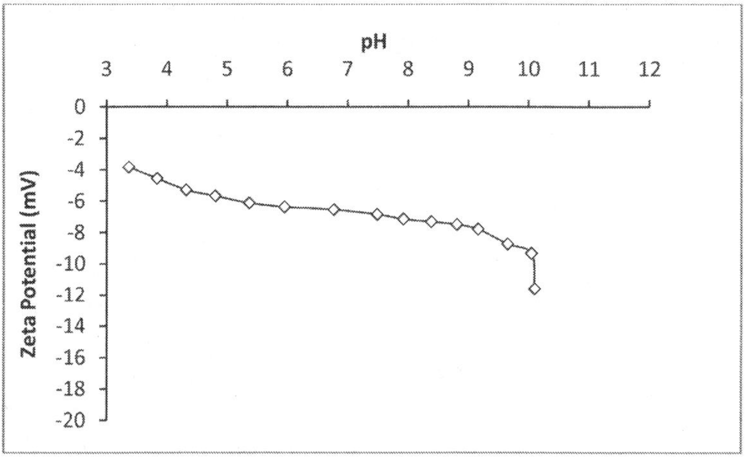

FIGURE 11.5(c) Zeta potential analysis at different pH of the CA membranes containing TiO_2 nanoparticles for the formulation T5.

6 is −6.356 mV and at pH 9 it was −10.78 mV. It is found that above the pH 6 and with increase in the value of pH the membrane surface becomes more negative. Further, at above pH 9 there is not much deviation of the membrane surface charge. While for cellulose acetate membrane TiO_2 nanocomposite T3 the value at pH 5 was −3.031 mV and at pH 7 it was −5.298 mV. The zeta potential value becomes more negative at pH 10, the recorded value was −5.918 mV. The zeta potential analysis for T5 was as follows at pH 3 the value recorded was −3.828 mV while at pH 7 it was −6.535 mV and at pH 10 the value of zeta potential was −9.332 mV.

Comparing the zeta potential obtained in this work with previous studies developed for CA membranes of reverse osmosis and nanofiltration [30–32], we may say that the incorporation of TiO_2 nanoparticles in the casting solution turn the surface of nanocomposite membranes less negative. The adsorption of ions will play an important role in the membrane nanocomposite performance [33–36].

11.3.4 PURE WATER FLUX OF TIO_2 NANOCOMPOSITE CA MEMBRANES

The membrane permeability is function of the pore size and pore distribution. As reported in the Table 11.1, the casting solutions compositions have different solvent ratios acetone/formamide that will result in membranes with different porosity. The membrane prepared from the casting solution with higher content in formamide (T1) will be more porous than the membrane prepared with a casting solution poorer in formamide (T5). Further, it is also important to note that addition of TiO_2 nanoparticles (encapsulated in PVP) makes the membrane more hydrophobic in nature and repels the water molecule.

The details of the membrane pure water flux at different pressures are reported in Figures 11.6a–11.6e and in Table 11.3

The T1 nanocomposite membrane, that result from casting solution with higher formamide content, presents the highest fluxes and at opposite site is the nanocomposite membrane T5 that was obtained from the casting solution with the lower formamide content. Through the slope of linear relation between the pure water flux and the pressure is possible to obtain the hydraulic permeability value for each membrane. The hydraulic pemeabilities presented for the nanocomposite membrane was 54.2, 47.6, 20.2 and 11.3 kg $m^{-2}h^{-1}bar^{-1}$, respectively for T1, T2, T4 and T5 nanocomposite membrane.

The hydraulic permeabilities reported in the literature for cellulose acetate membranes, prepared from casting solutions in the same range

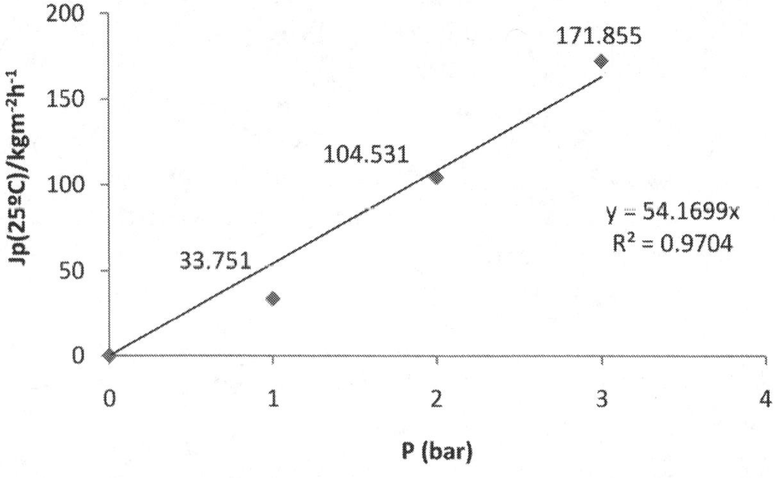

FIGURE 11.6(a) Effect of the pressure onto the water flux at the membrane composition (T1).

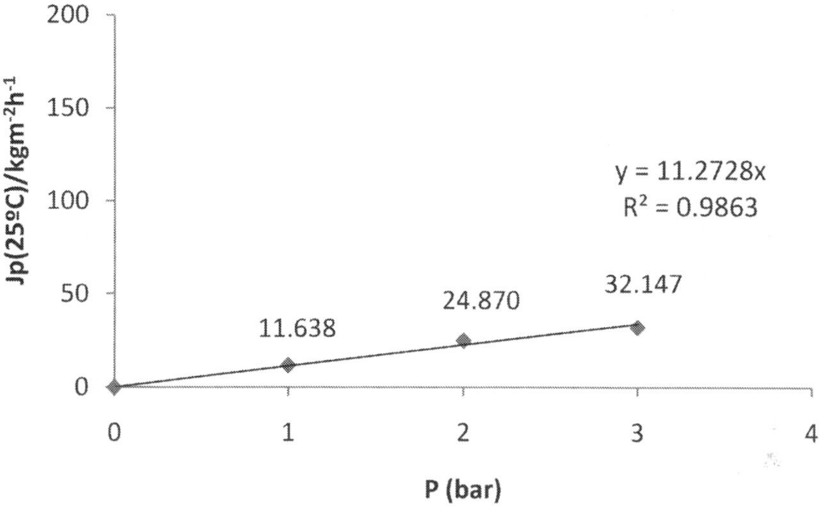

FIGURE 11.6(b) Effect of the pressure onto the water flux at the membrane composition (T2).

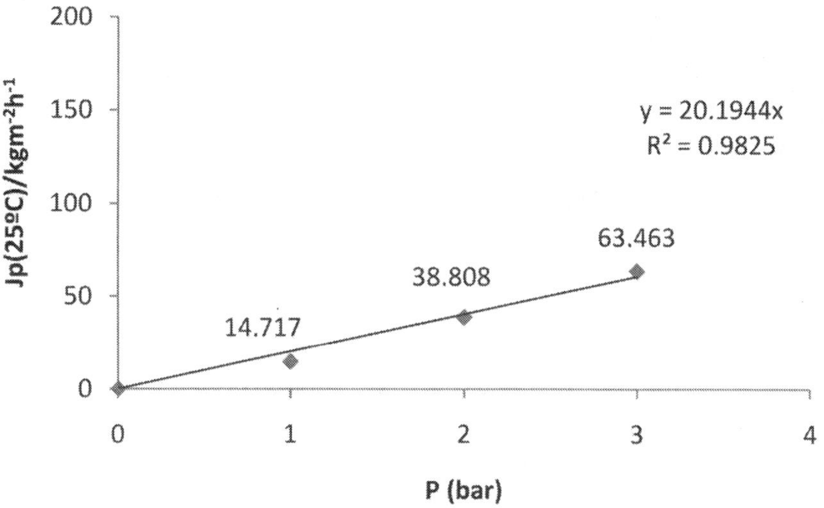

FIGURE 11.6(c) Effect of the pressure onto the water flux at the membrane composition (T4).

of the present ratios of acetone/formamide have values of 64.80, 39.96, and 8.46 kg $m^{-2}h^{-1}$ bar^{-1} for the correspondent T1, T3, and T5 membranes [33]. For the TiO_2/CA mixed asymmetric membranes prepared in this

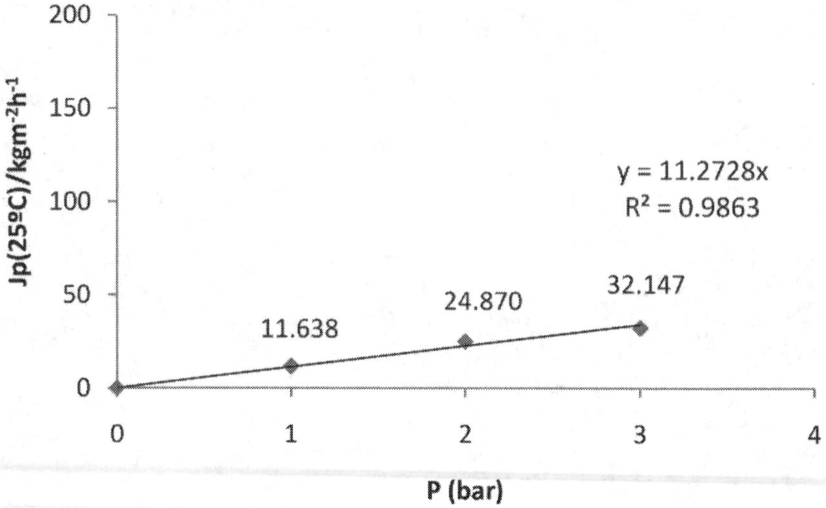

FIGURE 11.6(d) Effect of the pressure onto the water flux at the membrane composition (T5).

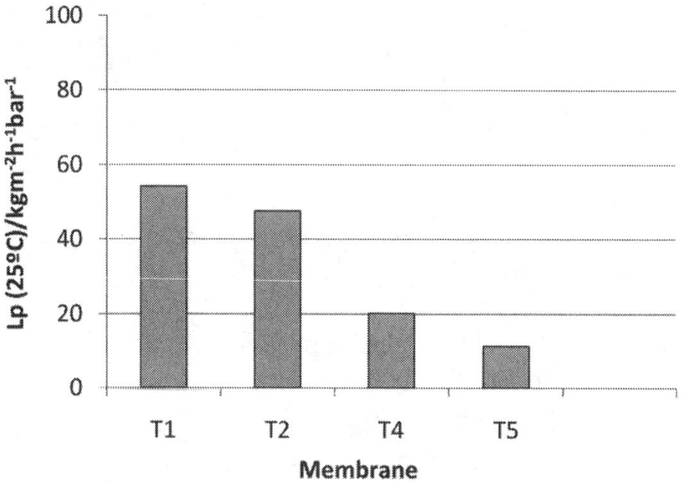

FIGURE 11.6(e) Comparison the effect of the pressure onto the water flux at a different membrane composition.

work presents an enhancement of the hydraulic permeabilities for the formulations T3 and T5, while for the T1 is observed a decrease in the hydraulic permeability value. Several works reports this enhancement in

TABLE 11.3 Permeation Results for Pure Water Fluxes (J_{pw}) at 1, 2 and 3 bar 25 C, Feed rate: 180 L/h), Hydraulic Permeabilities (Lp) for TiO_2/CA Nanocomposite Membranes

Membrane	P (bar)	J_{pw} (kgm^{-2}h^{-1})	Lp (kgm^{-2}h^{-1}/bar)
T1	1	33.8	54.2
	2	104.5	
	3	171.8	
T2	1	32.8	47.6
	2	94.4	
	3	148.3	
T4	1	14.7	20.2
	2	38.8	
	3	63.4	
T5	1	11.6	11.3
	2	24.9	
	3	32.2	

permeabilities for membrane containing TiO_2 nanoparticles that confer to the membranes antifouling characteristics [8, 37–39].

11.4 CONCLUSIONS

TiO_2 nanoparticles synthesis was successfully carried out using the sonochemical sol-gel approach. The particle size of the TiO_2 nanoparticle is near to 20–30 nm size. The shape is observed spindle in nature and contains both amorphous and crystalline phase. Different TiO_2 nanocomposite membranes are successfully prepared by addition of TiO_2 nanoparticles (encapsulated in PVP) into cellulose acetate/acetone/formamide casting solutions. It is observed that the TiO_2 nanoparticles have formed bond between the cellulose acetate and nanoparticles. The nanocomposite membrane becomes more hydrophobic in nature due to addition of TiO_2 nanoparticles. Further the addition of TiO_2 nanoparticles makes the membrane surface less negative when compared with results presented in literature for cellulose acetate membranes. The addition of nanoparticles also affect the porosity of the membrane and hence present an enhancement in hydraulic permeability.

ACKNOWLEDGMENTS

Shirish Sonawane wishes to acknowledges to the Erasmus Mundus Heritage Program for financial support to visit the Membrane Separation Group at the Department of Chemical Engineering at the Instituto Superior Técnico, Lisbon, Portugal.

KEYWORDS

- **active layer**
- **asymmetric composite membrane**
- **interaction of matrix and TiO_2 nanoparticles**
- **TiO_2 nanoparticles**
- **water permeation rate**
- **zeta potential**

REFERENCES

1. Baker, R. W., (2004). Membrane Technology and Applications, John Wiley & Sons: Chichester.
2. Drioli, E. Curcio, E., & Criscuoli, A. (2006). Membrane Contactors Fundamentals, Applications and Potentialities, Elsevier, Amsterdam, 41.
3. Norberta De Pinho, M., Matsuura, T., Nguyen, T. D. & Sourirajan, S., (1988). Reverse osmosis separation of glucose-ethanol-water system by cellulose acetate membranes, *Chemical Engineering Communications*, *64*, 113–123.
4. Arthanareeswaran, G., Thanikaivelan, P., Srinivasn, K., Mohan, D., & Rajendran, M., (2004). Synthesis, characterization and thermal studies on cellulose acetate membranes with additive, *Eur. Polymer J.*, *40*, 2153–2159.
5. Arthanareeswaran, G., & Thanikaivelan, P., (2010). Fabrication of cellulose acetate–zirconia hybrid membranes for ultrafiltration applications: Performance, structure and fouling Analysis, *Separation and Purification Technology*, *74*, 230–235.
6. Muddassir Ali, Muhammad Zafar, Tahir Jamil, & Muhammad Taqi Zahid Butt, (2011). Influence of glycol additives on the structure and performance of cellulose acetate/zinc oxide blend membranes, *Desalination*, *270*, 98–104.
7. Zafara, M., Ali, M., Maqsood Khan, S., Jamil, T., & Taqi Zahid Butt, M., (2012). Effect of additives on the properties and performance of cellulose acetate derivative membranes in the separation of isopropanol/water mixtures, *Desalination*, *285*, 359–365.
8. Abedini, R., Mahmoud Mousavi, S., & Aminzadeh, R. (2011). A novel cellulose acetate (CA) membrane using TiO_2 nanoparticles: Preparation, characterization and permeation study, *Desalination*, *277*, 40–45.

9. Jawalkar, J., More, P., Damkale, S., Kumarb, R., Yadav B., Vishwanath A., et al., (2009), Effect of Organic Chromophore on nano-sized TiO_2. Taylor and Francis. *International Journal of Green Nanotechnology: Engineering Materials, 1*, 40–50.
10. Shirsath, S. R., Pinjari, D. V., Gogate, P. R., Sonawane, S. H., & Pandit, A. B. (2013). Ultrasound assisted synthesis of doped TiO_2 nano-particles: Characterization and comparison of effectiveness for photocatalytic oxidation of dyestuff effluent. *Ultrasonics Sonochemistry Ultrasonics Sonochemistry, 20*(1), 277–286.
11. Vatanpour, V., Siavash Madaeni, S., Reza Khataee, A., Salehi, E., Zinadini, S., & Ahmadi Monfared, H., (2012). TiO_2 embedded mixed matrix PES nanocomposite membranes: Influence of different sizes and types of nanoparticles on antifouling and performance. *Desalination, 292*, 19–29.
12. Soo Lee, H., Joon Im, S., Hak Kim, J., Jin Kim, H., Pyo Kim, J., & Ryul Min, B., (2008). Polyamide thin-film nanofiltration membranes containing TiO_2 nanoparticles. *Desalination, 219*, 48–56
13. Sairam, M., Patil, M. B., Veerapur, R. S., Patil, S. A., & Aminabhavi, T. M., (2006). Novel dense poly (vinyl alcohol)-TiO_2 mixed matrix membranes for pervaporation separation of water–isopropanol mixtures at 30°C, *J. Membr. Sci., 281*, 95–102.
14. Yang, Y., Zhang, H., Wang, P., Zhang, Q., & Li, J. (2007). The influence of nano-sized TiO2 fillers on the morphologies and properties of polysulfone UF membranes, *J. Membr. Sci., 288*, 231.
15. Bendixen, N., Losert, S. Adlhart, C., Lattuad, M. & Ulricha, A. (2014). Membrane–particle interactions in an asymmetric flow field flow fractionation channel studied with titanium dioxide nanoparticles. *Journal of Chromatography A, 1334*, 92–100.
16. Yu, J. C., Yu, J. G., Ho, W. K., & Zhang, L. Z. (2001). Preparation of highly photocatalytic active nano-sized TiO_2 particles via ultrasonic irradiation. *Chem. Commun.*, 19, 1942–1943
17. Prasad, K., Pinjari, D. V., Pandit, A. B. & Mhaske, S. T. (2010). Phase transformation of nanostructured titanium dioxide from anatase-to-rutile is combined ultrasound assisted sol–gel technique, *Ultrasonics Sonochemistry, 17*, 409–415.
18. Prasad, K., Pinjari, D. V. Pandit, A. B., & Mhaske, S. T., (2010). Synthesis of titanium dioxide by ultrasound assisted sol–gel technique: Effect of amplitude (power density) variation. *Ultrasonics Sonochemistry, 17*, 697–703.
19. Shirsath, S. R., Pinjari, D. V., Gogate, P. R. Sonawane, S. H., & Pandit, A. B., (2013). Ultrasound assisted synthesis of doped TiO_2 nano-particles: Characterization and comparison of effectiveness for photocatalytic oxidation of dyestuff effluent. *Ultrason Sonochem, 20*(1), 277–286.
20. Kunst, B., & Sourirajan, S., (1974). An approach to the development of cellulose acetate ultrafiltration membranes. *J. Appl. Polym. Sci., 18*, 3423–3434.
21. Afonso, M. D., & De Pinho, M. N., (1990). Ultrafiltration of bleach effluents in cellulose production. *Desalination, 79*, 115–124.
22. Stamatialis, D. F., Dias, Cristina R., & Norberta de Pinho, M., (1999). Atomic force microscopy of dense and asymmetric cellulose-based membranes. *Journal of Membrane Science, 160*, 235–242.
23. Manuel Magueijo, V. Semiao, V. & Norberta de Pinho, M. (2006). Effect of membrane pore size and membrane-solute interactions on lysozyme ultrafiltration. *Material Science Forum, 514–516*(1), 1483–1487.

24. Sonawane, S. H., Antonine Terrien, Ana Sofia Figueiredo, Clara Goncalves, M. & Maria Norberta De Pinho, (2015). The role of silver nanoparticles on mixed matrix Ag/1 cellulose acetate asymmetric membranes, *Polymer Composites*, 38, 32-39.
25. Yang, Y. N., Zhang, H. X., Wang, P., Zheng, Q., & Li, J., (2007). The influence of nano-sized TiO_2 fillers on the morphologies and properties of PSF UF membrane, *J. Membr. Sci., 288*, 231–238.
26. Roelofs, K. S., Hirth, T., & Schiestel, T. (2010). Sulfonated poly(ether ether ketone)-based silica nanocomposite membranes for direct ethanol fuel cells. *J. Membr. Sci., 346*, 215–226.
27. Lu, Y., Sun, H., Meng, L., & Yu, S., (2009). Application of the Al_2O_3–PVDF nano-composite tubular ultrafiltration (UF) membrane for oily wastewater treatment and its antifouling research, *Sep. Purif. Technol. 66*, 347–352.
28. Wang, Y., Yang, L., Luo, G., & Dai, Y., (2009). Preparation of cellulose acetate membrane filled with metal oxide particles for the pervaporation separation of methanol/methyl tert-butyl ether mixtures *Chemical Engineering Journal, 146*, 6–10.
29. Chen, W., Su, Y., Zhang, L., Shi, Q., Jinming Peng, & Zhongyi Jiang, (2010). *In situ* generated silica nanoparticles as pore-forming agent for enhanced permeability of cellulose acetate membranes. *Journal of Membrane Science, 348*, 75–83.
30. Elimelech, M., Chen, W. H., & Waypa, J. J., (1994). Measuring the zeta (electrokinetic) potential of reverse osmosis membranes by a streaming potential analyzer. *Desalination, 95*, 269-286.
31. Childress, A. E., & Elimelech, M., (1996). Effect of solution chemistry on the surface charge of polymeric reverse osmosis and nanofiltration membranes. *J. Membr. Sci., 119*, 253-268.
32. Mukherjee, P., Jones, K., & Abitoye, J., (2005). Mukherjee, P., Jones, K., & Abitoye, J., (2005). *J. Membr. Sci., 254*, 303. *J. Membr. Sci., 254*, 303-310.
33. João Rosa, M., & de Pinho, M. N. (1994). Separation of organic solutes by membrane pressure-driven processes, *J. Memb. Sci. 89*, 235–243.
34. de Pinho, M. N., Matsuura, T., Nguyen, T. D., & Sourirajan, S. (1988). Reverse osmosis separation of glucose-ethanol-water system by cellulose acetate membranes, *Chem. Eng. Commun. 64*, 113–123.
35. Mukherjee, R., & De, S. (2014). Adsorptive removal of phenolic compounds using cellulose acetate phthalate–alumina nanoparticle mixed matrix membrane. *Journal of Hazardous Materials, 265*, 8–19
36. Takagia, R., Horib, M., Gotohc, K. Tagawab, M., & Nakagakid, M. (2000). Donnan potential and ζ-potential of cellulose acetate membrane in aqueous sodium chloride solutions, *Journal of Membrane Science, 170*(1), 19–25
37. Rahimpour, A., Jahanshahi, M., Rajaeian, B., & Rahimnejad, M., (2011). TiO_2 Entrapped nano-composite PVDF/SPES membranes: Preparation, characterization, antifouling and antibacterial properties. *Desalination, 278*, 343–353.
38. Bergamasco, R., da Silva, F. V., Arakawa, F. S., Yamaguchi, N. U., Reis, M. H. M., Tavares, C. J., et al., (2011). Drinking water treatment in a gravimetric flow system with TiO_2 coated membranes. *Chem. Eng. J., 174*, 102–109.
39. Abedini, R., Mousavi, M. S., & Aminzadeh, R., (2012). Effect of sonochemical synthesized TiO_2 nanoparticles and coagulation bath temperature on morphology, thermal stability and pure water flux of asymmetric cellulose acetate membranes prepared via phase inversion method. *Chem. Ind. Chem. Eng. Q., 18*, 385–398.

CHAPTER 12

A STUDY ON PERFORMANCE EVALUATION OF TERTIARY DYE MIXTURE DEGRADATION BY HYBRID ADVANCED OXIDATION PROCESS

BHASKAR BETHI and SHIRISH H. SONAWANE

Department of Chemical Engineering, National institute of Technology, Warangal, 506004, Telangana State, India, E-mail: shirish@nitw.ac.in, Tel.: +91-870-2462626

CONTENTS

ABSTRACT

In the present work, degradation of three commercial azo-dyes in the presence of hydrodynamic cavitation (HC) and hydrogen peroxide based hybrid advanced oxidation process (HAOP) has been investigated. The influence of quantitative addition of hydrogen peroxide (H_2O_2) for the degradation of dyes was examined. H_2O_2 has been used as an oxidizing agent to improve the degradation efficiency through cavitating device. UV-Visible scanning spectrophotometer analysis has been carried for the investigation of decolorization of mixture of dyes. It was observed that 39% of decolorization was obtained using HC alone and 25 % of decolorization was achieved using H_2O_2 alone. With the use of hybrid process 65 % of decolorization was obtained. The mineralization of dye solution was also evaluated by measuring the total organic carbon (TOC) of the dye solution. 28% of mineralization rate of tertiary dye mixture were attained using HC alone and it is increases to 51% by combining HC with 100 mmol/L of H_2O_2. Synergetic effect of HAOP based on kinetic data was also investigated for the degradation of dye solution in the presence of hydrodynamic cavitation and H_2O_2. Kinetic studies of the HAOP that followed the pseudo first-order reaction kinetics. HC + H_2O_2 have suggested that the combination of HC with other oxidizing agents is better than the individual processes for the degradation of dye solution. Synergistic effect of combined techniques has been found based on the rate kinetics of decolorization and degradation reaction.

12.1 INTRODUCTION

Water is the one of the resource and plays an important role in the human life to sustain on the earth. Even though it is also important resource in every chemical and non-chemical industry. Among chemical related industries, textile industry utilizes hues amount of water in the operations of dyeing and finishing [1]. As per the consideration of discharge volume of effluent and its composition, the wastewater produce from the textile industry is recognized as one of the major polluting source [2]. Total dyes which are produced throughout the world is abo 50–70% of azo class of dyes [3] and rest of the place occupied by the acid, reactive, metal complex dyes, etc. [3]. About 10–20 % of the total azo-dyes that are used for dyeing process in the textile industries were releasing in to the environment. It will create an immense effect on the aquatic and human life [2, 3]. Controlling and removal of those dye molecules is necessary before discharging into the environment. Due to the scarcity of portable water, the treatment and reuse of wastewater streams in many chemical industries became mandatory. Treatment of textile wastewater is considered as expensive and time consuming with biological treatment methods for the decolorization of textile industry wastewater [2]. There is need of time for developing alternative technologies and establishing very simple, economically feasible treatment technologies for the treatment of wastewater at on-site [4]. Up to now, various dye removal methods include biological processes, adsorption and chemical coagulation which do not show efficient dye degradation and create an ongoing waste disposal problem. The chemical oxidations of dye pollutants are usually effective towards the destruction of chromophoric structures of dyes.

In the past two decades, a novel technique called advanced oxidation process (AOP) has been investigated as a future technology for the degradation of organic pollutants in the wastewater. It is also reported that AOP is the one of the energy efficient and efficient degradation technology for the destruction of recalcitrant organic pollutants in wastewater. Some of the AOP's that are deeply investigated and reported in the literature it includes photocatalysis [5, 6], chemical oxidation [7, 8], electrochemical oxidation [9, 10], hydrodynamic cavitation [11]. Out of these AOP's hydrodynamic cavitation reported as an energy efficient treatment technique and due to its capacity to handle bulk volume. Hence, it can be easily to scale up for wastewater treatment. Literature on advanced oxidation processes also illustrates the use of single advanced oxidation process for the degradation of organic pollutants is not an efficient way for the complete mineralization of wastewater.

Few studies on degradation of mixture of dyes using photocatalysis technique, such as Bizani et al. [12] studied the photocatalytic degradation of two commercial azo-dyes (Red FNR and Cibacron Yellow FN2R) in the presence of TiO_2 suspensions as photocatalyst. They found that dyes were decolourized in the presence of nano TiO_2 (P-25) within 100 min of illumination. Palacio et al. [13] have studied the photocatalytic degradation of a mixture of six commercial azo-dyes, by exposure to UV radiation in an aqueous solution containing nano TiO_2 (P25). The optimum conditions for photocatalytic degradation they are achieved at concentrations of 0.5 g TiO_2/L and 50 mmol H_2O_2/L, respectively. Complete decolorization was attained after 240 min irradiation time.

Many studies were reported in the literature for the degradation of mixture of dyes using photocatalysis approach. Literature on mixture of dyes degradation using hydrodynamic cavitation is inadequate. The ability of degradation of organic pollutant in the HC is completely depends on the operating inlet pressure to the cavitating device and its cavitation number [14]. The mechanism involved for the degradation of organic pollutants in the hydrodynamic cavitation is mainly depends upon the two principles such as,

1. Generation of hydroxyl radicals (OH).
2. Thermal pyrolysis or cracking of organic molecules.

In the first case, the degradation of organic pollutants will occurs based on the generation of cavity bubbles, bubble growth and subsequent collapse of cavities in the recovered downstream pressure leads to the generation of •OH radicals. These •OH radicals are highly oxidative in nature and readily oxidize with organic pollutants at the bubble liquid interface [14, 15]. The degradation of organic pollutants in this case is completely depends on the reaction between •OH radicals and organic molecules in the bulk liquid.

In the second case, the degradation of organic molecules is possibly due to the thermal pyrolysis [14–16] or thermal cracking of molecules entrapped inside the cavity or located near the collapsing cavities. The principle of thermal cracking involves the breakage of higher molecular weight organic carbon to lower molecular weight organic carbon or short chain intermediate organic compounds.

Dye wastewater treatment using combination of HC + H_2O_2 has shown the attractive mineralization technique due to the addition of highly oxidative H_2O_2. Gore et al. [14] studied the degradation of reactive orange 4 dye using HC and in combination with other AOP's such as H_2O_2 and ozone. They found that the 15% of mineralization taking place using HC alone and increases to 32% by combining it with H_2O_2. Patil et al. [15] studied the degradation of imidacloprid by hydrodynamic cavitation based on different additives and combination with other processes such as Fenton's reagent ($FeSO_4$), advanced Fenton process (Iron Powder) and combination of $Na_2S_2O_8$ and $FeSO_4$. Degradation studies they were also carried out based on the use of hydrogen peroxide (20–80 ppm). Many studies reported the HC combined with H_2O_2 technique for the degradation of various organic pollutants such as single dye components and single pharmaceutical components, etc. It was observed from the earlier literature that insufficient literature and lack of research work is there on the degradation of mixture of components in wastewater using HC combined with other AOP's. The novelty of this work is to investigate the performance of HC combined with H_2O_2 for the degradation of mixture of three commercial azo-dyes. No degradation studies to the best of our knowledge have not been reported in the literature on HC combined with H_2O_2 in the case of degradation of mixture of three azo-dyes.

This case study reports the degradation of mixture of three commercial azo-dyes in the presence of hydrodynamic cavitation coupled with H_2O_2 and possible degradation mechanism of dye molecules in the presence of HC + H_2O_2. This work also reports mineralization rate of dye wastewater in the presence of hybrid process based on the TOC analysis.

12.2 MATERIALS AND METHODS

12.2.1 MATERIALS

The three commercial azo-dyes, which are used in this degradation study, namely methylene blue ($C_{16}H_{18}ClN_3SxH_2O$), brilliant green ($C_{27}H_{34}N_2O_4S$) dyes were procured from S. D. Fine Chemicals Ltd., Mumbai, India and crystal violet dye ($C_{25}H_{30}N_3Cl$) were purchased from Sisco Research Laboratories Pvt. Ltd. Mumbai, India. Hydrogen peroxide was purchased from Karnataka fine chem., Bangalore, India. All the chemicals procured were of analytical grade and were used in the degradation studies as it obtained from the supplier.

12.2.2 EXPERIMENTAL SETUP

Experimental setup used for the degradation of mixture of dyes (model pollutant) using hydrodynamic cavitation reactor coupled with H_2O_2 is shown in Figure 12.1. A storage/feed tank of 10 l volume used for holding and pumping of liquid dye solution through the cavitation device. The setup was arranged in a closed loop manner or recycle as shown in the Figure 12.1.

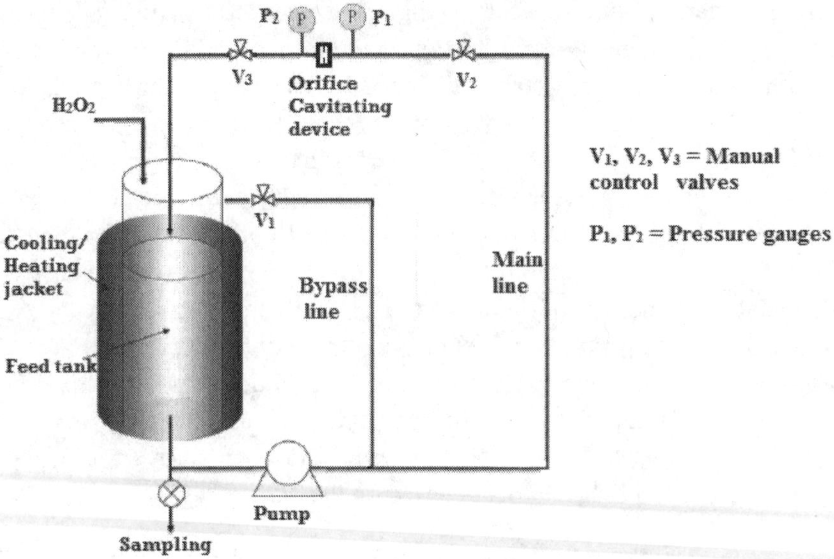

FIGURE 12.1 Experimental Setup of Hybrid AOP (HC + H_2O_2).

Which includes a feed tank, centrifugal pump (Power = 1.5 HP), pressure gauges and valves. Cooling jacket is provided to the feed tank in order to control temperature. Suction side of the pump is connected to the bottom side of the feed tank and the discharge line from the pump divided into two lines. The main line consists of an orifice is used as a cavitating device and second line is the bypass line, connected to the feed tank. Geometrical dimensions of orifice which is used for the degradation study was shown in Figure 12.2. A bypass line is provided to control the flow through the main lines. Control valves and

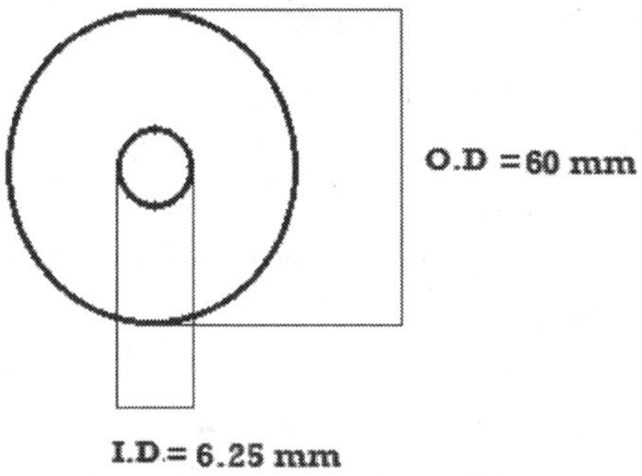

FIGURE 12.2 Geometrical dimensions of cavitating device (orifice).

pressure gauges are provided at appropriate places to control the flow rate through the lines containing the cavitating device (orifice) and to measure the fluid pressures, respectively. The material of construction of the entire system including cavitating device is mild steel.

12.2.3 ANALYTICAL PROCEDURE

Measurement of dye concentration for single dye solution was based upon their absorbance at their respective peaks. However, to measure the dye concentration of individual dyes in a solution of two dyes, values at single wavelength does not sufficient because response at each wavelength is affected due to the presence of other dyes. Whatever the concentration of CV, BG and MB dyes, the absorbance of mixture is always equal to the sum of absorbance of each dye indicating that these both dyes are not linked by chemical or physical bond. In other word no complexes are formed. Absorbance value of 3 dyes at (λ_{max}) of 590, 628, and 663 nm for solution of known concentration with varying dye ratios were considered for the calibration curve of mixed dyes. Maximum absorbance wave length for the above mixture of dyes was obtained at 619 nm. This wave length has been used to measure the concentrations of mixed dye from the calibration curve prepared for mixture of three dyes. The percentage decolorizations of mixture of dyes were calculated using the following equation.

$$\%\ \text{Decolourization} = 1 - \frac{[C_{\text{dye mixture}}]_f}{[C_{\text{dye mixture}}]_0} * 100 \qquad (1)$$

where, $[C_{(\text{dye mixture})}]_0$ is the initial concentration of dye mixture in mg/L; $[C_{(\text{dye mixture})}]_f$ is the final concentration of dye mixture in mg/L.

To know the complete mineralization of mixture of dyes, the TOC analysis were carried out using the Shimadzu TOC-L_{CPN} analyzer supplied by Shimadzu Japan. TOC-L_{CPN} analyzer mainly consists of three in-built stages for finding out the total carbon (TC) and inorganic carbon (IC), such as:

1. acidification;
2. oxidation; and
3. detection and quantification.

TOC-L_{CPN} Analyzer is works on the principle of high temperature catalytic oxidation (HTCO). Manually samples were injected into a HTCO combustion tube packed with platinum catalyst at 680°C in an oxygen rich atmosphere. The concentration of carbon dioxide generated was measured with a inbuilt non-dispersive infrared (NDIR) detector in the instrument. TOC can be calculated from the known values of total carbon and inorganic carbon. The following equation can be used to evaluate the TOC value.

$$\text{TOC} = \text{Total Carbon} - \text{Inorganic Carbon} \quad (2)$$

The percentage mineralization or degradation of mixture of dyes was calculated using the following equation.

$$\%\ \text{Degradation} = 1 - \frac{(\text{TOC})_f}{(\text{TOC})_i} * 100 \quad (3)$$

where, $(\text{TOC})_i$ is the initial TOC value of the dye solution in mg/L; $(\text{TOC})_f$ is the final TOC value of the dye solution in mg/L.

12.3 RESULTS AND DISCUSSION

12.3.1 DEGRADATION THROUGH HC ALONE

To study the decolorization and degradation of mixture of dye solution in the presence of HC, three commercial azo-dyes were used. Each dye consists of 200 mg/L of crystal violet (CV), methylene blue (MB) and brilliant green (BG), respectively were used for the preparation of 5 L dye stock solution. All the experimental runs were attempted with the tap water. Process operating parameters such as constant temperature (35°C), constant pressure (2 bar pump discharge pressure to the cavitating device) and normal solution pH 8.6 was used to investigate the decolorization and mineralization of tertiary dye mixture. Decolorization and degradation of dye solution was achieved by passing of dye solution through the cavitating device for 90 min in a recirculation mode. For every 15 min, samples has been collected and carried out the UV-Vis scanning spectrophotometer.

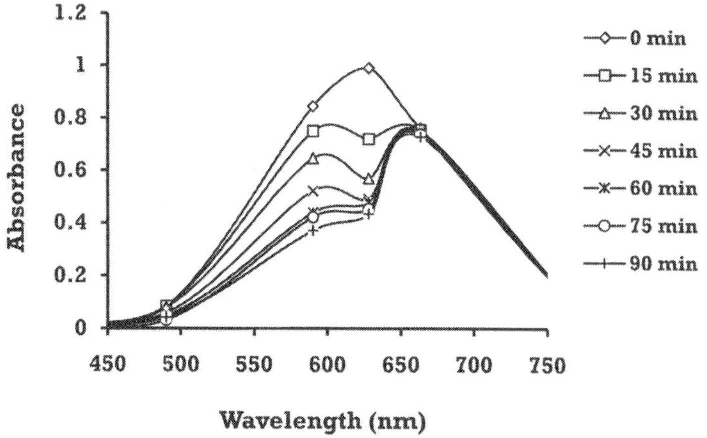

FIGURE 12.3 UV-Visible absorbance spectra of tertiary dye degradation using HC alone.

Figure 12.3 shows the UV-Vis absorbance spectra of mixture of three dyes for every 15 min of time interval over 90 min of study in the HC system, the profile of UV-Vis spectrum confirmed the dyes were decolorized for every 15 min of time interval. Concentration of mixture of dyes for every 15 min of time interval in the HC based decolorization study was shown in the Figure 12.4. Decolorization and degradation rate of tertiary dye mixture in the aqueous media were also studied in the presence of hybrid process. It was perceived that tertiary dye mixture in the aqueous media

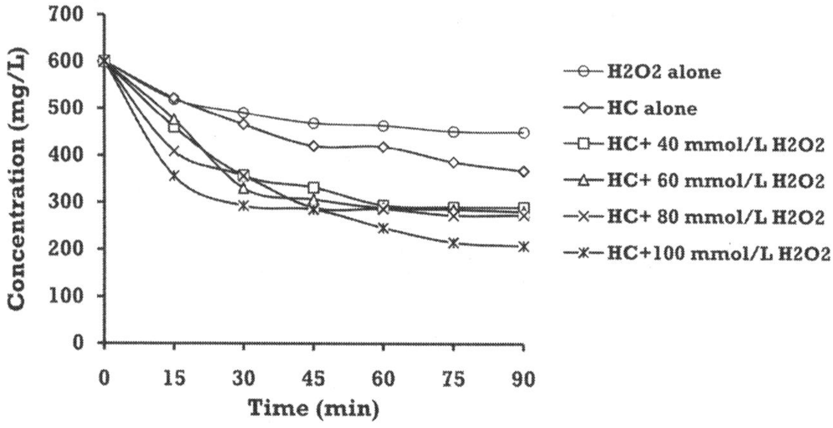

FIGURE 12.4 Decolorization performance of tertiary dye in various systems.

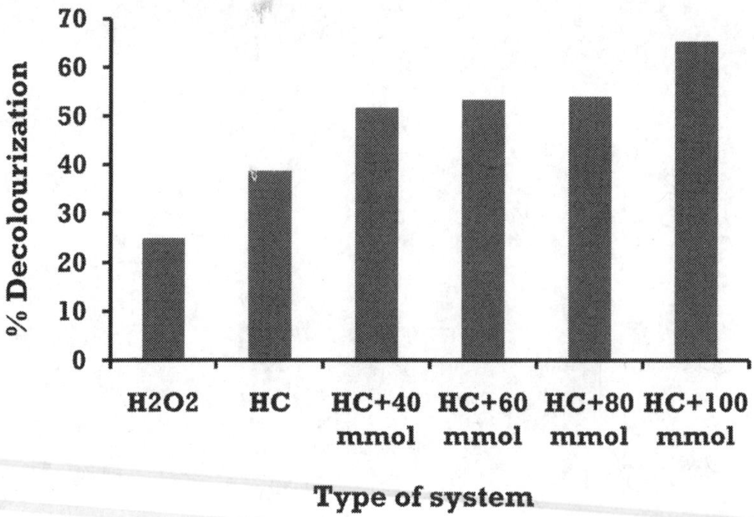

FIGURE 12.5 Percentage decolorization of tertiary dye in various systems.

was decolourized up to 38.8%. The decolorization data were obtained in this study were plotted and shown in the Figure 12.5 and the percentage of degradation of organic carbon was achieved up to 27.7 % in 90 min of degradation study has been confirmed from the TOC analysis and it was shown in the Figure 12.6.

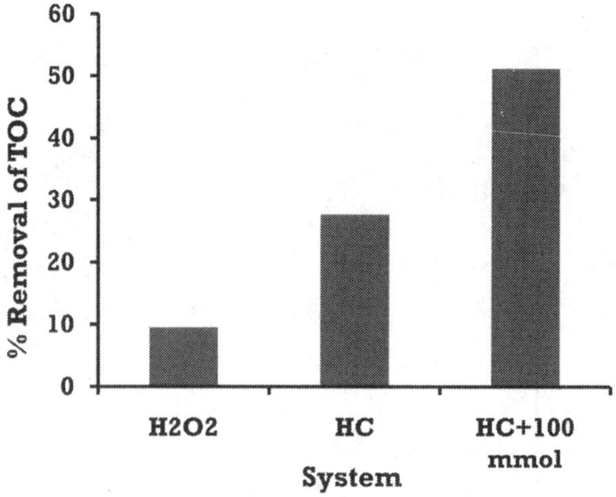

FIGURE 12.6 Percentage removal of TOC in various systems.

The hydrodynamic characteristic of the cavitating device (orifice) has been studied by measuring the flow rate in main line. The calculated C_v for an orifice at an inlet pressure of 2 bar is 0.017 C_v = 0.017) according to the following equation [14].

$$Cv = \frac{P_2 - P_v}{\frac{1}{2}\rho v_0^2} \tag{4}$$

where P_2 = recovered downstream pressure = 1.03 kg/cm^2; P_v = vapor pressure of the dye solution = 0.0573 kg/cm^2; ρ = density of the dye solution = 994 kg/m^3; v_0 = velocity of the dye solution at the throat of the cavitating constriction = 0.333 m/s.

12.3.2. DEGRADATION THROUGH HC + H_2O_2

Addition of hydrogen peroxide (H_2O_2) produce the additional supplement of •OH radicals to the •OH radicals generated by the HC, leads to the double the quantity of •OH radicals which enhance the rate of degradation of organic pollutant is extremely higher compared to individual advanced oxidation process (HC). H_2O_2 is known as a common oxidizing agent can be used for the treatment of wastewater due to its adequate oxidation potential of 1.78 V [14]. The phenomena of cavities generation, growth, and collapse creates the extreme surrounding conditions of high temperature (5000–1000 K) and pressure (1000 atm) [17], under these extreme conditions, if use H_2O_2 molecules can readily dissociate into •OH radicals [18–20]. Under the conventional stirred conditions (low temperature and pressure) the efficacy of H_2O_2 in oxidizing the organic pollutant is low due to poor dissociation of H_2O_2 into •OH radicals. The reaction mechanism for degradation of mixture of dyes using the hybrid process of HC and H_2O_2 is expected to take place like below.

$$H_2O_2 + HC \xrightarrow{35\,^{o}C,\ 2\ bar} 2 \bullet OH \tag{5}$$

$$H_2O + HC \xrightarrow{35\,^{o}C,\ 2\ bar} \bullet H + \bullet OH \tag{6}$$

$$\bullet OH + \bullet OH \rightarrow H_2O_2 \quad (7)$$

$$(CV+MB+BG) + \bullet OH + H_2O_2 \rightarrow \text{Intermediate compounds} + H_2O \quad (8)$$

$$\text{Intermediate compounds} + \bullet OH + H_2O_2 \rightarrow CO_2 + H_2O \quad (9)$$

Experiments were conducted using the hybrid process of HC and H_2O_2 at 2 bar inlet pressure and the solution pH of 8.6. 40, 60, 80, 100 mmol of H_2O_2 addition was carried out to study the decolonization and mineralization rate of dye mixture in the presence of hybrid process. Figures 12.7 and 12.8 shows the first order decolorization and degradation kinetics of mixture of dyes using the hybrid process, respectively. It was found that the decolorization and degradation rate increases with an increase in the molar concentration of H_2O_2. It was found that almost 65.35% of decolorization were achieved in 90 min. Figure 12.5 shows the percentage decolorization of tertiary dye over 90 min of reaction time using HC combined with H_2O_2 and 51% of degradation were achieved in 90 min has been found from the TOC Analysis. The percentage degradation of mixture of dyes in the presence of various molar concentrations of H_2O_2 coupled with HC was shown in the Figure 12.6.

12.3.3 SYNERGETIC EFFECT OF COMBINED TECHNIQUE

In order to know the synergetic effect of combined technique on degradation of tertiary dye mixture, individual experiments were carried out. To know the effect of H_2O_2 alone on tertiary dye mixture decolorization and degradation, experiments were carried out. It was observed that the tertiary dye mixture were decolourized up to 24.95% and degradation was taken place up to 9.58 %. Percentage decolorization of mixture of dyes using H_2O_2 alone with respect to time was shown in Figure 12.5 and the percentage degradation of mixture of dyes in the presence H_2O_2 alone over 90 min of degradation study was shown in the Figure 12.6. Synergetic effect of combined processes has been evaluated on the basis of reaction rate constant data of single and combined process. The synergetic index of the HC combined with H_2O_2 has been calculated to evaluate the efficiency of the combined process of HC and H_2O_2. The synergetic index (f) of HC combined with H_2O_2 has been calculated by using the following equation.

$$f_{(HC+H_2O_2)} = \frac{k_{(HC+H_2O_2)}}{k_{(HC)} + k_{(H_2O_2)}} \tag{10}$$

12.3.4 RATE KINETICS OF MIXTURE OF DYES DECOLORIZATION

Reaction kinetics based on the decolorization and degradation mechanism of organic pollutants in single and HAOP were studied (Figure 12.7). From the kinetics data it was observed that the decolorization and degradation reaction followed the pseudo first order reaction. Kinetic data which is obtained in this study also reveals the decolorization and degradation reactions of many organic pollutants in advanced oxidation processes followed a pseudo first order reaction [21, 22]. The rate constants of decolorization and degradation reaction of mixture of dyes will be calculate using the following equation.

$$\ln\left(\frac{C_0}{C}\right) = k \times t \tag{11}$$

where 'C' is the concentration of dye molecules present in mol/L, 'k' is the rate constant in min^{-1} and 't' is the time in minutes.

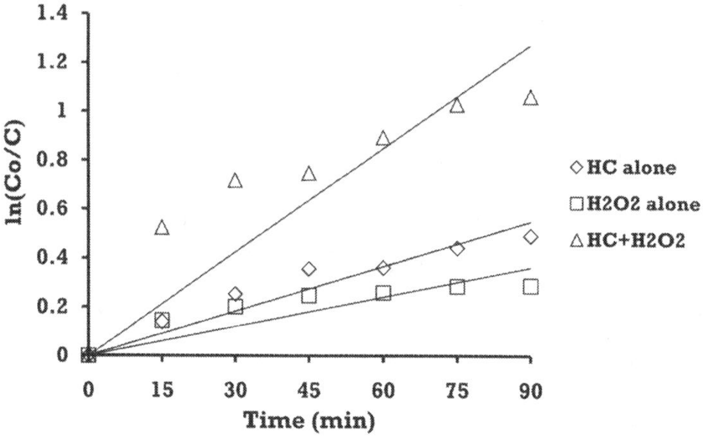

FIGURE 12.7 Rate kinetics of decolorization of tertiary dye mixture in various systems.

It was observed that degradation kinetic mechanism of mixture of dyes in hydrodynamic cavitation coupled with H_2O_2 followed the first order reaction and its conformed from the plot of ln(C_0/C) vs. time (t). The plot of ln(C_0/C) vs. time (t) shows a straight line and it was passed through the origin. Reaction rate constants (k) of mixture of dyes degradation in various systems, such as HC alone and HC + H_2O_2 has been reported in Table 12.1. Pseudo reaction kinetics of mixture of dyes degradation was shown the Figure 12.8.

12.4 CONCLUSION

The summery of this work concludes, a novel energy efficient technique HC + H_2O_2 was successfully applied for the degradation of mixture of azo-dyes. Decolorization and degradation of tertiary dye mixture is ineffective when the HC used alone. It was observed that 38.8% of decolorization and 27.7% of mineralization rate was achieved with the HC only. Extent of

TABLE 12.1 Rate Kinetics of Decolorization of Tertiary Dye Mixture in Various Systems

Method	K Value (min^{-1})	Synergetic coefficient
HC alone	0.006	—
H_2O_2 alone	0.004	—
HC + H_2O_2	0.014	1.4

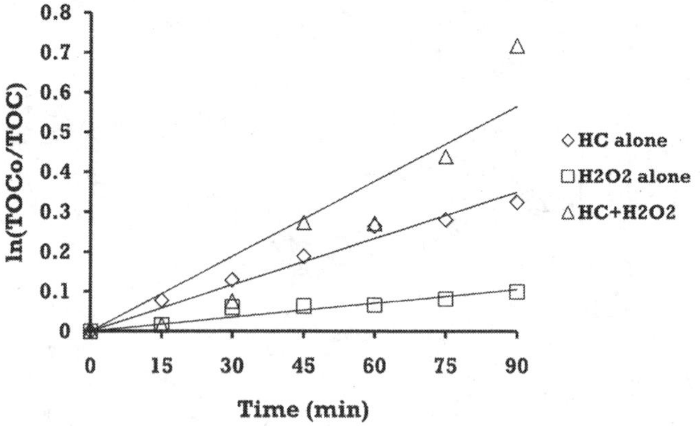

FIGURE 12.8 Rate kinetics of degradation of tertiary dye mixture in various systems.

TABLE 12.2 Rate Kinetics of Degradation of Tertiary Dye Mixture in Various Systems

Method	K Value (min^{-1})	Synergetic coefficient
HC alone	0.0039	—
H_2O_2 alone	0.0012	—
HC + H_2O_2	0.0063	1.23

decolorization and mineralization of mixture of dyes was achieved from 38.8 to 65.35% and 27.7 to 51%, respectively when the H_2O_2 was added. It was conformed that addition of H_2O_2 in the presence of HC made the process efficient way for the degradation of tertiary dye mixture in aqueous medium (Table 12.2).

ACKNOWLEDGMENTS

Shirish H. Sonawane acknowledges the Ministry of Environment and Forest (MoEF, Govt of India) for providing the financial support thorough Grant No: 10-1/2010-CT (WM).

KEYWORDS

- **hybrid advanced oxidation process**
- **hydrodynamic cavitation**
- **hydrogen peroxide**
- **mixture of dyes**
- **wastewater treatment**
- **decolorization**

REFERENCES

1. Anjaneyulu, Y., Sreedhara Chary, N., & Samuel Suman Raj, D., (2005). Decolorization of industrial effluents—available methods and emerging technologies: A review, *Rev. Environ. Sci. Biotechnol., 4*, 245–273.

2. Saharan, V. K., Pandit, A. B., Satish Kumar, P. S., & Anandan, S., (2012). Hydrodynamic cavitation as an advanced oxidation technique for the degradation of Acid Red 88dye, *Ind. Eng. Chem. Res. 51* (4), 1981–1989.
3. Madhavan, J., Grieser, F., & Ashok Kumar, M. (2010). Degradation of Orange-G by advanced oxidation processes, *Ultrason. Sonochem., 17*, 338–343.
4. Pankaj N. Patil, & Parag R. Gogate, (2012). Degradation of methyl parathion using hydrodynamic cavitation: Effect of operating parameters and intensification using additives, *Separation and Purification Technology, 95*, 172–179
5. Kansal, S. K., Singh, M., & Sud, D., (2007). Studies on photodegradation of two commercial dyes in aqueous phase using different photocatalyst, *J. Hazard. Mater., 141*, 581–590.
6. Zheng, H. L., Pan, Y. X., & Xiang, X. Y., (2007). Oxidation of acidic dye Eosin Y by the solar photo-Fenton processes, *J. Hazard. Mater., 141*, 457–464.
7. Daneshvar, N., Aber, S., Vatanpour, V., & Rasoulifard, M. H., (2008). Electro-Fenton treatment of dye solution containing Orange II: influence of operational parameters, *J. Electroanal. Chem., 615*, 165–174.
8. Ramirez, J. H., Duarte, F. M., & Martins, E. G., (2009). Modeling of the synthetic dye Orange II degradation using Fenton's reagent: from batch to continuous reactor operation, *J. Chem. Eng., 148*, 394–404.
9. Guivarch, E., Trevin, S., Lahitte, C., & Oturan, M. A., (2003). Degradation of azo-dyes in water by electro-Fenton process, *Environ. Chem. Lett., 1*, 38–44.
10. Vlyssides, A. G., Loizidou, M., Karlis, P. K., Zorpas, A. A., & Papaio Annou, D. (1999). Electrochemical oxidation of a textile dye wastewater using a Pt/Ti electrode, *J. Hazard. Mater., 70*, 41–52.
11. Virendra Kumar Saharan, Mandar P. Badve, & Aniruddha B. Pandit, (2011). Degradation of Reactive Red 120 dye using hydrodynamic cavitation, *Chemical Engineering Journal, 178*, 100–107.
12. Bizani, E., Fytianos K., Poulios, I., Tsiridis, V., (2006). Photocatalytic decolorization and degradation of dye solutions and wastewaters in the presence of titanium dioxide, *Journal of Hazardous Materials, 136*, 85–94.
13. Palácio, S. M., Quiñones, F. R. E., Módenes, A. N., Manenti, D. R., Oliveira, C. C., & Garcia, J. C. (2012). Optimised photocatalytic degradation of a mixture of azo-dyes using a TiO_2/H_2O_2/UV process, Water science & Technology, 65 (8), 1392–1398.
14. Gore, M. M., Saharan, V. K., Pinjari, D. V., Chavan, P. V., & Pandit, A. B. (2014). Degradation of reactive orange 4 dye using hydrodynamic cavitation based hybrid techniques, *Ultrasonics Sonochemistry, 21*, 1075–1082.
15. Pankaj N. Patil, Sayli D. Bote, & Parag R. Gogate, (2014). Degradation of imidacloprid using combined advanced oxidation processes based on hydrodynamic cavitation, *Ultrasonics Sonochemistry, 21*, 1770–1777.
16. Wang, X. K., Wang, J., Guo, P., Guo, W., & Wang, C. (2009). Degradation of rhodamine B in aqueous solution by swirling jet induced cavitation combined with H_2O_2, *J. Hazard. Mater., 169*, 486–491.
17. Didenko, Y. T., McNamara, W. B., & Suclick, K. S., (1999). Hot spot conditions during cavitation in water, *J. Am. Chem. Soc., 121*, 5817–5818.
18. Abbasi, M., & Asl, N.R., (2008). Sonochemical degradation of basic blue 41 dye assisted by nano-TiO_2 and H_2O_2, *J. Hazard. Mater., 153* 942–947.

19. Pang, Y. L., Abdullah, A. Z., & Bhatia, S., (2011). Review on sonochemical methods in the presence of catalysts and chemical additives for treatment of organic pollutants in wastewater, *Desalination, 277*, 1–14.
20. Merouani, S., Hamdaoui, O., Saoudi, F., & Chiha, M., (2010). Sonochemical degradation of Rhodamine B in aqueous phase: Effects of additives, *Chem. Eng. J., 158*, 550–557.
21. Gul, S., & Ozcan-Yildirium, O., (2009). Degradation of Reactive Red 194 and Reactive Yellow 145 azo-dyes by O_3 and H_2O_2/UV-C processes. *Chem. Eng. J., 155*, 684–690.
22. Wu, C. H., (2008). Effects of operational parameters on the decolorization of C.I. Reactive Red 198 in UV/TiO_2-based systems, *Dyes and Pigments*, *77*, 31–38.

INDEX

O

T

U

UV

V